Introductory Physics for Biological Scientists

Why do elephants have sturdier thigh bones than humans? Why can't ostriches fly? How do bacteria swim through fluids?

With each chapter structured around relevant biological case studies and examples, this engaging, full-color book introduces fundamental physical concepts essential in the study of biological phenomena. Optics is introduced within the context of butterfly wing coloration, electricity is explained through the propagation of nerve signals, and accelerated motion is conveniently illustrated using the example of the jumping armadillo. Other key physical concepts covered include waves, mechanical forces, thermodynamics, and magnetism, and important biological techniques are also discussed within this context, such as gel electrophoresis and fluorescence microscopy. A detailed appendix provides further discussion of the mathematical concepts utilized within the book, and numerous exercises and quizzes allow readers to test their understanding of key concepts. This book is invaluable to students aiming to improve their quantitative and analytical skills and understand the deeper nature of biological phenomena.

Christof M. Aegerter is a professor of physics at the University of Zurich. He has extensive research experience in biological and soft condensed matter physics, specifically bioimaging and the dynamics of biological growth. He currently teaches a course for first-year biology students, demonstrating introductory physics with examples from the biological sciences.

Introductory Physics for Biological Scientists

CHRISTOF M. AEGERTER

University of Zurich

CAMBRIDGE
UNIVERSITY PRESS

University Printing House, Cambridge CB2 8BS, United Kingdom

One Liberty Plaza, 20th Floor, New York, NY 10006, USA

477 Williamstown Road, Port Melbourne, VIC 3207, Australia

314–321, 3rd Floor, Plot 3, Splendor Forum, Jasola District Centre, New Delhi – 110025, India

79 Anson Road, #06–04/06, Singapore 079906

Cambridge University Press is part of the University of Cambridge.

It furthers the University's mission by disseminating knowledge in the pursuit of
education, learning, and research at the highest international levels of excellence.

www.cambridge.org
Information on this title: www.cambridge.org/aegerter
DOI: 10.1017/9781108525862

© Christof M. Aegerter 2018

First published 2018

Printed and bound in Great Britain by Clays Ltd, Elcograf S.p.A.

A catalogue record for this publication is available from the British Library.

Library of Congress Cataloging-in-Publication Data
Names: Aegerter, Christof M., author.
Title: Introductory physics for biological scientists /
Christof M. Aegerter, Universität Zürich.
Description: Cambridge, United Kingdom ; New York, NY : Cambridge University
Press, 2018. | Includes bibliographical references and index.
Identifiers: LCCN 2018018453 | ISBN 9781108423342 (hardback)
Subjects: LCSH: Physics.
Classification: LCC QC21.3 .A36 2018 | DDC 530–dc23
LC record available at https://lccn.loc.gov/2018018453

ISBN 978-1-108-42334-2 Hardback

Additional resources for this publication at www.cambridge.org/aegerter

Contents

Preface: How to Use This Book   *page* ix

1   Physics as a Basis for Describing Biological Systems   1
    1.1   Physical Effects in Biological Systems   1
    1.2   Physics-Based Methods for Investigating Biological Systems   11
    1.3   Acting and Thinking Like a Physicist: How to Gain Scientific Knowledge   16
    Quiz Questions   23

2   Errors, Units, and Scaling Laws   25
    2.1   Uncertainties and Error Propagation   25
    2.2   Units and Dimensional Analysis   36
    2.3   Scaling Laws   42
    Exercises   52
    Quiz Questions   58

3   Motions and Oscillations   61
    3.1   Describing Motions: Velocity and Acceleration   61
    3.2   Periodic Motions: Oscillations   67
    3.3   Describing Any Oscillation in Terms of Harmonic Ones: Fourier Series   73
    3.4   Motions in Two and Three Dimensions   76
    3.5   A Circular Motion Is an Oscillation   80
    3.6   Circular Motion as a Population-Dynamic System   83
    3.7   *Nonlinear Dynamical Systems in Ecology   85
    3.8   *Damped and Coupled Oscillations   88
    Exercises   95
    Quiz Questions   95

4   Resonances and Waves   98
    4.1   How Resonances and Waves Determine How We Interact
           with the Environment   98
    4.2   Forced Oscillations and Resonance   99
    4.3   One-Dimensional and Harmonic Waves   106
    4.4   Waves Are Transporting Energy   111
    4.5   The Physiology and Physics of Hearing   112
    4.6   Fourier Transforms   115

4.7   The Principle of Superposition   119
4.8   Wave Mechanics and Heisenberg's Uncertainty Principle   135
4.9   *The Doppler Effect   138
Exercises   140
Quiz Questions   143

5   Optics, Light, and Colors   151
5.1   How Light Interacts with Matter and What We Can Learn from This   151
5.2   Refraction and Reflection   152
5.3   Interference and Diffraction   154
5.4   Polarization   170
5.5   Geometric Optics   175
5.6   Absorption, Colors, and Fluorescence   203
5.7   *Absorption of Radiation   208
Exercises   211
Quiz Questions   213

6   Forces and Newton's Laws of Motion   217
6.1   Forces and Their Origin   217
6.2   Equations of Motion   227
6.3   Conservation of Momentum   239
6.4   Energy and Its Conservation   244
Exercises   254
Quiz Questions   257

7   Continuum Mechanics   260
7.1   Elasticity and Materials   260
7.2   Stress and Strain: Hooke's Law   263
7.3   Bending a Beam   271
7.4   Flowing and Shear Stress   273
7.5   Surface Tension   274
7.6   Fluids   282
7.7   Flow through Pipes and Blood Flow   293
7.8   *Cells and Tissues Are Neither Liquid nor Solid: Viscoelasticity   302
Exercises   305
Quiz Questions   308

8   Heat, Temperature, and Entropy   310
8.1   How the Interplay of Many Particles Leads to the Whole Being More
      than the Sum of Its Parts   310
8.2   Temperature and the Ideal Gas   316
8.3   Transport via Random Processes   321
8.4   Entropy and the Laws of Thermodynamics   333

8.5   *The Influence of Thermal Motion on Materials Properties   341
8.6   *Nonequilibrium Processes   349
Exercises   356
Quiz Questions   360

9   Electrical Charges and Currents   363
9.1   Electric Charges, Fields, and Potentials   363
9.2   Electric Fields and Potentials of Specific Charge Distributions 369
9.3   Molecular Interactions and Bonds   377
9.4   Electrical Currents   382
9.5   *Propagation of Nerve Signals   389
Exercises   395
Quiz Questions   398

10 Magnetism   401
10.1 Magnetic Fields in Biology   401
10.2 Properties of Magnetic Fields   402
10.3 Mass Spectrometry   407
10.4 *Nuclear Magnetic Resonance   408
Exercises   417
Quiz Questions   418

Appendix A   Mathematical Tools   420

Appendix B   Solutions to Quizzes   441

References   444
Index   447

Preface: How to Use This Book

This book grew out of an introductory lecture for first-year biology students at the University of Zurich that I have been giving for a couple of years. The entire material covered in the book I would teach in a year-long course (28 weeks of term) with three hours of lectures per week. Given the fact that physics is a prerequisite for biology students, which is usually not very popular, I have started to gradually change the standard curriculum of the introductory physics class by introducing the same concepts on biological examples and in this way also give insights to these biological problems that would not be covered in a biology class. In this way, the course is tailored to biology students in that the basic goal of the course actually is to gain an understanding of biological systems and common techniques in biology, which just happens to introduce all of the concepts usually also introduced in a basic physics course. This however does not mean that the physics is treated less rigorously or less quantitatively. In order to actually reach an understanding of the biological problems, a fully quantitative description of the processes and thus the introduction of the corresponding concepts need to be provided, which is why the course and the book are unapologetically quantitative where needed.

This is mirrored in the fact that biological research is currently becoming ever more quantitative, and as it happens, the methods used in such a quantitative treatment of biological problems are often more complex than what is usually treated in an introductory course. This is the reason that the present book contains advanced mathematical tools such as differential equations, Fourier-series, and statistics, which are used early on. All of these tools are, however, introduced in a very applied fashion on examples directly stemming from biology, such that their use can be immediately grasped. In addition, there is an appendix describing all of the basic concepts used, and this may be useful to turn back to or to study in advance to make sure all of the needed concepts are understood. One further reason for the use of mathematical descriptions throughout is that, when we are trying to understand the workings of magnetic resonance imaging (MRI) or how patterns form during embryogenesis, there is often a deeper level of understanding that we can gain by a mathematical treatment, such that we can make definite predictions. However, a qualitative understanding can be obtained without following any of the derivations, and often the result can even be guessed at by an understanding of the relevant parameters determining the behavior of a system. This can be seen throughout the book where in almost all instances a qualitative understanding of problems in terms of proportionalities and power-law dependencies is obtained from an understanding of the basic physics that determines the question at hand. In fact, the initial chapters introduce the tools of such a qualitative description in terms of scaling laws and dimensional analysis. Sections that go beyond the basic treatment, sometimes even of an introductory physics class for physics majors, are marked with an asterisk (*). However, given the nature of biological problems, these sections tend to be biologically even more interesting, concerning subjects such as population dynamics, blood flow, the conduction of nerve signals, or the properties of biomolecules or the cytoskeleton and electrophoresis. While the treatment of this needs some more advanced mathematical tools, these sections are still interesting to biologists for their conclusions regarding the biological problem at hand.

Given the importance of biological problems and physics-based methods used in biology, the order in which subjects are introduced does not follow the standard canon of an introductory physics class. Given the great importance of optical phenomena, in particular scattering and microscopy as biological tools, and vision, coloration, and navigation by the polarized sky as biological problems, we start with a discussion of waves, oscillations, and optics after an introductory chapter laying the foundations of why physics is important and how quantitative tools can help making predictions. This grasps the imagination of biology students from the beginning. Therefore, when it is time to discuss mechanics, heat, and electrical phenomena, the students already know that this is going to be important. In addition to biological problems, some important biological techniques, such as gel electrophoresis, fluorescence microscopy, MRI, and mass spectrometry are discussed in detail in the context of the corresponding physics concepts, in order to get biology students to understand the basic principles behind them and therefore make them appreciate what these experimental methods can and cannot do. In places, the text goes further than a usual introductory physics course, where it is biologically relevant. Thus there is an introduction to statistical physics, soft matter, viscoelasticity, nonequilibrium and nonlinear dynamics, as well as pattern formation, all of which have direct applications in biology. These typically are the final sections with an asterisk of the corresponding chapter on heat, mechanics, elasticity, or fluid dynamics. In all of this, it is important to not only study the theory but to solve problems oneself. Therefore, every chapter closes with problem sets, including basic qualitative questions and more advanced problems directly describing an interesting biological question. Solutions for these problems are available for download at www.cambridge.org/aegerter

1 Physics as a Basis for Describing Biological Systems

You may be asking yourself, why a book on physics for biologists? Why do I as a biologist need to know anything about physics? The fact that many universities have a physics class as a requirement for life science students does not actually answer this question, but possibly makes you think that there might be a reason. We will try to elucidate this reason or these reasons, for there are several, in this introductory chapter, where we'll look at the history of biology as well as a few specific biological problems. The short summary of this answer is:

- Principles from physics and physical effects are of great importance in understanding how biological systems behave and which constraints evolution has to fulfill in shaping organisms.
- Experimental methods developed in physics have time and again revolutionized the practice of biological research and biology itself and are still doing so today.
- Physics is a prime example for scientific and quantitative thinking, which is the endeavor that turns scholarly research into science.

This introductory chapter will make all of these point more specific and give some detailed accounts of where physical principles can lead to a better understanding of biological systems or where physics-based methods have yielded novel insights into biology. Finally, we'll see how the methods of physics of quantitatively analyzing a problem and finding abstract toy models for real-life situations can give deep insights into how nature works as well as what science actually can (and cannot) achieve. Apart from these methods of a quantitative description of nature, the other main aspect of this book is to give you some of the basic physics principles behind biological phenomena – from the molecular to the organismal scales.

1.1 Physical Effects in Biological Systems

Physical effects can have a major influence on the shapes and forms, the development, and the behavior of biological systems. This is because animals, plants, and all living beings adapt to their environment by the means of evolution. This environment, however, is determined by the laws of physics – and these already existed before there were biochemical entities, let alone living beings. In the following, we shall briefly look at four examples, which we discuss in more detail in later chapters, in which physical circumstances require a certain kind of biological system. Such examples will continue

Figure 1.1   The butterfly *Morpho menelaus* captivates with its clear blue color (left). The origin of this coloration from a physical point of view is investigated in this section. When adding alcohol to the wing, it turns green and reverts to blue when the alcohol has evaporated (right). This indicates that the blue is not due to a pigment.

to concern us in the rest of the book, and we will use them to highlight various physical concepts.

1.1.1 The Coloration of Butterfly Wings

The South American butterfly *Morpho menelaus* is noted for its bright blue color (see Figure 1.1). However, blue pigments are extremely rare in nature, and also the blue color of *Morpho menelaus* is not caused by a pigment. We can make this plausible by placing a drop of alcohol on the wing. We see that the color changes to a clear green, which should not happen with a pigment. The effect is reversible. As soon as the alcohol evaporates, the butterfly is blue again. We get another hint on the color's origin if we repeat the same experiment with a drop of water. Here nothing happens; the water pearls off the surface of the wing. This is due to the greater surface tension of water and means that there are very small structures on the butterfly wing scales, which can be filled by alcohol, but not by water. You can read more on surface tension later in Section 7.5.

In order to see these structures, however, we need a microscope, one with a resolution that is better than the wavelength of the light, because the relevant structures are of roughly this size. Figure 1.2 shows a schematic of the nanometer-sized structures found on the cross-section of the scales on a wing of *Morpho menelaus* when imaging it with an electron microscope. In this cross-section, it is noticeable that layers of wing material (cuticular structures) form in highly ordered structures. These layers have a period, which corresponds approximately to a quarter of the wavelength of blue light (blue light has a wavelength of about 480 nanometers [nm], red one of 630 nm). What does this have to do with color?

When light impinges on the different layers, a part of this light is reflected and another part is transmitted. This is like a glass pane where we mainly see the reflected part of the light from inside at night and the transmitted part of the light from outside during the day. How much of the light passes through and how much is reflected depends on the refractive index difference of the two materials, and we will look at this quantitatively in Section 5.2.

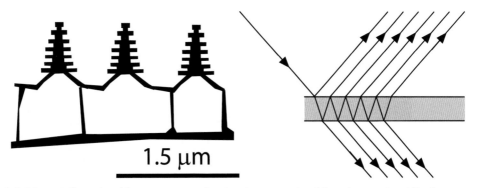

Figure 1.2   Left: Schematic illustration of the nano-structures found on the cross-section of the scales on a wing of *Morpho menelaus*. These terraced structures effectively form several thin layers of chitin approximately 100 nm thick and also roughly 100 nm apart. The reflection of light on these layers forms the secret of the blue color. Right: Light paths reflected on a thin layer have different lengths depending on which boundary they are reflected from. If this difference corresponds to half the wavelength of light, constructive interference occurs.

For a glass–air interface, one obtains that approximately 4% of the light is reflected. As it happens, this is similar on wing scales, since the refractive index of the cuticular structures is approximately equal to that of glass (about 1.5). The situation is shown schematically on the right of Figure 1.2 for one such layer.

The light that passes through the first boundary meets a second boundary at the other end of the layer, where it again can be either be reflected or transmitted. The light reflected there will finally arrive at the same position as the light that was reflected already at the first boundary – that is, at least if it is transmitted again at the upper boundary. Now we have the situation that two beams of light originating from the same place also arrive at the same place, but they have taken different paths for this. In this case, they can do something special due to the wave nature of the light: they can interfere. This means that depending on what the path-length difference between those two beams is, they can either intensify or even cancel out completely. To decide which of these happens, it is important whether the path-length difference is half of a wavelength (where the light beams will cancel out) or an entire wavelength (where the light beams will amplify). Therefore, depending on how thick the layer is (light reflected on the bottom will have passed through the layer twice), light of different colors, or wavelengths, will either cancel out or amplify. This can be seen not only in butterfly scales, but also directly in the coloring of a soap film. Soap films are very thin layers filled with water, and their color actually directly indicates their thickness. As we have said previously, the reflected part of the light is very small for a single layer; we need to stack several layers on top of each other to obtain an intense reflection. This will also lead to a reduction of the reflection for those colors that cancel each other out, and the reflection spectrum (or color dependence) will become increasingly sharp, appearing only at wavelengths, which are about four times the thickness of the layers. We will discuss in detail why this is so in Section 5.3.3. For the thicknesses of the butterfly structures, this is shown in Figure 1.3 for an increasing number of layers (up to 10) and the obtained sharp reflection band corresponds to the bright blue we observe.

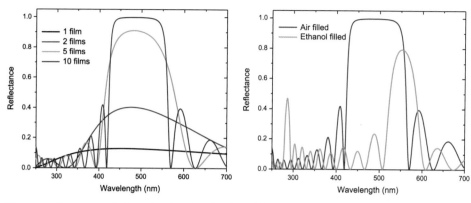

Figure 1.3 When several layers are stacked one after the other, the effect of the individual layer is strengthened (left) (Filmetrics, 2017). This is shown here by the color dependence (or spectrum) of the reflection for different layers chitin with a thickness of 80 nm at a distance of 125 nm. Already with five layers, only a very narrowly limited area of wavelengths is well reflected. These layers and their spacing correspond well to those observed on a butterfly wing scale, and the observed peak in reflection corresponds to a blue color. On the right, a simulation of a stack of 10 layers is shown, once when air is filling the voids between the layers and once when this is filled by ethanol. As can be seen, the peak in reflection shifts in wavelength to around 530 nm, corresponding to green.

But why does the wing turn green when alcohol is added? As previously mentioned, the alcohol wets the nano-structures of the butterfly wing. This means that we no longer have a structure consisting of layers of air and cuticular structures, but one in which alcohol and cuticular structures alternate. Since alcohol has a refractive index of about 1.3, the path difference of the differently reflected light beams changes (it is lengthened relative to the wavelength). This implies that the distance between the layers no longer corresponds to the wavelength of blue light, but to that of green light (see Figure 1.3, on the right).

This principle of coloring is, however, by no means limited to butterflies. Almost all shades of blue in nature and many shades of green are made by such structural effects, e.g., all colors of peacock feathers, the green of the feathers of ducks, the colors of the armor of many beetles or the colors of the scales of many fish, and so on. Due to the fact that the path-length actually changes when the angle of reflection of the light changes, you can easily check which colors are likely to be so iridescent. If the shade of the color changes with viewing angle, the origin of this color is probably due to interference.

It is even possible that animals control their coloring by controlling the corresponding nano-structures themselves. This leads, for example, to the adaptability of lizards, chameleons, and squid, which can be very impressive.

1.1.2 Navigation of Ants

Let us take the desert ant *Cataglyphis* as another example. In an impressive series of experiments over several decades, Rüdiger Wehner from the former Zoological Institute at the University of Zurich has extensively studied the navigation behavior of these insects and described the various navigation mechanisms in detail. Desert ants are well suited as an

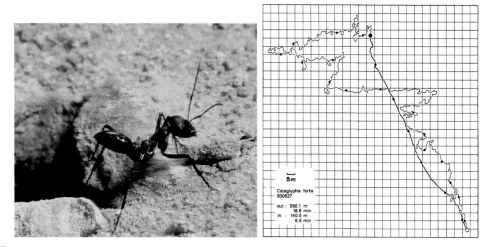

Figure 1.4 Left: The desert ant *Cataglyphis*. Right: The path of a *Cataglyphis* to the food source and back again. Beginning at nest (open circles), the ant looks for food and goes on a tortuous path until it finds the food source (filled circles). From this point onward, it directly moves toward the nest and goes straight back. Figures from Wehner (1982, 2000); Wehner and Wehner (1990), used with permission from EMH Swiss Medical Publishers Ltd and Taylor & Francis Ltd, respectively.

experimental system because, unlike bees, they move on a surface and are easy to observe in their natural habitat (the experiments were carried out in Algeria and Namibia), i.e., their paths can be recorded very well (see Figure 1.4 on the right).

For example, Rüdiger Wehner and his coworkers put "sunglasses" on the ants, which consisted of a resin that predominantly filtered out the blue to ultraviolet light from the spectrum of the white light coming from the sun. These ants with "sunglasses" no longer found their way home. A microscope image of an ant's eye with and without this filter is shown in Figure 1.5. There, it can also be seen that the filters can be removed again, showing that one can examine whether the effect is reversible, i.e., whether the ants find their way home again when the sunglasses are removed (which they do).

The experiments with the ants wearing "sunglasses" are summarized on the right side of Figure 1.5. For very many different ants, the directions of departure after they have found the food source are shown (see also Figure 1.4). The upward direction in the graph represents the direct line to the nest. On the right side, the paths chosen by ants without filters are shown, while on the left side those with filters are shown. As is clearly visible, almost all ants without filters reliably find their way home, whereas most of the ants with filters get lost.

What is so special about blue or ultraviolet (UV) light? Well, the sky is blue (and actually also ultraviolet, which we cannot see), so maybe it has something to do with some pattern in the sky that we normally cannot observe, but the ants can. As can be seen in Figure 1.6, the light from the sky is polarized. This means that the electromagnetic waves that make up the light have a certain direction of vibration, depending on where they come from in the sky. Polarization will be discussed in greater detail in Chapter 5 concerned with optics. At the

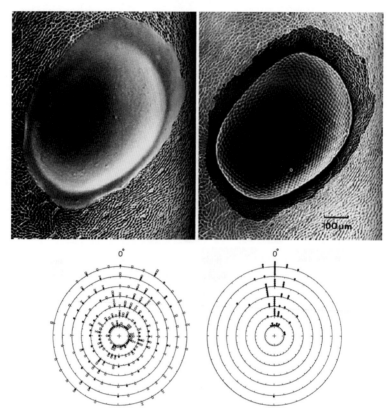

Figure 1.5 Top: Ant's eye with and without "sunglasses." Bottom: Direction of the ants after finding food relative to the direction toward the nest (0 degrees). The ants with "sunglasses" (green) do not find their way home anymore. Figures from Wehner (1982, 2000), used with permission from EMH Swiss Medical Publishers Ltd.

moment, we only need to know that the polarization corresponds to a certain direction and that it does have a certain pattern in the sky. If one makes the experiment in Figure 1.6 at different times, one can see that the polarization pattern changes together with the position of the sun, but is actually constant relative to the sun. Specifically, the sky is always polarized in the direction that makes a right angle to the direction pointing toward the sun.

To find out why this could be so, we carry out an experiment. Since we have no sky and no sun in our laboratory, we illuminate a water tank from one side. We now add scatterers to the water, for instance add a few drops of milk to a bucket, and observe two things. First, the transmitted light becomes reddish and increasingly so with increasing scatterer concentration, and second, the scattered light, i.e., the light at an angle to the transmitted light, becomes blueish. Again this increases with increasing scatterer density. You can try this at home by shining white light into a glass of water with a few drops of milk dissolved in it. If you look at the glass, such that you do not have the light source in sight (preferably it is at a right angle to where you look from), the water appears blue. If you look directly at the light source though the glass, the light appears red or orange. With this, we have

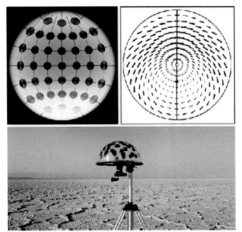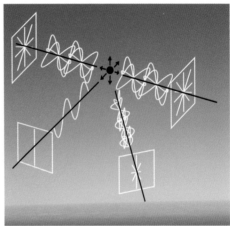

Figure 1.6 Picture of the sky with a fish-eye lens such that the whole sky is imaged (below). At different points, polarizers are attached which can indicate the direction of the polarization (upper left). Top right: The polarization pattern of the sky when the sun is at an angle of about 60 degrees. The length of the bars indicates the degree of polarization, and the direction of the lines is that of the polarization. On the far right: Schematic representation of the polarization of the light by scattering. The light is completely polarized only in the direction perpendicular to the incident light. This leads to the polarization pattern of the light in the sky. Figure on the left from Wehner (1994, 2000), used with permission from EMH Swiss Medical Publishers Ltd.

made an experimental model for the sky in the color of the light that is scattered and at the same time we have a model for the sunset in the transmitted light that looks reddish. The white, incident light contains all the wavelengths of the visible spectrum, but the blue portion, which corresponds to shorter wavelengths, is scattered more strongly, such that the scattered light appears blue to us, just as the sky appears blue. The transmitted light lacks the blue component and therefore appears red. But we actually wanted to explain the polarization pattern with our experiment. For this purpose, take a polaroid filter (or polarizing filter), as you may have in your sunglasses at home, and hold it in front of the direct transmitted light. While the light will become a little dimmer due to the absorption of the filter, there will be no change if you turn the filter at varying angles. The transmitted light is not polarized. If, on the other hand, you hold the filter in front of the scattered light (the one that appears blue), rotating the filter will show large variations in intensity, with the glass becoming dark after turning the filter by 90 degrees. The physical interpretation is that the light that is scattered at right angles must be fully polarized. The same thing happens in the sky. Here, the air molecules (but also dust particles and other impurities) scatter the sunlight. This means that there is light that hits our eyes from other directions than the sun because it has been scattered. If this scattering is at a right angle, then the light reaching us is completely polarized (see Figure 1.6, far right). Therefore, the direction of the polarization is at a right angle to the direction of the sun, as we have seen. It follows that a *Cataglyphis* can read the direction to the sun on any part of the sky if it can determine the polarization of the sky. However, this light needs to be scattered, and this only happens for blue or UV light, i.e., this signal is only available in blue (or UV) light. So if you remove

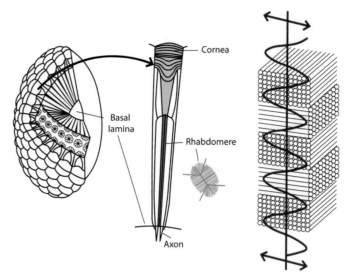

Figure 1.7 The structure of an ant's eye. Inside the facet eye are the visual cells. The light-sensitive part (the rhabdom) has a special nano-structure, which arranges the light-absorbing rhodopsin in two perpendicular directions. This is done in the microvilli (right) forming the rhabdomeres. The mutually perpendicular parts are antagonistically transmitted into the brain so that a polarization measurement is carried out. The size of the structures is adapted to the wavelength of blue to ultraviolet light in order to obtain the largest possible signal by constructive interference.

the blue light (filter it out with "sunglasses"), the *Cataglyphis* is missing the compass and it does not find its way home anymore.

There remains the question how the ants can see this polarization pattern of the sky. We humans cannot do this without technical aids (the Vikings used sunstones to navigate via the polarized sky). For this purpose, we look at the structure of the ant eye more closely, as shown in Figure 1.7. In the facet eye of the ant, the light receptor molecules are present in ordered structures (the microvilli). As a result, the molecules can only be excited by one direction of oscillation of the light, and thus the degree of polarization and its direction can be seen by the eyes. Since, in particular, the blue and UV light shows a polarization, the distance of the arrangement of microvilli actually corresponds precisely to the wavelength of UV light. This implies that only the blue and UV light reaches the receptor molecules. This is precisely the same interference effect as we have just seen in the butterfly wings.

1.1.3 Propagation of Nerve Signals

Whenever we interact with the world, we use our nervous system, be it to hear, see, or use muscles. Signals from our sensory organs are translated to nerve signals and action signals from our brain are transported to the muscles via nerve cells. All of these signals are transported in a fraction of a second. How these electrical signals are transported in the nervous system is something that we will study in Chapter 9. In doing so, we will deal with the conduction of electrical currents in general, but in the course of this, we will see

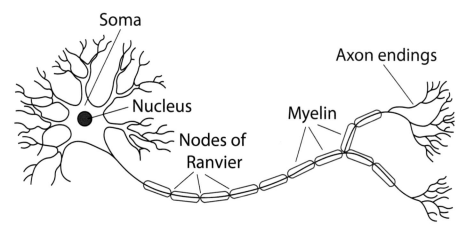

Figure 1.8   Schematic representation of a nerve cell with a myelinated axon for better nerve conduction.

that that if the axons were used as a normal cable, they would not be able to transport the nerve signals with the necessary speed. In fact, it would take several seconds for your brain to send a signal to your hands and reaction times would be very different. In this context, we will look at two different solutions that nature has found for this. One of them is indicated in the schematic structure of a nerve cell shown in Figure 1.8: our treatment of electrical currents will show that a thicker insulation layer around the axon of the nerve cells will result in a much faster transport. This solution is found in many neural systems where the axons are surrounded by a myelin sheath produced by the Schwann cells. The other solution of nature is to use voltage-dependent ion channels for transporting charges in and out of neurons.

1.1.4 Bone Structure and Allometric Scaling Laws

In Section 2.3, we will deal with scaling laws, i.e., the description of how certain properties of an object depend on its size. Such effects are also very frequent in nature and are of great importance in evolution. For example, you may be interested to know how large a land animal can actually become. This depends, among other things, on how strong the bones have to be in order to carry the animal's weight. Since the load-bearing capacity of the bones depends on their cross-sectional area and hence the cross-sectional area of the animal's legs rather than its weight; the thigh bones of larger animals have to become ever more massive in order to carry the load (see Figure 1.9 on the left). The limiting size is then given by the fact that a leg cannot consist solely of bone or cannot be wider than it is long.

On the other hand, size can also be an advantage. The energy consumption of an animal (its metabolic rate) increases less rapidly with the size than the weight. This is due to the fact that a large part of the energy consumption is used to maintain the body temperature since body heat is constantly lost. This heat loss happens through the surface, as we will see in Section 8.3.6 on heat conduction. Thus metabolic rate is proportional to the body surface.

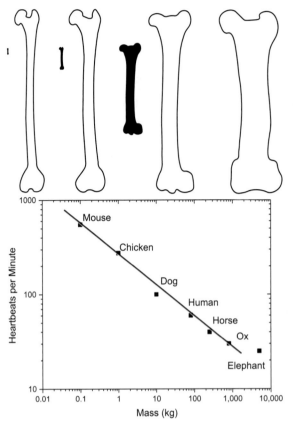

Figure 1.9 Top: Thigh bones of animals of different sizes ranging from mouse (left) to fox, horse, and elephant (right) spanning several orders of magnitude in weight. More massive animals have more massive bones due to the nonlinear increase in weight relative to bone strength. Bottom: The heart rate of different animals depending on their mass. Small animals have a much higher metabolic turnover. Data are from www.msdvetmanual.com/appendixes/reference-guides/resting-heart-rates.

Because the relative energy consumption decreases with size, predominantly large animals are found in the vicinity of the poles. The exceptions, which normally confirm the rule, are then sufficiently well-insulated animals, i.e., those with a relatively thick coat. For fun, one can calculate, for example, that a mouse, which has the same energy consumption per kilogram of body weight as a cow, should have a coat of 20 centimeters (cm) thickness in order to maintain its body temperature in our latitudes. Turning this around, a cow with the same metabolic rate per kilogram as a mouse would have a body temperature in excess of 100°C, i.e., its blood would be boiling.

The fact that the relative energy consumption depends on the size also has the direct effect that the heart rate decreases with increasing mass. After all, the heat lost in metabolic rate comes from the combustion of oxygen that is transported in the blood. This dependence of heart rate on mass can also be empirically verified, as shown in bottom graph in Figure 1.9, where the number of heartbeats per minute is plotted for different animals.

1.2 Physics-Based Methods for Investigating Biological Systems

But physics and physical effects are not only important to understand the properties of biological systems, they are also important in the development of measurement techniques that are widely used in biological research. Hence, let us look at some standard experimental techniques in biology and chemistry and briefly discuss how they depend on physical principles. We will cover some of them in detail in later chapters after having laid the proper foundations of the corresponding physics.

1.2.1 Microscopy

A first major leap in biological research came from the development of the microscope. Only in this way was it at all conceivable to find out that there are microscopically small organisms. Likewise, we would otherwise not know anything about the construction of living beings from different cells or about the construction of cells. Many of these findings are based on the light microscope, but this has a limited spatial resolution due to the wave nature of the light. We will deal extensively with this resolution limit in later chapters of the book. However, in the last 10–15 years, there have been many developments in the field of microscopy, which in turn are about to revolutionize biological investigations. These include, in particular, fluorescence microscopes, which allow smaller structures to be examined, as well as a three-dimensional examination. These are, e.g., confocal microscopes or the STED (STimulated Emission Depletion) microscope (see Figure 1.10).

Further findings, such as the structure of butterfly wings or the ant's eye, as we have seen previously, need microscopes with very high spatial resolution, comparable or even better than STED. Given their longer existence, mainly electron microscopes have been used in the past. We will, however, not deal with these further.

1.2.2 X-Ray Diffraction

Another way to increase the resolution compared to optical microscopy is to use "light" with a much smaller wavelength. This "light" is also called X-rays, whose wavelength is roughly comparable to the size of an atom. The diffraction of waves on objects of similar size will be treated in Chapter 5, on optics, and Section 4.7, on acoustics. When one sends X-rays through ordered matter, such as a crystal, regular patterns of X-ray spots are formed that give information about the order of the object being studied. Mathematically, the pattern of the X-ray reflections is given by a Fourier transformation of the arrangement of the atoms or molecules, but we will treat this only qualitatively. One of the most famous examples of an X-ray diffraction pattern from biology is given in Figure 1.11. This is the diffraction pattern of a DNA taken by Rosalind Franklin and Raymond Gosling, on the basis of which Francis Crick and James Watson recognized the double helix structure and the base-pairing of DNA. For this, it was very helpful that

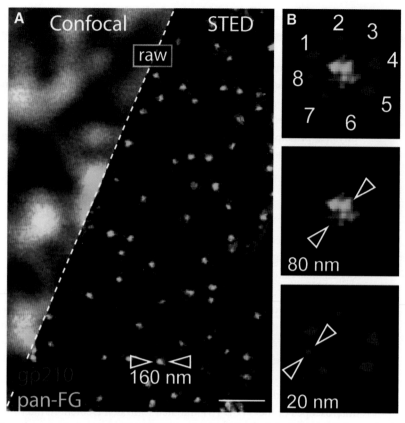

Figure 1.10   Comparison of a confocal microscope with a STED (STimulated Emission Depletion) microscope. Due to the clever selection of the illuminating light, a fluorescent dye is excited only in a very small range. This allows an optical microscopy with extremely good resolution down to 10 nm or less. Figure from Göttfert et al. (2013), used with permission.

Crick became thoroughly acquainted with Fourier transformations in his initial career as a physicist.

X-rays can also be used in a form of microscopy. Here, the large range of X-rays in matter is of advantage and allows one to obtain three-dimensional tomograms of optically opaque materials. As an example, Figure 1.12 shows the X-ray tomogram of the pupa of a fruit fly *Drosophila melanogaster*, with a light image for comparison.

1.2.3 Nuclear Magnetic Resonance

From the study of the magnetic properties of atomic nuclei, the method of nuclear spin resonance has emerged. We shall also deal with this later. It has been shown that the chemical properties of the atoms have a slight influence on the magnetic properties of the nuclei, which can be measured very accurately. It is thus possible to investigate the chemical properties of substances by measuring the nuclear spin resonance. At the

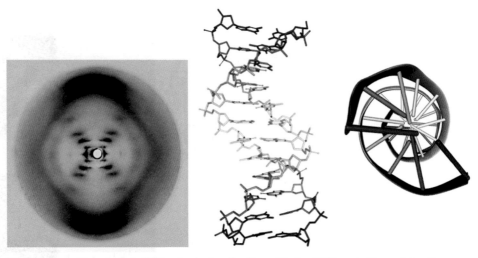

Figure 1.11 Left: Diffraction pattern of oriented DNA molecules from Franklin and Gosling (1953), used with permission. The characteristic arrangement of the diffraction in an X allows the identification of the structure with a helix. Further details of the diffraction pattern (such as a missing spot on the X) led Watson and Crick to conclude that the structure of B-DNA consists of a double helix (right). Data from Drew et al. (1981).

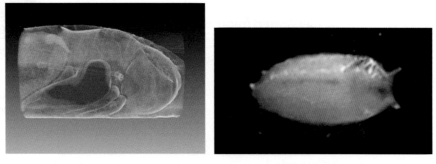

Figure 1.12 Left: X-ray tomogram of a *Drosophila* in the pupal stage. Right: To compare, the optical image of such a pupa.

same time, the chemical properties and bonds determine the structure of molecules, with the result that nuclear spin resonance is also used to study the molecular structure of, e.g., proteins that cannot be crystallized and therefore cannot be examined with X-ray diffraction (see the preceding section). An example of the thus determined structure of a cytochrome is given in Figure 1.13 on the left.

However, nuclear spin resonance can be used not only with molecular resolution, but also on the millimeter scale. In this case, use is made of MRI, which allows three-dimensional, noninvasive imaging. The basic principle of imaging, however, is different from that used for molecular structure determination. While we will not be able to discuss structure determination in detail, we will be discussing MRI as well as the fundamental principles of magnetic resonance in Section 10.4.

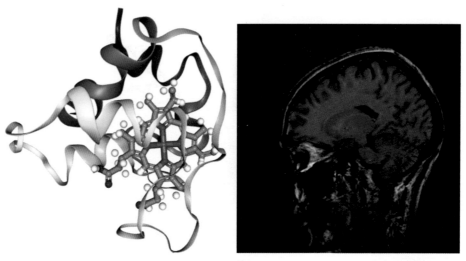

Figure 1.13   Left: Reconstruction of a cytochrome from nuclear spin resonance spectroscopy. Data from Detlefsen et al. (1991). Right: Section through a human head in a magnetic resonance image.

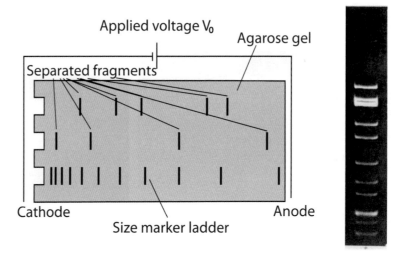

Figure 1.14   Schematic setup for gel electrophoresis and a typical electrophoresis gel. The labeled bands correspond to different lengths of pieces of DNA.

1.2.4 Electrophoresis

Electrophoresis is an integral part of today's molecular biology research. It allows to separate certain sections of genes, depending on how large they are (see Figure 1.14). Likewise, proteins can be separated by electrophoresis because of their size. The method is based on two physical properties of the molecules investigated. On the one hand, their electrical charge, which ensures that the molecules are transported through the agarose gel by an applied electric field and, on the other hand, the mobility of these long-chain molecules in the disordered gel. To describe this mobility, it is important to understand how long-chain molecules such as DNA or denatured proteins behave. We will look at this in detail when we study the thermal movement of molecules, and we will see in Section 8.5.5 that the mobility of the molecules is strongly influenced by this. This will allow us to consider how the separation of molecules works and where its limits lie.

1.2.5 Mass Spectrometry

Like electrophoresis, mass spectrometry is a technique for the separation of molecules or molecular components by their size (or mass). Especially for proteomics, mass spectrometry has become a standard tool in the last 10–15 years. The molecules are first converted into charged ions in an ion source and then accelerated in an electric field before they pass through a homogeneous magnetic field perpendicular to $\vec{v}$. Due to the Lorentz force (see Section 10.2.3), such a particle will make a circular motion in the magnetic field with a radius solely given by mass to charge ratio. Hence, depending on the mass of the particle, it will exit the magnetic field at different distances from the source. Moving a detector along this distance (see Figure 1.15) and measuring the particle flux as a function of position (twice the radius of curvature) then gives a mass spectrum of the substance which is analyzed and hence yields insights into its structure and composition (see Figure 1.16).

There are, however, many different types of mass spectrometers with different principles. Let us now discuss a second example: charged particles carry out a circular orbit with a very definite frequency in a magnetic field, the cyclotron frequency $\omega = \frac{q}{m}B$, so when B is known, this frequency directly gives the inverse mass of the particle. The movement of such a charged particle corresponds to an electric current that can be measured. This current oscillates correspondingly at the same frequency, that is, to determine the distribution of the masses, one must determine which part of the measured oscillation is at which frequency.

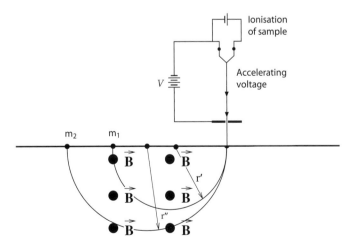

Figure 1.15   Schematic design of a mass spectrometer.

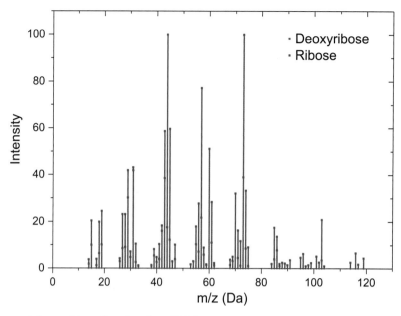

Figure 1.16 Mass-Spectra of ribose and deoxyribose. Data from SDBS, 2017.

This can be done mathematically with a so-called Fourier analysis of the signal in order to determine the different frequencies. We will deal with this in the description of waves.

1.3 Acting and Thinking Like a Physicist: How to Gain Scientific Knowledge

On the way to gaining knowledge, **the observation, the phenomenon** in nature, always comes first. Every thunder crack is preceded by a flash, the rainbow is only visible with the sun in the back, blue butterflies are green when alcohol comes to the wing, and so on. The general method of science can maybe be summarized as follows:

Observations of nature lead to concrete experiments that reproduce and at the same time abstract the phenomenon. For this, models are created that determine the level of abstraction and thus the essential parameters of the phenomenon to be explained. The concrete experiment consists of a quantitative measurement, which faithfully reproduces one or more properties of the phenomenon. To realize an experiment, models must be created that describe the abstract phenomenon. Different models are combined into a larger theory. These theories make new predictions, which can be experimentally tested (falsified).

Let's now look at each of the steps in this process in detail.

1.3.1　Making an Abstract Image of the World

Consider any phenomenon is that is interesting enough to be studied more deeply. For this purpose, one has to try to get to the bottom of the nature of this phenomenon. However, to do this, it may be necessary to distance oneself very far from the actual phenomenon. Only this makes it possible to have a controllable, abstract, or simpler version of the phenomenon, where repeatable and quantitative experiments can be performed. If, for example, we want to investigate the phenomenon of the rainbow, we should be able to get rid of the weather conditions as far as possible. In the end, we do not want to be dependent on the whims of the weather when studying the phenomenon. In other words, we will investigate a simplified, abstract version of a rainbow when we send white light through a single, large "water drop," i.e., a water-filled spherical container. While one might think that this has little to do with a rainbow, we have actually created a model system where we can carry out experiments and that is simple enough to be described mathematically. We do see, however, that in such an experimental situation we do observe what looks like a rainbow and we can control the relevant parameters much better in our model system (see Figure 1.17). You should be aware that the term "model system" is used somewhat differently in physics when compared to biology. The essential property of a physical model system is that it only contains the relevant parts of the phenomenon and therefore discards all the unimportant parts. Thus, the model rainbow is not an actual rainbow (for example, it does not need to be raining in order for us to observe it), but it describes the intensity distribution, color sequence, angular size, and polarization of the phenomenon and it does so in quantitative measurements. By the way, the abstraction of the rainbow that we have just seen belongs to one of the earliest surviving investigations of this method of modern science. The monk Dietrich von Freiberg made this experiment a little more than 700 years ago!

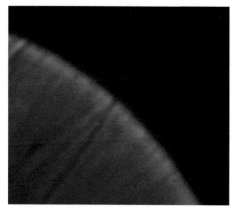

Figure 1.17　Rainbows in the field and in the lab. The color, size, and polarization of the rainbow in the lab is identical to the one in the field.

1.3.2 Quantifying Predictions and Measurements

The main reason for the preceding abstraction is that you can do experiments that are easy (or at least possible) to control and where you can make clear measurements where you know their uncertainties. Only in the case of a quantitative measurement can you actually determine whether one quantity depends on another one and whether this dependence is relevant or whether it might as well not be present given the uncertainty. Without the initial abstraction of the problem, it would be difficult to find a connection at all, since under certain circumstances too many parameters *could* influence the measurement or the phenomenon and thus the uncertainty of the measurement could not be determined. But if one has restricted oneself to the essential parts of the problem and determines their dependencies quantitatively, one finds laws that are unchanged when replacing unimportant variables. Since these laws are based on quantitative measurements, they can be characterized by functional dependencies, that is, by mathematical functions, and hence we obtain mathematical laws of nature. For example, the metabolic rate of animals always depends on their mass regardless of which species or genus they belong to. Acceleration of a mass by gravity on earth is always given by Newton's gravitational law (a constant divided by the distance to the center of the earth squared), whether the mass is an apple on a tree or the moon. Likewise, the pressure of a gas at the same temperature is always inversely proportional to its volume, i.e., it is described by Boyle's law or the ideal gas law. This is the case whether we consider oxygen, nitrogen, or helium. This also means that if we know these mathematical relationships between variables, we can also use them to make further predictions, or to see how other combinations of parameters influence a result. Therefore, mathematics, or more precisely the description of the functional dependencies of the various parameters, is an enormously important tool in the natural sciences.

1.3.3 Models and Hypotheses

In certain cases, a quantitative description of an abstract property already constitutes a law of nature. One example is Newton's law of gravity or the conservation of energy. Frequently, however, one also wants to understand a mechanism for how such a quantitative relationship between abstract properties comes about. To this end, hypotheses are combined with previous findings to make a model, which is intended to describe the experimental facts due to underlying processes. An example of such a model is the statistical description of a gas and the corresponding interpretation of thermodynamic properties, such as the temperature, from the distribution of the velocities of the atoms or molecules. In this way, the previous ideal gas law can be established microscopically and we "understand" the law by virtue of the mechanism that we have put forth to obtain it.

The necessary hypotheses for such a model in turn are predictions, which should be experimentally verifiable. On the other hand, a good model not only makes a prediction for exactly those experimental facts we want to describe, but also for a number of other experiments, and that number ideally is as large as possible. These may have already been executed or they may not. In any case, what we would like is for such models to coherently and consistently describe different phenomena. To take up the previous example of the ideal gas law, the description of the temperature from the statistical properties of the atoms contains not only the laws of gases, but the whole of thermodynamics, i.e., the nature of heat, entropy, and irreversibility. Thus, the laws of thermodynamics, in particular the concept of entropy, can also be obtained from a statistical point of view, which was championed by Boltzmann and which we will look at in detail later on. In addition, this description provides a mechanism for diffusion (i.e., Brownian motion) and chemical kinetics. Thus a model for one set of observations can have very far-reaching consequences in other places.

In biology, Darwin's theory of evolution presents such an example. What Darwin wanted to describe was a mechanism for the observation of species formation. This mechanism is variation through mutation, heredity of this variation, and finally selection. Various hypotheses were necessary for this mechanism. On the one hand, Darwin had to postulate how traits are inherited, as well as a sufficient variation through mutations. He postulated a blending of the characteristics as a hereditary mechanism, which would reduce the variation and therefore would actually not be compatible with selection later on. After a few generations, there would be no traits left to vary and select on. A different mechanism of heredity was described by Gregor Mendel at about the same time, finding the inheritance of two copies of discrete features with dominant and recessive traits. This mechanism agreed with Mendel's quantitative determinations of the frequency of different traits in experiments with peas. This mechanism of heredity allows for a continuation of variation and therefore for selection at later stages, but remained unknown to Darwin. However, in spite of the fact that the proposed mechanism for inheritance in Darwin's version is wrong, the great merit of the theory of evolution is that it recognizes the importance of heredity and variation at all. Also, evolution actually shows that a blending inheritance of the traits must be false, since otherwise evolution and speciation would not be possible. In addition, since the selection does not take place at the level of heredity, evolution also has far-reaching consequences for developmental biology, since the development must be controlled in such a way that an individual capable of selection emerges with the inherited properties. In this development, nothing should go wrong and the changes in the development plan must be evolutionarily stable. The far-reaching influence of evolution in biology goes so far that Theodosius Dobzhansky said, "Nothing in Biology makes sense, except in the light of evolution." Thus evolution is a really important theory. However, the influence goes even further – the theory of evolution implies the hypothesis that the earth must have a certain age for the mechanism to function at all, which has given it a great influence on the development of geology, for instance. We will return to this in Section 8.3.6.

1.3.4 Falsification

We have just seen that some aspects of Darwin's proposed mechanism for evolution were actually wrong. This is not a problem at all and does not single out this theory. Any theory is actually wrong. Scientifically, theories and their predictions can only be refuted, or **falsified**. If an observation does not match with the prediction of a theory, that theory must be false. In order to draw such a conclusion, however, the prediction as well as the measurement must be quantitative, otherwise they cannot be compared with one another. In addition, the basic assumptions of the model behind the theory should be explicit, since the error is probably there and knowing the assumptions can therefore lead the way to a more correct notion. If, however, the prediction is correct, this can only be determined within the accuracy of the measurement. It could be that, on closer scrutiny, with better precision, the prediction deviates from the observation. Or it could be that the prediction coincides with the experiment only accidentally.

> Theories can therefore always only be falsified and never verified. A scientific proof is a contradiction in terms!

The fact that theories can only be wrong in principle is not to say that all theories are equally wrong. Take, for example, the two assertions (or theories): "The earth is a disk," and "The earth is a sphere." Both are wrong! But the disc-shaped earth certainly corresponds much less to reality than the spherical one. But it is not just how much a theory is quantitatively wrong. A theory that has survived many attempts at falsification before a special case has been found to be an exception, and thus has been disproved, is certainly much more trustworthy than one that has been directly and repeatedly rebutted. If a theory has not been checked very often, you need other criteria for its trustworthiness. These criteria are also given by our idea of the world that it is (hopefully) described by simple and few laws. Thus, for example, if two theories describe the same facts and both have not (yet) been falsified, it is advisable to prefer the one whose basic principles are simpler (or which has fewer basic assumptions). This is also known as Occam's razor and is a good rule of thumb for judging different theories. In the end, with fewer basic assumptions, the probability that one of these foundations is falsified is smaller. This does not, however, mean that the simple solution is always the right one. Too simple a solution typically quickly shows its flaws and then there actually is a difference to be made between the theories. Occam's razor is only valid when the competing theories give the same results for what has been actually measured. Finally, in addition to fewer basic assumptions of a given theory, one would like to have as few natural laws as possible and a theory covering as large a range of phenomena as possible, therefore has a preference.

In summary, we have greater trust in a theory under the following conditions:

- The more often it makes predictions that are correct (or more precisely, not wrong) within the uncertainty of the measurement
- The more wide-ranging the subjects are, where the theory makes correct predictions
- The simpler its basic assumptions are

However, as stated previously, the latter two (and especially the last one) of these points are characterized by human prejudice. However, in the course of the history of science, they have proved to be not too wrong, so they can be used as faithful guides and rules of thumb, at least in physics.

1.3.5 Reproducibility

Up to now, we have tacitly assumed an essential property of science as given, which, however, is anything but trivial: **reproducibility**. This means that all scientific results and conclusions should be independent of the person making the observation or drawing the conclusion. Any scientist who carries out the same experiment or calculates the same theoretical derivation should come to the same conclusion. Not only other scientists should obtain the same result in repeated observation, but you should as well.

If experiments are not reproducible, this is a very good indication that the system has not been understood well enough and the abstraction of the phenomenon has to be changed because essential parts are missing. In this context, it is important to know what it means that the same result occurs. Since measurements always have an uncertainty, a measurement can, of course, only be reproduced within this uncertainty. Therefore, it is extremely important that measurements are always given with their uncertainty, otherwise it is not possible for other scientists to check the reproducibility, and the measurement is actually scientifically meaningless. How these uncertainties are assessed will be discussed in detail in the next chapter.

1.3.6 Approximations and Uncertainties

In order to decide whether a statement is falsified, its prediction must be incompatible with an experiment. However, there are no infinitely accurate measurements; they are always fraught with uncertainty (also called errors). On the other hand, also the predictions are almost always uncertain. After all, the predictions are based on previous measurements of

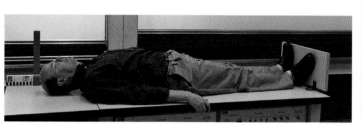

Figure 1.18 Measuring one's height while laying down gives a different result from when standing up. The difference of about 2 cm can be guessed at from the uncertainties of different types of height measurements.

natural constants, which are also error-prone. So if we make predictions from a theory, we do not have to perform the corresponding calculations with absolute accuracy. It makes no sense to calculate the prediction more precisely than its uncertainty, just as it makes no sense to specify the result of a measurement more precisely than the corresponding uncertainty. If we know what the uncertainties are, be it in the measurement, or in the quantities going into our calculation, we can often be a lot quicker and make quantitative estimates rather than accurate calculations. As long as we know the error that we make in such an estimate, we can still compare this prediction very well with the experiment and decide whether the theory is meaningful or not.

In order to be able to make such estimates, we must know something about the accuracy of the measurement we perform. As an example, consider the assertion that people are shorter when standing than when laying down. Such lengths are difficult to determine with a better accuracy than about a millimeter. On the other hand, if the effect were very large (e.g., 10 cm or larger), this would be visible to the naked eye. Therefore, since there is the assertion, but you can't see it with the naked eye, we expect the difference in lengths to be somewhere between the limits of these accuracies, i.e., somewhere between a millimeter and 10 cm. Given this range, a reasonable estimate is the geometric mean, so that we expect a difference of about 1 cm. When we actually measure this difference experimentally, we find a difference of about 2 cm, so the estimate based on the accuracies of different measurements is not too bad (see Figure 1.18).

In addition, we need to know something about the law we are investigating. When we investigate an exponential process, we can allow for fewer errors than when we are

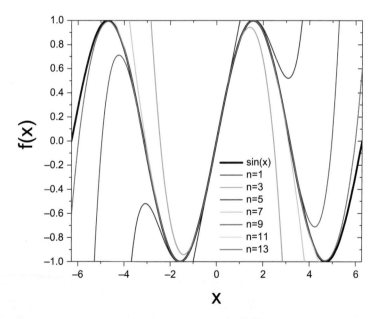

Figure 1.19 Successive approximations of the sine using a Taylor expansion. Apart from the sine itself, the first seven terms of the polynomial are shown. The more terms that are taken into account, the wider the range of applicability of the approximation. Even the first term only gives a reasonable approximation for angles smaller than 30 degrees.

investigating a linear process. This means we have to look at how the accuracy in a parameter of a function determines the accuracy of the function itself. This is known as error propagation, and we will deal with this in detail in the next chapter.

Mathematically, the most important tool here is the expansion of a function as a polynomial. In a small enough domain, we can describe every function as a linear function, thus making the extrapolation of the uncertainty for each dependency analogous to that of a linear function. As an example, the successive expansion of the sine is shown in Figure 1.19 in ever-higher terms of its Taylor polynomial.

Quiz Questions

1.1   Scientific statements

Which of the following qualifies as a scientific statement (i.e., is falsifiable)?

A Humans will never be able to set foot on the moon.
B Some laws that describe nature cannot be found scientifically.
C In a different galaxy, there may be fundamentally different laws of nature at play.
D The earth is actually only 4,000 years old. Any other estimates of its age are due to the fact that the initial conditions were set such that a different result is obtained.

1.2   False statements

All of the following statements are false (have been falsified). However, some are more wrong than others. Which ones are the most wrong?

A Inheritance follows Mendel's laws.
B The earth is a sphere.
C The earth is a flat plate.
D The active substance in a homeopathic remedy becomes more potent the more strongly diluted it is.
E The resulting acceleration of an object is given by the total applied force divided by the mass of the object.
F The total mass of substances in a chemical reaction remains constant.

1.3   Taylor expansion

What is the first-order Taylor expansion of a bell-curve $\exp(-x^2)$ around zero? In other words, how can you approximate the function when x is small and you only consider linear changes from zero?

A 1
B $1 + x$
C x
D $1 - x$
E $1 - x^2/2$

1.4 Taylor expansion 2

What is the first-order Taylor expansion of $ln(1 + x)$ around zero? In other words, how can you approximate the function when x is small and you only consider linear changes from zero?

A x

B $1 + x$

C $1/(1 + x)$

D $1 - x$

E $1 + x/2$

2   Errors, Units, and Scaling Laws

As we have indicated in the introductory chapter, one of the most fundamental things one does in a quantitative science is to perform measurements and obtain functional relationships between variables in this way. Here, we will look at all the steps involved in this properly. Starting with how measurements are necessarily uncertain and how these uncertainties are quantified, we will go on to see that all measurements have to be done for physical quantities, which have units by which they are measured. This is not only important in measuring these quantities, but can also be of tremendous help in obtaining physical relationships between variables, since in the end, a single quantity can only have a single type of unit associated with it. Therefore, different ways of describing the same quantity have to give this same unit. This is known as dimensional analysis and will be treated in detail in Section 2.2.2. Finally, the functional relationships one finds between natural quantities are typically given by power laws. How these are specified and derived, we will see in Section 2.3 on several biological examples.

2.1   Uncertainties and Error Propagation

As an example for a measurement, consider Figure 2.1. The length of a board is measured with a meter stick. To get the length of the board, we have to estimate it given the readings on the meter stick. The more closely spaced they are, the more precisely we can read the length of the board. Basically, if we take the closest reading to the actual length of the board as our measurement, the value we obtain changes with the type of meter stick used.

The result of such a measurement is given by the number read from the scale as well as the corresponding unit: For the example of the board, using 10 cm marks (the middle photo in Figure 2.1) this gives the following:

$$x = 0.6 \text{ m}$$

As we have indicated, this is not the actual length of the board. The result only indicates an approximation of the length. We do not actually know the real length. The difference between the actual length and the result of the reading of the scale is called **error of the individual measurement**. We will also denote this by the Greek letter sigma, σ, in the

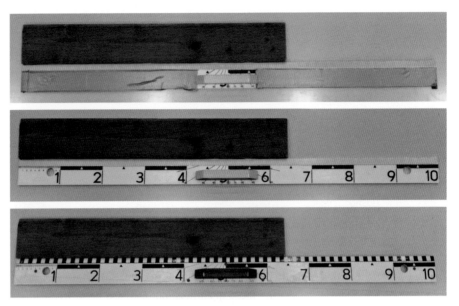

Figure 2.1   Measuring the length of a board with different rulers.

following. In the example of the length of the board, this error will be roughly given by a few centimeters, given that the scale has marks at a 10 cm interval. So while we can estimate the length and its uncertainty with some confidence, we do not actually know either the exact error or the true value.

How to obtain an **estimate** of the true value and its measurement error given the measurement apparatus as well as a series of measurements is the object of error calculation. The probable measurement error can be estimated from the nature of the scale as we have seen in the particular example of the division in the scale, as well as the **scattering from many measured values**, if the scale is not the limiting factor. If a measurement is repeated several times, the mean value of all measured values $\langle x \rangle$ is certainly a better estimate than a single measurement would be.

A proper result of a measurement must always indicate an estimated measurement uncertainty:

$$x = (64 \pm 2) \text{ cm}$$

or

$$x = 64(2) \text{ cm}$$

In this second representation, the number in brackets corresponds to the uncertainty in the last significant digit of the measurement result. This representation is particularly useful because it naturally limits the precision of the quoted value to its uncertainty.

There are no absolutely accurate measurements – in no science. Each measurement is subject to uncertainty. A measurement without errors makes no sense. Likewise, stating the measured value to a higher precision than the uncertainty makes no sense.

2.1.1 Systematic and Statistical Uncertainties

When we look at how we measured the length of a board, we could get an intuitive feel for a fundamental distinction between two different types of measurement errors. Depending on how we use the scale and what the scale actually is, we get very different results on repeated measurement. This is summarized by the following distinction between systematic and statistic uncertainties:

- **Systematic measurement errors** are those parts of the error that remain the same under repeated measurement. That is, the deviation is the same every time, both in size and direction (sign). In the preceding example, reading the scale such that we always take the closest mark as the measured value gives a large systematic error if the density of marks is too low, i.e., we would read 60 cm with 10 cm marks on all repeated measurements. Other systematic errors are reading a scale from an incorrect angle, i.e., the parallax error, or an incorrectly calibrated scale, where a meter stick is not actually a meter long. The better you know and understand your system, the smaller the systematic error becomes. For instance, a calibration of the scale before a measurement ensures that there is no systematic calibration error. Reading the scale from the proper angle ensures that there is no parallax error, etc. If one knows their cause, systematic measurement errors can in principle be avoided or corrected. However, this correction can never be determined completely due to various uncertainties. For instance, even if you have calibrated your scale before a measurement, the scale to which you calibrated may have a small error or in the time between the calibration or the measurement settings may have changed a little. Therefore, there is always an *uncertainty* associated with systematic errors that cannot be reduced by more frequent measurement, but only by a more careful or clever measurement setup. The smaller a systematic error is, the more accurate a measurement is.
- **Statistical errors**, on the other hand, are due to inherent fluctuations in measuring variables. Repeated measurements under identical conditions do not always produce the same measurement value. For example, if we measure a distance with a very accurate scale, we do not always position the zero at exactly the same location, but at random positions in close vicinity of actual zero. Similarly, if a lot of observers measure the run time of Usain Bolt with stopwatches, each one will obtain a slightly different value as long as the watches indicate with enough precision, say, one-thousandths of a second. The average value of all these measurements, however, becomes more precise the more measurements are performed. One can treat these errors with the methods of

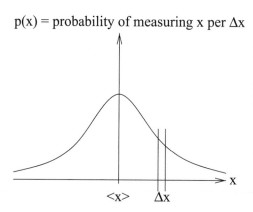

Figure 2.2 Left: Schematic of the probability distribution of a measurement. Right: Histogram of many measurements of beads falling down a pin board leading to a peaked distribution.

statistical mathematics as we will do later in this chapter. Statistical measurement errors are unavoidable. However, with a higher number of measurements, the effects of such statistical errors can be reduced. The smaller the statistical error of a measurement, the more precise it is.

2.1.2 Distribution of Statistical Uncertainties

Let us consider the measurement of some physical quantity, for example, a position x. We call the true value of this position (that we do not know, remember) $\hat{x}$, while the result of a single measurement is called x_i. If the measurement is reasonably performed and able to actually measure the position, the probability of measuring a certain value x_i should be greater the closer this value x_i is to that of $\hat{x}$. Given this reasonable assumption, we can qualitatively draw the probability $w(x)$ of measuring x, indicated in this section.

However, it is often easier and more comparable to experiments to work with the so-called probability density $p(x)$, which is defined such that $p(x)\Delta x$ corresponds to the probability that the measured value lies between the values of x and $x+\Delta x$ (see Figure 2.2). This corresponds to the determination of a histogram from a large number of measurements. In principle (mathematically), we can make this interval arbitrarily small, at which point one writes dx instead of Δx for an infinitesimally short interval. This means that for a continuous probability distribution $w(x)$, the probability density becomes $p(x) = \frac{w(x)}{dx}$. Since a measurement has to yield some result, the sum of all the probabilities must be equal to one. With infinitesimal intervals, this sum becomes an integral and we can write for the probability density:

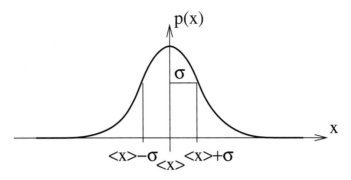

Figure 2.3 A Gaussian bell curve.

$$\int\limits_{-\infty}^{+\infty} p(x) \cdot dx = 1 \tag{2.1}$$

There are a few quantities that we can determine from such a probability density that can be useful in describing the corresponding experiments and measurements. Basically, we will use these in the following to describe uncertainties and measurement values. They are the following:

- **Mean value** $= \langle x \rangle = \int p(x) \cdot x \cdot dx$
- **Variance** $= \sigma^2 = \int p(x) \cdot (x - \langle x \rangle)^2 \cdot dx = \langle x^2 \rangle - \langle x \rangle^2$
- **Standard deviation** $= \sigma = \sqrt{\sigma^2} =$ mean error of a single measurement

These quantities actually fully describe the probability density if $p(x)$ is a Gaussian bell curve (see Figure 2.3), which we will see shortly is commonly observed for all kinds of randomly fluctuating quantities. In that case also, $\langle x \rangle = \hat{x}$. This Gaussian distribution is given by the following:

$$p(x) = \frac{1}{\sqrt{2\pi\sigma^2}} \cdot e^{-\frac{(x-\langle x \rangle)^2}{2\sigma^2}} \tag{2.2}$$

The full width at half the height (FWHM) of this curve is roughly given by FWHM = 2.35 σ, i.e., σ is a good measure for the width of the curve. The area under the curve $p(x)dx$ gives the probability of finding a measurement within the interval dx. If we use a finite interval as opposed to the infinitesimally small dx, we find, for instance, that 68% of the measurements will fall within the range of $[\langle x \rangle - \sigma, \langle x \rangle + \sigma]$. In an interval $[\langle x \rangle - 2\sigma, \langle x \rangle + 2\sigma]$, the so-called 2σ interval, we'll find about 95% of all measurements and in the 3σ interval, there are 99.7% of all the measurements. The probability of being outside of such an interval is usually called the p-value. In the biomedical literature, a p-value of 0.05 is usually used to denote whether a result is significantly different from the expectation. This thus corresponds to a measurement falling outside of the 2σ interval. Given these numbers, it makes sense to use the standard deviation σ as a measure of the error of the individual measurement. We'll make this more quantitative in the next section.

2.1.3 Estimation of the True Values

As mentioned, in practice neither $\hat{x}$ nor σ are known beforehand. Actually, the probability density $p(x)$ is not really known beforehand. We thus have to try to estimate them using measurements that we have made. From a large number of measurements, we can form a histogram of all the obtained values, which gives an approximation of the probability density for discrete, finite intervals. From this, one can obtain discrete versions of the previous definitions of the mean and the standard deviation (using sums instead of integrals), which in turn we can use as estimates for the true value and the deviation from this true value. Given a series of individual measurements $x_1, x_2, \ldots, x_n$, which vary only statistically, we thus obtain the following:

- Estimate the true value $\hat{x}$ from the average of the single measurements:

$$\langle x \rangle = \frac{1}{n} \sum_{i=1}^{n} x_i \tag{2.3}$$

- Estimate the deviation from the true value via the standard deviation:

$$\sigma = \sqrt{\frac{1}{n-1} \sum_{i=1}^{n} (x_i - \langle x \rangle)^2} \tag{2.4}$$

Sometimes the so-called root mean square (RMS) is used as a rough estimate for σ, which for a large number of measurements n is very close to the estimated standard deviation:

$$r.m.s. = \sqrt{\frac{1}{n} \sum_{i=1}^{n} (x_i - \langle x \rangle)^2} \tag{2.5}$$

The main difference of the standard deviation and the RMS can be seen when considering a single measurement. In that case, the average of the measurements is the measurement itself, such that the RMS turns out to be zero. However, if we have only carried out a single measurement, we have no way of knowing the variability of this measurement and the error is certainly not zero. Thus the RMS in this case cannot represent a reasonable estimate of the error. This is, however, the case for the standard deviation, since for a single measurement $n - 1$ is also zero and the standard deviation thus yields $\frac{0}{0}$, which in fact can yield any value (mathematically speaking, the value is not defined) just as is the case for the variability, which can have any value if we have only carried out a single measurement.

The more measurements one makes, the better these estimates should become, i.e., for $n \to \infty$, $\langle x \rangle \to \hat{x}$.

But how fast do we improve the accuracy of the measurement? If we make several series of measurements, with n measurements each, we can determine $\langle x \rangle$ and σ for every one of these and we will get different values for each of them. Given all of these different averages and standard deviations, we can use a crucial piece of statistical mathematics, the central limit theorem, to learn more. The central limit theorem tells us that the distribution of these estimated mean values, $\langle x \rangle$, always approaches a Gaussian distribution for a large number of measurements, with a mean value of $\hat{x}$ and a variance of σ^2/n. The amazing thing about this is that this is also true if the distribution of individual measurements x_i is not Gaussian. This theorem thus allows us to estimate the error of the estimated mean value, where we obtain $m = \sigma/\sqrt{n}$.

Instead of "estimated error," we often also say **uncertainty**.

To make a long story short, if we determine a quantity with a sequence of n measurements with the results x_i we can give the estimates for the true value as well as the statistical error the following values:

Result of a measurement:

$$\langle x \rangle \pm m \tag{2.6}$$

Here $\langle x \rangle$ is the mean of the measurements, i.e., the estimation of the true value obtained from the measurements x_i

$$\langle x \rangle = \frac{1}{n} \sum_{i=1}^{n} x_i \tag{2.7}$$

and m is the error of this mean, obtained via a determination of the standard deviation of the measurements x_i

$$m = \sigma/\sqrt{n} = \sqrt{\frac{1}{n(n-1)} \sum_{i=1}^{n} (x_i - \langle x \rangle)^2} \tag{2.8}$$

An error should only be quoted with *one*, at the very most two, significant digits. *The level of accuracy, i.e., the number of significant digits has to be the same in the error and the value of the measurement.*

It makes no sense to specify the measured value more precisely than its uncertainty! In order to force this in a natural notation, one can indicate the error within brackets behind the last significant digit of the result. So, for example, $x = 1.573 \pm 0.004$ is written as $x = 1.573(4)$. In this way, there is no incentive to give more significant digits than the error allows for.

While the error m of the average value is getting smaller with a higher number of samples, the value of the standard deviation σ remains the same. The probability distribution of the measured values does not change with an increased number of measurements! It is just that we can determine the average with higher precision when we have more statistics.

It is often useful to give the **relative**, estimated error r, which is defined by the ratio of the absolute error and the mean value: $r = \frac{m}{\langle x \rangle}$.

For example, if we measure a length with the result of $x = 100(1)$ cm, i.e., we have an absolute error of $m = 1$ cm, then we have a relative error of $r = 1/100 = 0.01$ or otherwise put 1%. We will see why the relative error is often more useful when we treat error propagation in the next section.

When plotting data, we can indicate the uncertainty of a measured value by error bars, which are either vertical or horizontal lines emanating from the mean value, where the length of the bar away from the mean corresponds to the uncertainty or the error.

2.1.4 Error Propagation

Very often, however, there is no direct measuring stick for the quantity we want to measure. What we have to do then is to measure several other quantities (for which a scale exists) and then to calculate the value of the quantity of interest from these other measurements. To determine how the errors of these measurements determine the error of the quantity of interest is the objective and purpose of error propagation. To see how this works and give a feeling of where the final recipe comes from, we will consider two examples in which we measure two quantities to determine another, derived quantity.

As a first example, we again consider the measurement of the length of a board. When we measured the length of the board, we pushed the beginning of the board to the zero line of the scale. This actually also constitutes a measurement. In general, we always measure the length of a section via the measurement of the start and end positions, L_1 and L_2. The length of the line is then given by $\Delta L = L_2 - L_1$. Both position measurements have a probability distribution, $p_{L_1}(x)$ and $p_{L_2}(x)$. It is obvious that the probability distribution of ΔL must depend on these two distributions, but first we have to think about how exactly they have to be combined. If we measure the length of the section, i.e., the difference in positions, then our variable is going to be $x = \Delta L$, then we must have measured $L_1 = y - x$ and $L_2 = y$, where y can in principle be any position. The probability of having made these two measurements is given by the product of the probabilities of each of these measurements, i.e., $p_{L_1}(y-x) \cdot p_{L_2}(y)$. Since this has to result in a value of $\Delta L = x$ for all values of y, we get the probability distribution of ΔL by summing over all the values for y, i.e., by integrating over y. Therefore,

$$p_{\Delta L}(x) = \int p_{L_1}(y - x) \cdot p_{L_2}(y) dy \qquad (2.9)$$

Mathematically, this is called the convolution of the two distributions p_{L_1} and p_{L_2}. In case $p_{L_1}(x)$ and $p_{L_2}(x)$ are Gaussian distributions, one can directly calculate the result of this convolution. One obtains another Gaussian distribution, where the mean is $\langle \Delta L \rangle = \langle L_2 \rangle - \langle L_1 \angle$ and the standard deviation is as follows:

$$\sigma_{\Delta L}^2 = \sigma_{L_1}^2 + \sigma_{L_2}^2. \qquad (2.10)$$

This means that the errors of the measurements of the initial and end position are added quadratically to give the error of the length of the section (see Figure 2.4).

As a second example, consider the measurement of speed. To measure speed, there is no simple scale for direct measurement. But one can measure a speed by measuring the time, Δt, taken for something to move a certain distance, Δx. Then $v = \frac{\Delta x}{\Delta t}$ is the (average) speed you are looking for. So, if we specify a certain distance, which we can determine very precisely, the uncertainty in the measurement of the speed depends only (or mainly) on the uncertainty of the measurement of time. But how does it depend on this? For slow objects,

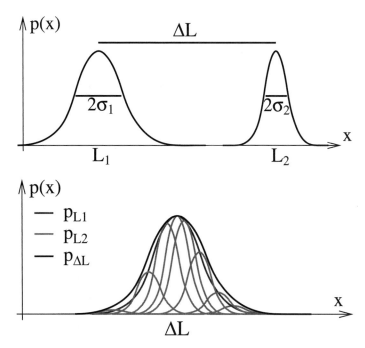

Figure 2.4   Combining probability distributions of two measurements using a convolution.

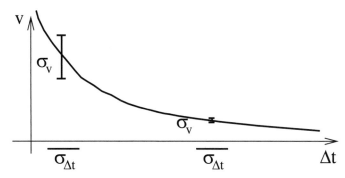

Figure 2.5   Dependence of the measured speed on the flight time for a given distance. For fast objects, the required accuracy in timing becomes increasingly higher.

such as a model plane, you can determine a speed by setting a distance and measuring the time with a stopwatch, i.e., a relatively poor accuracy in time is sufficient for acceptable accuracy in speed. For fast projectiles, such as a bullet shot from a gun, however, this is not possible, and we need a much better accuracy. This can be seen graphically in the representation in Figure 2.5. Here the speed v is shown as a function of the measured time Δt, given a fixed distance Δx.

If we have an uncertainty in Δt, which is an error $\sigma_{\Delta t}$, this corresponds to an uncertainty in v, i.e., an error σ_v. We can read this from the graph when we look at the difference in

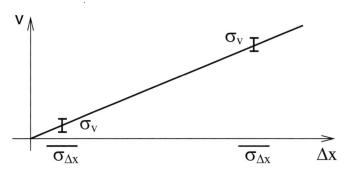

Figure 2.6 Dependence of the measured speed on the distance for a given flight time. Here the uncertainty in the speed is dependent only on the time interval.

v that comes from having different times. Formally written, this becomes $\sigma_v = v(\Delta t + \sigma_{\Delta t}) - v(\Delta t - \sigma_{\Delta t})$. As long as these differences are small enough, we can use a Taylor expansion for v in powers of Δt and only keep the linear term. Then we can write this via the variation (or the partial derivative) of v with Δt:

$$v(\Delta t \pm \sigma_{\Delta t}) \simeq v(\Delta t) \pm \frac{1}{2}\frac{\partial v}{\partial \Delta t}\sigma_{\Delta t} \tag{2.11}$$

If we put this into the relationship for the uncertainty of v, we get the following:

$$\sigma_v = \frac{\partial v}{\partial \Delta t} \cdot \sigma_{\Delta t} = \frac{\Delta x}{\Delta t^2} \cdot \sigma_{\Delta t} \tag{2.12}$$

To be complete in our discussion of the example of a speed measurement, we should also take into account that the distance measurement has an error as well. This error can then also influence the final uncertainty. If we plot the speed as a function of the distance Δx at fixed Δt, we obtain Figure 2.6. Here we can also read off the uncertainty in v given by $\sigma_{\Delta x}$. Since in this case the slope is constant, we obtain the uncertainty of the speed independently of the interval in Δx as follows:

$$\sigma_v = \frac{\partial v}{\partial \Delta x} \cdot \sigma_{\Delta x} = \frac{1}{\Delta t} \cdot \sigma_{\Delta x} \tag{2.13}$$

Both of these contributions must now be added to obtain the final uncertainty, similar to what we have done previously in the measurement of a section. Also, the statistical errors in distance and time will be distributed according to a Gaussian, which means that we have to add their contributions in squares and obtain for the total uncertainty in the speed:

$$\sigma_v^2 = \left(\frac{\partial v}{\partial \Delta x}\right)^2 \cdot \sigma_{\Delta x}^2 + \left(\frac{\partial v}{\partial \Delta t}\right)^2 \cdot \sigma_{\Delta t}^2 \tag{2.14}$$

This is the law of error propagation according to Gauss.

This can also be done for general dependencies of one variable on many others. Given a function $f = f(x_1, x_2, ..., x_n)$ depending on n different quantities, the uncertainty of this function σ_f is given by the errors σ_i of the different quantities x_i via propagation of these errors using the following:

$$\sigma_f^2 = \left(\sigma_{x_1}\frac{\partial f}{\partial x_1}\right)^2 + \left(\sigma_{x_2}\frac{\partial f}{\partial x_2}\right)^2 + \left(\sigma_{x_3}\frac{\partial f}{\partial x_3}\right)^2 + \cdots + \left(\sigma_{x_n}\frac{\partial f}{\partial x_n}\right)^2 \tag{2.15}$$

With this law of Gaussian error propagation, we can also understand the statement that the error of the mean of n measurements is given by $\sigma/\sqrt{n}$ if σ is the error of a single measurement. The average of a quantity $\langle y \rangle$ is given by $\langle y \rangle = \frac{y_1+y_2+\cdots+y_n}{n}$. Thus we have $\frac{\partial \langle y \rangle}{\partial y_i} = \frac{1}{n}$, because the different measurements are independent of each other. If every single measurement has an error of σ, which is true by definition, we obtain with error propagation

$$\sigma_{\langle y \rangle}^2 = \sum_{i=1}^{n}\left(\frac{\partial \langle y \rangle}{\partial y_i}\right)^2 \cdot \sigma^2 = \sum_{i=1}^{n}\left(\frac{\sigma}{n}\right)^2 = \frac{\sigma^2}{n}, \tag{2.16}$$

or just the expression we had simply stated previously as derived from the central limit theorem.

For the example of measuring a speed, we have a function f given by $f = v = \frac{\Delta x}{\Delta t}$. So we need the derivatives for propagating the errors: $\frac{\partial v}{\partial \Delta x} = \frac{1}{\Delta t}$ and $\frac{\partial v}{\partial \Delta t} = -\frac{\Delta x}{\Delta t^2}$. This yields the following:

$$\sigma_v^2 = \left(\frac{1}{\Delta t}\right)^2 \cdot \sigma_{\Delta x}^2 + \left(-\frac{\Delta x}{\Delta t^2}\right)^2 \cdot \sigma_{\Delta t}^2 \tag{2.17}$$

This can now be used to determine the error of the velocity measurement from the measurements of Δx and Δt as well as their respective errors. As noted earlier, it may be convenient to use the relative errors in calculations. This is such a case, and the relationship between the errors of v and those of Δx and Δt will be much easier if we consider relative errors. If we transform the preceding equation by using the fact that $v = \frac{\Delta x}{\Delta t}$, we get the following:

$$\sigma_v^2 = \left(\frac{\Delta x}{\Delta t}\right)^2 \cdot \left(\frac{\sigma_{\Delta x}}{\Delta x}\right)^2 + \left(\frac{\Delta x}{\Delta t}\right)^2 \cdot \left(\frac{\sigma_{\Delta t}}{\Delta t}\right)^2 = v^2\left(\left(\frac{\sigma_{\Delta x}}{\Delta x}\right)^2 + \left(\frac{\sigma_{\Delta t}}{\Delta t}\right)^2\right) \tag{2.18}$$

or by dividing the whole equation by v^2 and inserting the definition of the relative error $r_v = \sigma_v/v$, we finally obtain the following:

$$r_v^2 = r_{\Delta x}^2 + r_{\Delta t}^2 \tag{2.19}$$

In this case, therefore, the *relative* errors add quadratically.

We will see in the next two subsections that physical quantities are very often related to other physical variables in the form of power laws (the speed is an example of this). In such relationships, the relative error is also very useful. Thus, if the quantity f is given by $f(x, y, z) = x^a y^b z^c$, then the derivatives in the error propagation law become $\frac{\partial f}{\partial x} = ax^{a-1}y^b z^c = \frac{af}{x}$, $\frac{\partial f}{\partial y} = bx^a y^{b-1}z^c = \frac{bf}{y}$ and $\frac{\partial f}{\partial z} = cx^a y^b z^{c-1} = \frac{cf}{z}$. Inserting these into the law of propagation of errors, we then obtain the following:

$$\sigma_f^2 = f^2 \cdot \left(\left(\frac{a\sigma_x}{x} \right)^2 + \left(\frac{b\sigma_y}{y} \right)^2 + \left(\frac{c\sigma_z}{z} \right)^2 \right) \tag{2.20}$$

or written using relative errors:

$$r_f^2 = a^2 r_x^2 + b^2 r_y^2 + c^2 r_z^2 \tag{2.21}$$

In the previous example of speed, we had two variables, $x = \Delta x$ and $y = \Delta t$, whose exponents therefore were $a = 1$ and $b = -1$. In this case, thus, $a^2 = b^2 = 1$, and the relative errors are added quadratically, as we have already seen for the special case.

2.2 Units and Dimensional Analysis

All that we have looked at in the preceding section concerning measurements was based on the fact that we compare a quantity with a scale for this quantity. This scale specifies the unit of this quantity of interest. One can invent several different units for the same quantity and for instance a different system of units is used in the United States compared to the rest of the world. Just because we have different systems of units, however, the quantity itself cannot change. This means that the systems of units must be able to be transferred into one another without influencing the described quantity. The consequence of this fact when thought through mathematically is that the units of derived quantities are always linked to the basic units via power laws. Therefore, every physical quantity that is influenced by other quantities will be so according to a specific set of power laws. This forms the basis of dimensional analysis, which is an extremely powerful tool, not only to check calculations for consistency, but also to "derive" laws of physics by virtue of the fact that they have to describe quantities with the proper dimensions. Before we describe this in detail, however, let us first define the units we actually will use throughout the rest of the book to describe all the variables we encounter. This is the international system of units.

2.2.1 SI Units

The basic units thus define the system of units, since the unit of every quantity will depend on these units in the end. In our previous example of velocity, the unit of velocity is determined when the unit length and the unit of time are given. In the international system units (SI), seven basic units are defined, from which the units of all other variables can be derived.

The homepage of the International Bureau of Weights and Measures (BIPM) in Paris (www1.bipm.org/en/si/) shows the current definitions of the basic SI units:

Length: 1 meter (m) is the distance that light travels in vacuum during the period of 1/299,792,458 seconds.

Mass: 1 kilogram (kg) is the mass of the Pt-Ir cylinder of the Bureau International des Poids et Mesures in Sèvres, after applying the standardized cleaning procedure. This will be changed in 2018.

Time: 1 second (s) is the time taken for 9192631770 oscillations of a transition between two different energy levels of the isotope ^{133}Cs.

Electrical current: 1 ampère (A) is the strength of a constant current that flows through two different conductors in a vacuum, which are straight, infinitely long, and of negligible diameter, when they create a force of 2×10^{-7} Newton per meter in a distance of 1 m.

Temperature: 1 Kelvin (K) is 1/273.16 part of the thermodynamic temperature of the triple point of H_2O.

Quantity of matter: 1 mole (mol) is the quantity of matter that contains the same number of molecules as there are carbon atoms in 12 grams (g) of carbon of the isotope ^{12}C.

Light strength: 1 candela (cd) is the light emitted by 1/600000 m^2 of the surface of a black body at the freezing temperature of platinum at a pressure of 101325 N/m^2.

We will not treat the unit candela any further; it is only of concern for photographers and lighting technicians, since it depends on what we humans perceive as visible light. It therefore cannot be determined from natural constants and thus is specific to human endeavors. Also, the preceding definitions are about to change, and the kilogram should soon be no longer defined by a specific object kept in a vault in Paris, but by some combination of natural constants, similar to the definition of the meter. In the newly envisions definitions, which should come into effect sometime in 2018, not only the speed of light would be defined, but also Planck's constant, the Boltzmann constant, Avogadro's number, and the elementary electric charge. This will actually mean that all units except the second end up being defined via prescribed natural constants.

For other physical quantities than those concerning the base units, other SI units can be directly derived from the basic units. The special feature of the SI basic units is that all possible derived variables are obtained by combinations of integer powers from the basic units. Table 2.1 gives some examples of physical quantities with derived units, some of which also have their own respective names (but not their own definition).

In order to indicate particularly large or small values, prefixes to the units are used, which signify a certain power of 10, as indicated in Table 2.2.

2.2.2 Dimensional Analysis

Since every physical quantity must always be connected to a unit, it is also possible to deduce the relation between different quantities from the knowledge of their respective units. If a combination of different quantities is to give a different physical quantity, then this combination of the units of these quantities must also have the unit of the quantity sought. Since we cannot add up different units (for this, it would have to be the same unit), the different units must be multiplied by one another, or, more precisely, be connected by power laws, as discussed earlier. In this way, we can often determine the functional

Table 2.1 Some derived SI-units

Quantity	Relationship	Unit	In basic units	Symbol
Speed	Distance/time		$\frac{m}{s}$	v
Acceleration	Speed/time		$\frac{m}{s^2}$	a
Force	Mass*acceleration	Newton	$N = \frac{kgm}{s^2}$	F
Momentum	Mass*speed		$\frac{kgm}{s}$	p
Density	Mass/volume		$\frac{kg}{m^3}$	ρ
Energy	Force*distance	Joule	$J = Nm = \frac{kgm^2}{s^2}$	E
Power	Energy/time	Watt	$W = \frac{J}{s} = \frac{kgm^2}{s^3}$	P
Pressure	Force/area	Pascal	$Pa = \frac{N}{m^2} = \frac{kg}{ms^2}$	p
Electrical charge	Current*time	Coulomb	$C = As$	Q
Electrical potential	Energy/charge	Volt	$V = \frac{W}{A} = \frac{J}{As} = \frac{kgm^2}{s^3A}$	U or V
Electrical resistance	Voltage/current	Ohm	$\Omega = \frac{V}{A}$	R

Table 2.2 Prefixes for SI units

Factor	Prefix	Symbol	Example (lengths)
10^{24}	yotta	Y	Size of the observable universe: 130 Ym
10^{21}	zetta	Z	Distance to the Andromeda galaxy: 20 Zm
10^{18}	exa	E	Distance to the center of the milky way: 250 Em
10^{15}	peta	P	Distance to the next-nearest star (proxima centauri): 40 Pm
10^{12}	tera	T	Distance to pluto: 6 Tm
10^9	giga	G	Distance to the sun: 150 Gm
10^6	mega	M	Radius of earth: 6 Mm
10^3	kilo	k	Height of Mount Everest: 9 km
10^2	hecto	h	Highest trees in the world: 1 hm
10^1	deka	da	Length of a Brachiosaurus: 3 dam
10^{-1}	deci	d	Width of a hand: 1 dm
10^{-2}	centi	c	Width of an index finger: 2 cm
10^{-3}	milli	m	Width of a human hair: 0.1 mm
10^{-6}	micro	μ	Wavelength of visible light: 0.5 μm
10^{-9}	nano	n	Radius of a DNA molecule: 1 nm
10^{-12}	pico	p	Radius of a hydrogen atom: 50 pm
10^{-15}	femto	f	Radius of a proton: 1 fm
10^{-18}	atto	a	Smallest length measured: 1 am
10^{-21}	zepto	z	
10^{-24}	yocto	y	

relationships that we have identified with physical laws in Section 1.3.2, without going through lengthy derivations or considering basic axioms. We'll now see how this is done.

First, we offer a word on notation. The unit of quantity x will be denoted by $[x]$ in the following. Thus if, for example, x is a length, then $[x] =$ m, i.e., we use the meter as a unit of length (we are using SI units). To give an example of a derived quantity, if we want to obtain a velocity with $[v] =$ m/s and we have at our disposal the values of a mass ($[m] =$ kg), a time ($[t] =$ s), and a distance ($[x] =$ m), we get a value with the right unit only if we take $[x/t] = [x]/[t] =$ m/s $= [v]$. The unit of mass does not appear in this equation. Since the physical quantities must be connected in the same way as the units, we can conclude that $v \propto x/t$. In other words, we have found that the mass does not influence the value of the speed we obtain, even though it was given as one ingredient of the problem!

This is, for instance, useful in the quick checking of the result of a calculation. Suppose you have a mass ($[m] =$ kg), an acceleration ($[g] =$ m/s^2), and a distance ($[h] =$ m) and should have calculated an energy ($[E] =$ J $=$ kgm^2/s^2) from these variables. You obtain $E = mgh$. Could this be correct? Let's check the units. If the result is correct, then we must have $[E] = [mgh] = [m] \cdot [g] \cdot [h] =$ kg $\cdot$ m $\cdot$ m/s^2 $=$ kgm^2/s^2 $=$ J. Therefore, yes, this result could be correct indeed (and actually it corresponds to the potential energy of height in a constant gravitational field)!

But you not only can use it to check if the result is correct, but also to possibly correct a calculation where something went wrong. Suppose you wanted to obtain a force ($[F] =$ N $=$ kgm/s^2) from a mass ($[m] =$ kg), a speed($[v] =$ m/s), and a distance ($[r] =$ m). You obtain $F = mv/r$. Let's again check the units: $[m][v]/[r] = \frac{\text{kg} \cdot \text{m}}{\text{m} \cdot \text{s}} = \frac{\text{kg}}{\text{s}}$ and $[F] =$ kgm/s^2 $= \frac{\text{kg}}{\text{s}} \frac{\text{m}}{\text{s}} = [mv/r] \cdot [v]$. Otherwise put, to obtain the right result, you're still missing a speed in your result. A possibly correct result would therefore have been $F = mv^2/r$, which would have the right units (and actually corresponds to the force required to keep something in circular orbit at constant speed).

Time for Free Fall

We can go even further than this and "derive" physical laws using just the units. For instance, if we would like to determine the falling time T of an object, or the period of a swinging pendulum, we can study the problem and identify important parameters, that could possibly influence this time. One such parameter is certainly the falling distance L (or the length of the pendulum). Another one might be the acceleration due to gravity g, and maybe also the mass m of the falling object (or the bob in the pendulum). Given these different parameters, we now assume that T is given by a power law determined by different exponents for these three parameters:

$$T = AL^a \cdot g^b \cdot m^c \tag{2.22}$$

Here, A is some number that we do not know. The important part is that A does not have a unit and therefore does not correspond to any physical quantity; it is simply a number. Of course, for a proper calculation the value of A will be important, but if we are only interested in how the physical parameters (that can actually be changed) influence the result, we do not have to worry about the value of A, but rather we are interested in the exponents of the power laws. These we can determine when looking at the units, where A will drop out of the equation, because it does not have any units. In the case of the time for free fall, we thus obtain the following:

$$[T] = [L]^a \cdot [g]^b \cdot [m]^c \tag{2.23}$$

Alternatively, if we insert all of the units directly, we obtain the following:

$$m^0 \cdot s^1 \cdot kg^0 = s = m^a \cdot (m/s^2)^b \cdot kg^c = m^{a+b} \cdot s^{-2b} \cdot kg^c \tag{2.24}$$

Because the basic units m, s, and kg do not depend on each other (they are independent base units, after all), the power laws in the preceding equation must be fulfilled for each basic unit separately. This means that the exponents for every unit must be the same on both sides of the equation. This gives three different equations for the three different basic units. For the meter, we get $a + b = 0$; for the second, we get $-2b = 1$; and for the kilogram, we get $c = 0$. We can solve these equations and obtain for the different exponents: $c = 0$, $b = -1/2$ and $a = 1/2$. Therefore, we have a mathematical description of the falling time as follows:

$$T = AL^{1/2} \cdot g^{-1/2} \cdot m^0 = A\sqrt{L/g} \tag{2.25}$$

This means that the mass of the object does not determine the falling time! This is something that we did not assume. In addition, we see that the fall time increases as the square root of the falling distance. This means that if we want to double the falling time, we have to quadruple the falling height. We can measure this experimentally and we see that this is indeed the case (within the experimental accuracy). In this experiment, we can also determine the numerical prefactor and see that in this case it matches the measurement errors with $A = \sqrt{2}$.

The same reasoning can also be used to determine the period of a pendulum, that is, a mass oscillating on a piece of thread of a given length. Again, the mass of the object is not important, which you can test by studying pendulums of equal length, but with very different masses, where you indeed find the same period. By changing the length of the pendulum, you can also determine the prefactor A, which here will be different from the case of free fall. Experimentally one finds a value compatible with 2π.

In general, it would be possible that the approach we followed in equation 2.22 is not correct. It might be that the quantity (in this case, the time for free fall) is given by the sum of two terms of power laws or something more complicated. However, also in this case, we could learn useful things from dimensional analysis, because even in that case, it is necessary that the two terms have the same unit (in the example, both would have to be a time). One would then have to determine from the given factors two quantities with different prefactors, both of which have the right unit. The complication is that one does not know what the prefactors are, and therefore one does not know which of the two terms is more important. But even so, interesting things can be learned, as we will see for the case of air resistance very shortly.

Maximum Walking Speed

The dependence of the oscillation period on the length of the pendulum can also be used to understand a phenomenon in nature that you may have noticed before, but maybe not

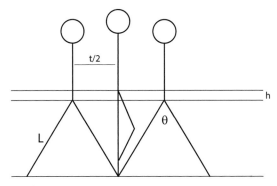

Figure 2.7   A schematic step sequence. During the time of a step, the human performs a pendulum movement on the leg.

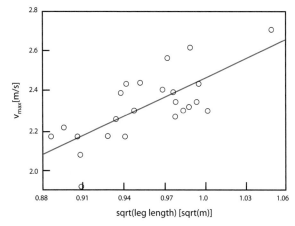

Figure 2.8   The maximum walking speed for people of different sizes. On the horizontal axis, the square root of the length of the leg, in meters, is plotted. The linear relation in the data corresponds to the scaling law $v_{max} \propto \sqrt{L}$. Data from Webb, 1996.

thought too much about. When you look at small and large animals, it is noticeable that small animals (such as mice) run very frequently, while large animals (such as elephants) never (or hardly ever) do so. Why could this be?

When walking (as opposed to running), the resting leg is simply changed from one leg to the other (see Figure 2.7). In the time between the change of the legs, the body performs an inverted pendulum motion, which produces the distance covered. That is, the time between two steps T has to be given by the period of the pendulum, and hence by the length of the leg L via $T = A\sqrt{L/g}$. The walking speed is then obtained by the ratio of step length and the time between steps T. However, the step length is again given by some percentage of the length of the leg itself; it is proportional to L, such that when walking one cannot exceed a speed of $v = A'\sqrt{L \cdot g}$. The constant A' is a number depending only on the opening angle of the legs, so it is approximately the same for all animals. This means that the maximum speed that an animal can still walk at depends only on the length of the leg and, to be more quantitative, is proportional to the square root of the length of the leg. This can also be seen in Figure 2.8, where the maximum walking speed for people of different sizes is plotted as a function of the square root of the lengths of their legs, showing a linear relationship.

Figure 2.9   Different kinds of flyers – natural and man-made.

So if two animals walk at about the same speed, the larger one can still be below its maximum walking speed and therefore walk, while the small one is already above it and therefore has to run.

2.3 Scaling Laws

As we have seen in the last section for walking speed, it can happen that there is a physical quantity in a relation, which we cannot change, e.g., some constant of nature. In the example of walking speed, this was the acceleration due to gravity, g. In that case, it can make sense to only look at the parameter(s), which actually can be changed to see how a system behaves in different situations. So in the example of the walking speed, we wanted to see how the walking speed depends on the animal type (or on the length of the animal's legs). The resulting dependencies are then still power laws, but the prefactors now are physical quantities and therefore do have a unit. Such relationships are called a scaling law. In the preceding example, this means that the maximum walking speed increases with the square root of the leg length, i.e., $v_{max} \propto \sqrt{L}$. This is such a scaling law. In contrast to what we have seen with dimensional analysis, it is important to note that the proportionality does not imply that the units are the same on the right and on the left. This is because we now allow natural constants of any kind (which do have units) as factors of proportionality.

2.3.1 Flying Speeds and Masses

In the following, we will look at the origin of a few scaling laws, as well as how to find them empirically. Consider Figure 2.10, which shows the flight speeds of different aircraft and flying animals as a function of their mass. The type of the flying object varies enormously and can go from a fighter jet to a fruit fly, where the masses can differ by a factor of 100 billion (or 10^{11}, i.e., 11 orders of magnitude). It is not surprising, then, that there is not much to be seen in the figure, because nearly all of the data points correspond to objects much smaller than the fighter jet.

But if we plot the masses on a logarithmic scale, that is, we plot the logarithm of the masses as in Figure 2.11, this can be circumvented. This is due to the property of the

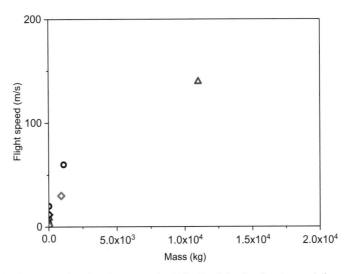

Figure 2.10 Flight speeds of different animals and airplanes at sea level. The size of the aircraft varies greatly from a fruit fly to a jumbo jet. It is very difficult to see any regularity from such a linear recording.

logarithm that the logarithm of a product is the sum of its logarithms. The following equation expresses this:

$$log(x \cdot y) = log(x) + log(y) \tag{2.26}$$

That is, when we plot the mass on a logarithmic scale, we record the data such that an increase by a certain factor (e.g., a doubling) always corresponds to the same linear distance on the plot. This means that we treat all mass scales equally and only the relative differences are of interest to us. A linear scale would mean that the same distance always corresponds to the same absolute increase. That is, if two sizes are relatively similar, but both have a large value, they will be plotted far apart on a linear scale, while they are close on a logarithmic scale. On the other hand, two values that are a large factor apart but both a small absolute value are close together on a linear scale and far apart on a logarithmic one. As we have seen in Figure 2.10 showing airspeeds as a function of mass, a linear scale is not very helpful for the comparison of strongly different values. On the simple logarithmic representation (Figure 2.11), we can now see clearly that there are many data points and that the data follow a certain course. The speeds generally increase with increasing mass.

However, a quantitative relationship is still difficult to detect, since the speeds also cover a large range of two to three orders of magnitude. If we now apply a logarithmic scale to the speeds as well (Figure 2.12), then we find a clearer relation between the mass and the speed, which to a good approximation corresponds to a straight line, or in other words, a linear relation in the logarithms of speed and mass. This we can summarize in a simple equation:

$$log(v) = a \cdot log(M) + b \tag{2.27}$$

But what does this connection mean? If we apply the fundamental property of the logarithm to a power law, we can write $log(x^a) = a \cdot log(x)$. We can thus rewrite the relation we found

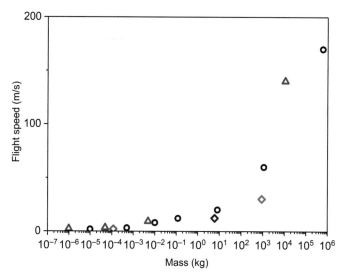

Figure 2.11 The data from Figure 2.10, where the masses are shown on a logarithmic scale. Despite the representation over many scales, no quantitative regularity can be seen.

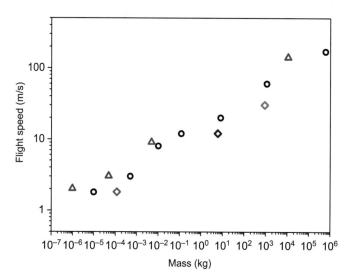

Figure 2.12 The data from Figure 2.10 on a double logarithmic scale. In this representation, a linear relation between the flight speed and the mass is discernible. This corresponds to a power law.

from the graph as $\log(v) = \log(M^a) + b = \log(10^b \cdot M^a)$. In other words, we obtain a scaling law between the flight velocity and the mass as follows:

$$v \propto M^a, \tag{2.28}$$

if we ignore the prefactor 10^b, which is some physical quantity (i.e., has units), but is nevertheless constant with regard to changes in mass.

Furthermore, this tells us that the exponent of the scaling law is given directly by the slope of the straight line in the double logarithmic plot. From the graph, we can thus directly determine the exponent of this scaling law by reading off the slope of the straight line. For the data given, this yields a slope of about $a = 1/6$, with an uncertainty of about 10%. What we have found from the data here, empirically, is that there is a scaling law between the speed and the mass of a flying object where the speed depends on the mass to the power of $a = 1/6$. The data do not tell us why this should be so, but we can still read this quantitative relationship off the graph with the data. Can this scaling law be explained is some way and not just be stated empirically? What do we need to know about the flight speed?

If something is supposed to fly, then it must have lift, F_L. If it is flying at a constant speed, the lift must match the weight, F_G. So in principle we would have to deal with what forces act on a flying object, which we will only discuss in more detail in Chapter 6. We can, however, use dimensional analysis to obtain the information we need.

Which quantities determine the lift? The lift is essentially a form of friction from the airflow, which is characterized by the properties of the air (the density $[\rho] = \text{kg/m}^3$ and the viscosity $[\eta] = \text{kg/(m} \cdot \text{s)}$) as well as of the object experiencing the lift (its speed $[v] = \text{m/s}$ and size $[L] = \text{m}$). From the point of view of dimensional analysis, we now have four variables (ρ, η, v, L) but only three independent units (m, kg, s), which means that the system of equations that we obtain from dimensional analysis will not have a unique solution. Therefore, we cannot describe the system as a simple power law of these four variables.

We can, however, assume that, as described previously, the force is given by a sum of two (or more) forces, each depending on three of these quantities. We then have to experimentally determine the ratios of the two dimensionless prefactors for different situations. So we take the two possible trios of quantities (ρ, L, and v) as well as (η, L, and v) of the given parameters and determine the relationship for a force for each of these. The general approach is thus as follows:

$$F = A\rho^a v^b L^c + B\eta^d v^e L^f \tag{2.29}$$

where A and B are dimensionless prefactors (as is always the case in dimensional analysis), whose relative magnitude can shift the weight given to one of the two forces to a certain extent.

For the first term, we obtain a lift force

$$F_{L,A} \propto \rho^a v^b L^c \tag{2.30}$$

where we want to determine the exponents needed in order to match up the units on both sides. This yields the following:

$$[F_{L,A}] = N = \frac{\text{kgm}}{\text{s}^2} = \text{kg}^1\text{m}^1\text{s}^{-2} = [\rho]^a[v]^b[L]^c = \frac{\text{kg}^a\text{m}^b\text{m}^c}{\text{m}^{3a}\text{s}^b} = \text{kg}^a\text{m}^{b+c-3a}\text{s}^{-b} \tag{2.31}$$

This gives three equations for the three exponents, which are $a = 1$, $b = 2$, and $1 = b + c - 3a = c - 1$, which together with the first two equations directly give $c = 2$. Thus we obtain a relationship for the lift force as follows:

$$F_{L,A} \propto \rho L^2 v^2 \tag{2.32}$$

This means that we have found that the lift increases with the square of the speed and also increases with the area of the wings.

We can proceed in exactly the same way for the second force in our general ansatz and obtain the following:

$$[F_{L,B}] = N = \frac{\mathrm{kgm}}{\mathrm{s}^2} = \mathrm{kg}^1 \mathrm{m}^1 \mathrm{s}^{-2} = [\eta]^d [v]^e [L]^f = \frac{\mathrm{kg}^d \mathrm{m}^e \mathrm{m}^f}{\mathrm{m}^d \mathrm{s}^d \mathrm{s}^e} = \mathrm{kg}^d \mathrm{m}^{e+f-d} \mathrm{s}^{-(d+e)} \tag{2.33}$$

which yields $d = 1$, $e = 1$, and $f = 1$, or in other words, $F_{L,B} \propto \eta L v$, which together with the previous derivation gives the general form of what the lift force is expected to be as follows:

$$F_L = A\rho L^2 v^2 + B\eta L v \tag{2.34}$$

Which of the two types of lift dominates the process cannot be determined from this analysis. This will depend on the values of the two prefactors A and B as well as the four variables L, v, ρ, and η. How these four variables determine which of these forces dominates can be further simplified and summarized into a single number, namely the ratio of the two forces $\frac{F_{L,A}}{F_{L,B}} = Re = \frac{v\rho L}{\eta}$. This is called the Reynolds number and determines at which combination of size, speed, and materials properties force $F_{(LA)}$ becomes more important than force $F_{(L,B)}$ or vice versa. We will deal with this in more detail in Section 7.6, where we look at fluid dynamics.

But let us look at the problem of lift a little more closely. For this purpose, we define a dimensionless coefficient that corresponds to the lift. This means that we have to divide the lift F_L by one of the forces found using dimensional analysis. This gives the lift coefficient $c_L = \frac{F_L}{\rho v^2 L^2}$, or in other words:

$$c_L = A + B\frac{\eta}{\rho L v} = A + B/Re \tag{2.35}$$

If we look at this relationship, we see that for very large Reynolds numbers, the second term becomes ever smaller, irrespective of the values of A and B. Therefore, for very large Reynolds numbers, we have $c_L \simeq A$ and the force is given by $F_{L,A}$. On the other hand, if the Reynolds number is small, the second term becomes increasingly large and we have $c_L \simeq B/Re$, meaning that the lift is given by $F_{L,B}$. Where this transition is depends on the ratio of A and B, which we cannot obtain from dimensional analysis and which we have to determine experimentally. This experiment is shown in Figure 2.13 for a cylinder, and we can see that values of $A = 0.6$ and $B = 24$ actually describe the experiment very well with the predicted dependence. This then also gives a transition from one type of friction to the other somewhere between $Re \simeq 100 - 1,000$. Now we can use this knowledge to determine which regime is relevant for the lift of our flying objects, and since the viscosity of air is quite small, any of the sizes and speeds in the many flying objects we have looked

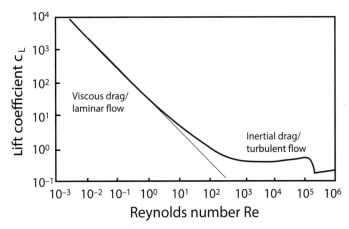

Figure 2.13 Schematic illustration of the lift coefficient ($\frac{F_L}{\rho L^2 v^2}$) as a function of the Reynolds number. The illustration is plotted on a double logarithmic scale. At low a Reynolds number, the lift coefficient varies as $1/Re$, corresponding to viscous flow, whereas over a large range of high Re the lift coefficient is constant, which means that the lift depends on ρ, v, and L only.

at correspond to a Reynolds number above the transition. Therefore, the lift is given by $F_L \propto \rho L^2 v^2$.

To come back to the question of the relationship between airspeed and mass of the flying object, we must now recall that the lift must correspond to the weight. That is, $F_L = F_G$ and because the weight is given by the mass, $F_G = Mg \propto M$, we obtain

$$M \propto L^2 \cdot v^2 \qquad (2.36)$$

The other parameters (air density ρ and gravitational acceleration g) are the same for all the different airplanes, and therefore we ignore them in the relation that should give us the scaling law dependence between mass and speed. We can see that we are almost there, since we have a relationship between mass M, speed v, and size L at the moment. For the final result, we need to get rid of the size L, i.e., we need to find a connection between the object's size and its mass. The mass of the flyer is directly proportional to its volume V, which itself is given by the size cubed:

$$M \propto V \propto L^3 \qquad (2.37)$$

We get this, for instance, from the units for volume and length, but we certainly know it to be true in case that the length in all three spatial directions changes in the same way, i.e., if all differently sized airplanes look about the same when we enlarge or reduce them homogeneously. So we obtain $L \propto M^{1/3}$ and therefore $M \propto M^{2/3} \cdot v^2$ or $M^{1/3} \propto v^2$, which is exactly the scaling law that we have previously found by purely looking at empirical data:

$$M^{1/6} \propto v \qquad (2.38)$$

In this derivation, we have made the explicit assumption that all airplanes look the same. This is actually not quite true. A glider plane has a different shape from a jumbo

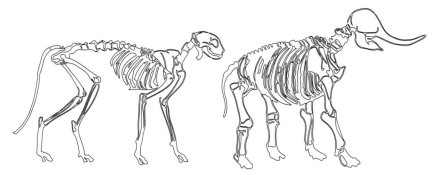

Figure 2.14   The skeletons of two differently sized animals (a cat an an elephant) plotted on the same scale. The elephant (right) is clearly identifiable as the larger animal by its "more massive" leg bones.

jet, an eagle looks different from a scaled-up hummingbird. When you look closely at the data in Figure 2.12, you can actually see those differences in the various airplanes. The maneuverable ones with small wings have higher speeds relative to the curve, whereas the gliders with long wings are slower. That is, the derivation of the scaling law positing that $v \propto M^{1/6}$ gives us not only this expectation, but also tells us under what circumstances we can expect which deviations. This is because in order to derive a quantitative relationship, we must make our assumptions explicit, and if these assumptions are not met, we will find deviations from the law.

2.3.2 Bone Shape and Size

If we look at the changes in shape, which we have seen in the different birds in the previous section, we may also ask ourselves whether these show a dependence on the size. Consider the two skeletons in Figure 2.14. Both are artificially drawn to the same size, but nevertheless it is recognizable which animal is larger. On the right is the skeleton of an elephant, on the left that of a cat. The thigh bones of the elephant are much more massive than those of the cat, which means they are thicker in proportion to their length. Here we have a clear example that animals do not look the same when the size is changed. This fact has already attracted the attention of Galileo Galilei, who made a scaling argument to explain this fact. The essential point of such an argument must be that the different directions can not be treated equally. Galilei looked at the load on a leg bone, i.e., the weight of the animal. As in the case of the airplanes, we have a load that grows with the length cubed $F_G \propto L^3$, where L is some characteristic length of the animal, e.g., the length of the thigh bone. On the other hand, a bone can withstand only a limited mechanical stress. We'll treat mechanical stress and elasticity in Section 7.1, but let it suffice that mechanical stress is a loading force divided by the loaded cross-sectional area A. The stressed cross-section is therefore given by the bone thickness squared, d^2, such that the mechanical stress is given by $\sigma = F_G/A \propto L^3/d^2$. Since the maximum mechanical stress that a bone can withstand is approximately the same for all animals (bone material is basically the same

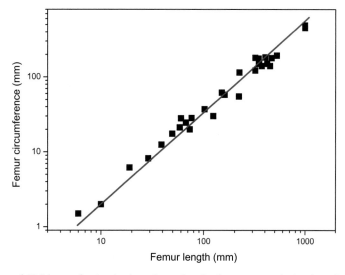

Femur circumference (mm) vs Femur length (mm)

The circumference of thigh bones of various land vertebrates (ranging from mouse to elephant) as a function of their lengths. The representation is using double-logarithmic scales and the line points to a power law with an exponent of 5/4.

in all vertebrates), Galilei has therefore concluded that bones must have a specific type of scaling connecting their lengths and widths given by $L^3 \propto d^2$ or

$$d \propto L^{3/2} \tag{2.39}$$

This is a quantitative prediction that can be checked empirically: if one considers the femurs of many different animals and measures their diameter (or the circumference, which is measured more reproducibly) as a function of the bones' length, we should find a power law with an exponent of 3/2, i.e., a straight line with a slope of 3/2 on a double logarithmic plot of the data. These measurements are shown in Figure 2.15. You can see that there actually is a straight line, i.e., a power law dependence, but the slope is not 3/2. Rather, we find a slope of about 5/4 with an accuracy of about 5%, so this exponent can be determined exactly enough to say that the measurement is incompatible with the expectation of 3/2, thereby falsifying the theory.

So Galilei was wrong! His error was that the pressure load on the bones is not actually the limiting factor for bone stability. Since the general argument makes sense and we actually do find a scaling law between bone thickness and bone length, we can try to improve the prediction by considering which load actually would pose a limitation for bone stability. When we look at the stresses on a femur, i.e., when and how a bone breaks, we see that we had better consider the bending or buckling of the bone. We will do this in detail in Section 7.3, where we will find that the force required to bend a round pipe (diameter d, length L) is given by $F_B \propto d^4/L^2$. This force must correspond to the load, such that the limit is $L^3 \propto d^4/L^2$, or in other words:

$$d \propto L^{5/4} \tag{2.40}$$

which is precisely the relationship we had found empirically. The fact that we find no linear relationship between bone diameter and bone length also tells us that land animals can only reach a certain size, since at some point the bone will become wider than long it is long. Dinosaurs are already getting close.

How the thigh bone of the larger animal knows that it must become more massive to ensure stability is an interesting biological question. From the bones of astronauts or paraplegics, we know that a load is actually necessary for bones to grow at all. Wolff's (empirical) law states that when mechanical stresses exceed a certain limit, bone growth is induced, and conversely when the stress is below another limit, bone is degraded. This is a very elegant regulation mechanism, which leads to the shape of bones, which is exquisitely adapted to the stresses the animal incurs, in animals of all sizes. For instance, one can use this in anthropology to determine which of our ancestors walked upright, since this incurs very different loads and thus leads to a very different shape of the skeleton around the hips. This regulation mechanism also leads directly to the formation of hollow bones, since in the interior the loads are always very small. However, the actual regulatory mechanism, i.e., how the bone "measures" the applied load, is still not known!

2.3.3 Metabolic Rates

So far, we have not looked at the accuracy of the exponent's definition in detail but have been content with a direct determination, which then could be "explained" by some model for the scaling law. However, the fluctuations of the data always allow for the possibility of several fitting lines (with different slopes) that are compatible with the data. This can be important if different models predict different exponents, as we have seen in the

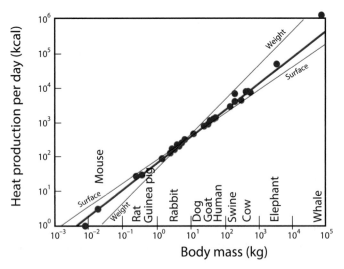

Figure 2.16 The metabolic rate of different animals depending on their mass. An exponent of 0.75(7) is determined mainly by the data points at very high masses (whale and elephant). Data from Kleiber, 1947.

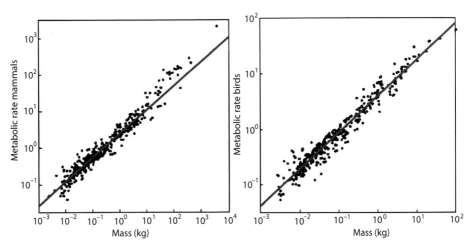

Figure 2.17 The metabolic rate of different mammals and birds depending on their mass. In both cases, the data are in good accordance with a power law with an exponent of 2/3. Data from Dodds et al., 2001.

preceding section for the bones. Due to fluctuations, it may be that the data do not actually allow differentiation between the two models. One such example is the dependence of the metabolic rate of an animal on its mass. Here, historical data (see Figure 2.16) suggest an exponent of 3/4, however, the uncertainty of this exponent here is about 10%. This means that the data are not completely incompatible with an exponent of 2/3, using the 2 sigma rule. Such an exponent would be physically more sensible, since the main part of the metabolic rate is needed to compensate for heat losses (see Section 8.3.6 for more details). This heat loss is given by the animals' surface area, that is, by the mass to the two-thirds power. Newer and more precise determinations have actually produced an exponent of 0.68(2) for warm-blooded animals (see Figure 2.17).

2.3.4 Flightless Birds

We can take together two of the scaling laws we have just discussed to find out why ostriches and other very large birds cannot fly. What we will need for this are the scaling of flight speed with mass and that of metabolic rate with mass. This is because the flight speed is directly related to the power needed to sustain flight. The power is given by the energy used per unit of time, hence it has to be given by the force used for propulsion times the speed, $P_{needed} = F_L \cdot v = Mgv$. This means that the relative power needed, i.e., the power per kilogram is directly given by the flight speed, such that we know how this relative power needed increases with the mass of the flying object: $P/M_{needed} = gv \propto M^{1/6}$. This actually means that bigger and bigger airplanes get more and more costly to fly.

On the other hand, from the scaling of the metabolic rate, we get that the power an animal can deliver increases with mass more slowly than the mass itself. The power delivered by the muscles is proportional to the metabolic rate, $P_{delivered} \propto B \propto M^{2/3}$, and hence the relative power delivered, i.e., the power per kilogram, actually decreases with mass as

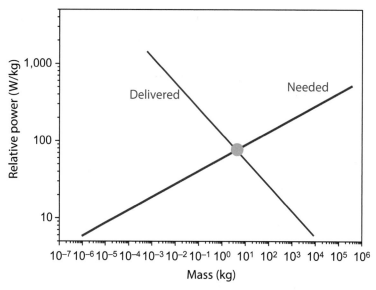

Figure 2.18 The scaling laws of flight speed with mass and metabolic rate with mass directly indicate the power needed for flight and the power delivered by muscles. For masses above a certain size, the power delivered is smaller than the power needed, hence birds above this mass, such as ostriches, will not be able to fly.

$P/M_{delivered} \propto M^{-1/3}$. In other words, larger and larger animals have less and less power to actually put out.

Therefore, there must be a point where the power needed for flight exceeds the power delivered by the animal, since the increasing needed power has to cross with the decreasing delivered one at some point. Thus the scaling arguments show us that there has to be a maximum size for flying animals, but we do not actually know what this limit is. In order to do this, one has to actually take into account the prefactors of all of the scaling laws in question and then set the needed and delivered relative powers equal, i.e., $P/M_{needed} = P/M_{delivered}$ or $AM^{1/6} = A'M^{-1/3}$, which gives $M^{1/2} = A'/A$ or $M = (A'/A)^2$. We could, however, also read the crossing point directly from a graph of the two powers in Figure 2.18, where we see that the result will be somewhere in the range of 5–20 kg. Therefore, human-powered flight in the same way that birds fly will be impossible. The only way around this is to decrease the flying speed of a human-powered flyer to bring it into the range of power delivered by a human, which as we have seen can only be done by changing the geometry of the flyer and increasing its surface area greatly relative to its volume.

Exercises

2.1   Error propagation
You have determined the diameter of a cell to be 10.0(5) μm using a microscope.

(a) What is the surface area of the cell and its uncertainty?

(b) What is the volume of the cell and its uncertainty?

Assume that the cell is spherical.

2.2   Absolute and relative errors

You have measured the pressure at the bottom and the top of a tower (with this, you will eventually be able to determine the tower's height, as we will see in later chapters). At the bottom, you determine a pressure of $p_0 = 1010(1)$ Pa and at the top a pressure of $p_h = 1000(1)$ Pa.

(a) What is the relative accuracy of your pressure determination?

(b) What is the absolute error in the difference of the pressures at the top and the bottom $\Delta p = p_0 - p_h$?

(c) What is the relative error in the difference of the pressures at the top and the bottom $\Delta p = p_0 - p_h$?

2.3   Data analysis

Suppose you wanted to measure the speed of a model glider plane, by determining its flight time for a given distance several times using a stopwatch. You obtained the following results for the flight time Δt in seconds: 1.20; 1.16; 1.23; 1.06; 1.12; 1.14; 1.05; 1.28; 1.15; 1.18; 1.07; 1.11; 1.14; 1.17; 1.10.

(a) What possible sources of errors could you have in such a measurement, both for the time measurement as well as the distance measurements? Which of these are systematic errors and which of these are statistical errors? What estimate would you set for these different sources of errors?

(b) Determine the average of the preceding measurements.

(c) Determine the standard deviation of the measurements as well as the error of the average from (b), i.e., the standard deviation divided by the square root of the number of measurements. Compare these to the estimate of the sources of errors you obtained in (a).

(d) With the results from (b) and (c), as well as the flight distance of 4.8(1) m (the error should roughly correspond to your result in (a)), determine the flight speed of the plane and its uncertainty.

2.4   Absolute and relative errors 2

Suppose you have to pipette an amount of 275 μl for some biochemical protocol. In your lab, you have two pipettes with maximum volumes of 250 μl and 5 μl respectively. The settings on the pipettes are such that the volume can be varied from 10% of the maximum volume up to the maximum volume. The quoted uncertainty by the seller is 1% for both pipettes, except at the minimum volume, where the quoted uncertainty is 5%.

(a) What is the error in the total volume that you pipette if you use only the 250 μl pipette, using the maximum volume once and the minimum volume once?

(b) What is the error in the total volume that you pipette if you use only the 250 μl pipette, pipetting 150 μl once and 125 μl once?

(c) What is the error in the total volume that you pipette if you use only the 5 μl pipette 55 times with the maximum volume?

Always determine the relative as well as the absolute error.

2.5 Error propagation 2

In developmental biology, the building of spatial gradients using "morphogens" is of great importance in the determination of spatial information and patterning within developing organisms. In this, the concentration of such a morphogen determines whether or not certain other genes are transcribed or not using some threshold level. In order for such a process to be developmentally robust, this concentration gradient needs to be determined with good accuracy. As we will see in later chapters, the concentration dependence of the maternal morphogen Bicoid in an early *Drosophila* embryo follows an exponential distribution given by $c_{Bcd}(x) = c_0 e^{-x/\lambda_{Bcd}}$, where the different parameters are determined by the diffusion and breakdown properties of the protein. The figure does, however, also show that there is a certain degree of variation in this concentration gradient.

(a) In the preceding expression, x is the distance along the anterior–posterior axis of the embryo, and $\lambda_{Bcd} = 120$ μm is the distance after which the concentration has dropped to $1/e$ (roughly equal to 0.37) of the initial concentration c_0. From data, the uncertainty of λ and c_0 can be determined, yielding $\sigma_\lambda = 20$ μm and $\sigma_{c_0}/c_0 = 0.05$. How do these uncertainties correspond to an accuracy in positioning in the embryo? Determine the uncertainty of the concentration for a position at $x = 200$ μm, both with absolute and relative error.

(b) Now let us consider the function of the morphogen. Which position corresponds to a threshold concentration c_T, initiating the transcription of a protein? For this purpose, solve $c_{Bcd}(x_T) = c_T$ for x_T and determine the uncertainty in x_T. For a numerical result, assume that $c_T = 0.2c_0$.

(c) In (b), we have assumed the threshold concentration c_T to be exactly given. However, there is also a level of uncertainty involved in this. Do the same as in (b), but taking into account the uncertainty in c_T, i.e., σ_{c_T}.

(d) If we are normalizing concentrations to the initial value c_0, we eliminate the uncertainty of c_0, but increase the uncertainty in c_T. Again determine the uncertainty in x_T if c_T is a given fraction of c_0 and compare this with the result from (b).

2.6 Dimensional analysis

(a) We will see that the diffusivity of a substance is measured in m^2/s (this actually means that equal distances are not traveled in equal times). Similarly, we will see that the mobility of a substance (determining the speed obtained for a given force) is measured in m/(N s). What is the unit of the ratio of diffusivity and mobility? Which physical property might this correspond to?

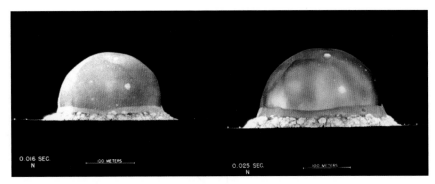

The shock wave of the first atomic explosion in Alamogordo at two different times.

(b) We have seen what the drag force in turbulent flows is using dimensional analysis. Do the same for viscous drag. Now the drag force depends on the viscosity η (unit Pa s), the area A (unit m^2) and the speed v (unit m/s).

2.7 Dimensional analysis 2

Figure 2.19 shows the shock wave of the first atomic explosion in Alamogordo at two different times. The time resolution in both cases is probably about 1 ms. From these images, you can determine the energy released in the explosion. For this purpose, proceed as follows:

(a) Determine the energy E given a shock wave at position R at time T traveling in air of density $\rho = 1.0(1)\mathrm{kg/m^3}$.

(b) For both images you can thus determine an energy release. For this purpose, assume that the dimensionless prefactor in your result in (a) is one.

(c) Determine the error of the released energy, given the uncertainties in R, T, and ρ. Use error propagation on the result of (a) for this purpose.

(d) Determine the errors for your results in (b) after estimating the errors of the quantities involved from the images.

(e) Compare the variation of your results in (b) with the estimate of the corresponding uncertainties.

2.8 Power laws and logarithmic scales

Figure 2.20 shows the dependence of a molecule's molecular weight on its size on a double logarithmic scale. determine the slope of the straight line. What can you learn from this? Give a physical reason why the exponent should have the value that you have found.

2.9 Power laws and logarithmic scales 2

Figure 2.21 shows data for the circumference of thigh bones of different land animals as a function of their lengths on a double-logarithmic scale.

(a) Determine the slope of the straight line. What can you learn from this?

(b) What is the uncertainty in the determination of the exponent in (a)? Consider the line with maximum and minimum slope that is still compatible with the data for an estimate of the uncertainty.

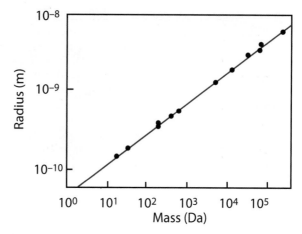

Figure 2.20 The dependence of a molecule's molecular weight on its size on a double logarithmic scale.

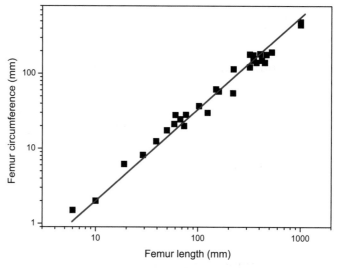

Figure 2.21 Data for the circumference of thigh bones of different land animals as a function of their lengths on a double-logarithmic scale.

(c) Does the uncertainty depend on whether you suppose that the uncertainty of the measured circumference has a constant absolute value irrespective of its size or a constant relative value? If so, how? Which of these two options is more probable? What is the uncertainty of the exponent now?

2.10 Power laws and logarithmic scales 3

Figure 2.22 shows data for the number of blood vessels of different types on their size on a double-logarithmic scale.

(a) Determine the slope of the straight line. What can you learn from this?

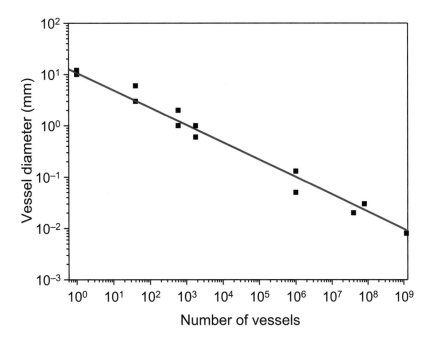

Figure 2.22 Data for the number of blood vessels of different types on their size on a double-logarithmic scale.

(b) What is the uncertainty in the determination of the exponent in (a)? Consider the line with maximum and minimum slope that is still compatible with the data for an estimate of the uncertainty.

(c) If you consider that the number of blood vessels should have an uncertainty given by a relative error of $1/\sqrt{N}$, what uncertainty of the exponent do you obtain?

2.11 Scaling laws and the body mass index

(a) How do length, L, and mass, M, scale in fish? What do you expect based on geometry? Does this make life easier on a fish market? What can you learn from the constant of proportionality?

(b) How about humans? What relation between length, L, and mass, M, do you expect based on geometry? What can you learn from the constant of proportionality?

(c) The body mass index (BMI) is defined as the ratio of the mass to the length squared, $BMI = M/L^2$. Why does this not correspond to the expectation from (b)? If the BMI is a good measure for the average population, what do you expect to be different between short and tall people?

(d) Let's determine a good measure empirically. In Figure 2.23, length and mass of many people are plotted on a double-logarithmic scale. Determine the exponent in $M \propto L^\alpha$ that best describes these data. Also determine an uncertainty for your exponent.

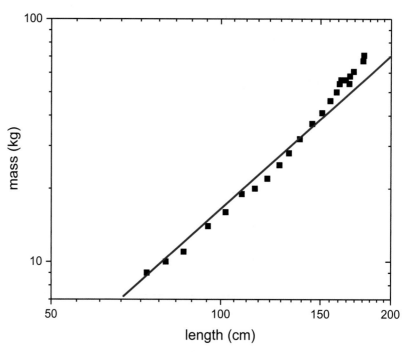

Figure 2.23 Length and mass of many people plotted on a double-logarithmic scale.

Quiz Questions

2.1 Error propagation

You have determined the side length of a cube to be $\ell = 1.00(1)$ m. What is the uncertainty of the volume of this cube ($V = \ell^3 = 1$ m³)?

A. 0.01 m³
B. 0.02 m³
C. 0.03 m³
D. $\sqrt{3} \cdot 0.01$ m³
E. 0.1 m³

2.2 Error propagation 2

You have determined the side lengths of a cube to be $a = b = c = 1.00(1)$ m. What is the uncertainty of the volume of this cube ($V = abc = 1$ m³)?

A. 0.01 m³
B. 0.02 m³
C. 0.03 m³
D. $\sqrt{3} \times 0.01$ m³
E. 0.1 m³

2.3 Error propagation 3

What is the error of x/y, $\sigma_{x/y}$, if x and y have errors of σ_x and σ_y?

A. $\sigma_{x/y} = \sigma_x + \sigma_y$

B. $\sigma_{x/y} = \sigma_x/\sigma_y$

C. $\sigma_{x/y} = \sqrt{\sigma_x^2 + \sigma_y^2}$

D. $\sigma_{x/y} = \sqrt{\sigma_x^2/\sigma_y^2}$

E. $\sigma_{x/y} = (x/y)\sqrt{\sigma_x^2/x^2 + \sigma_y^2/y^2}$

2.4 Scaling laws

What do you obtain if you plot a straight line $y = a \cdot x$ on a double logarithmic scale?

A. A straight line with slope a

B. A straight line with slope 1

C. A straight line with slope 0

D. A straight line with slope -1

E. No straight line

2.5 Logarithmic scales

What do you obtain if you plot an exponentially increasing function $N(t) = N_0 \cdot exp(\lambda t)$ on a single-logarithmic scale?

A. A straight line with slope 1

B. An exponential

C. A constant

D. A straight line with slope λ

E. A straight line with slope $1/\lambda$

F. A logarithmically increasing function

2.6 Scaling laws 2

What do you obtain if you plot a parabola $y = a \cdot x^2$ on a double-logarithmic scale?

A. A straight line with slope a

B. A straight line with slope 1

C. A straight line with slope 2

D. No straight line

2.7 Scaling laws 3

What do you obtain if you plot a straight line $y = a \cdot x + b$ on a double-logarithmic scale?

A. A straight line with slope a

B. A straight line with slope 1

C. A straight line with slope b

D. A straight line with slope 0

E. No straight line

2.8 Units

Dimensional analysis can be very useful in checking calculations quickly in terms of whether they correspond to a physical situation. We use kg, m, and s as units of length, mass, and time, respectively. Then the units of force (F) and energy (E) are kg m/s^2 and kg m^2/s^2 respectively. Which of the following relations is possible based on units (x, v, a, t, and m are place, speed, acceleration, time, and mass, respectively)?

A. $F = ma$

B. $x = at^3$

C. $E = \frac{1}{2}mv$

D. $E = m \cdot a \cdot x$

E. $v = \sqrt{Fx/m}$

F. $a = v^2/x$

G. $x = v/a$

H. $x = vt$

I. $x = vt \sin t$

J. $x = vt \sin vt$

K. $x = (at^2) \sin(vt/x)$

L. $a = x^2/v$

3 Motions and Oscillations

In many fields of science, we are interested in how different things move. These movements can vary widely from things we've seen before, such as the search pattern of *Cataglyphis*, the movement of DNA in an electrophoresis gel, the flying of a bumblebee or a hummingbird, our legs while walking, to the jumping of a grasshopper, and can also comprise inanimate matter, such as electrons in a particle accelerator, the motion of the sun around the center of the galaxy, or a pendulum swinging back and forth. The most fascinating thing is that all of these movements can be described with the same mathematics and are therefore subject to the same laws. With all of these, we ask ourselves, when is the thing whose movement we want to describe where? Again, what we are interested in is effectively a functional dependency, namely, the position of the object as a function of time. What this dependence looks like determines how we can describe or perceive the motion. To describe the movement, different quantities are required, which we will introduce in detail in this chapter. In particular, these are the changes in the position and, in turn, the rate of change of these changes. These will be called velocity and acceleration. Through these two parameters, we can describe a very large class of motions, as we have already described in Section 1.3.2. In addition, we will see later in the treatment of forces that the acceleration plays a very special role in physics because it is directly given by (proportional to) the acting forces. In order to describe all of these motions, we first have to set the stage and discuss what we mean by velocity and acceleration and how they are related to the motion of the object. For the time being, we restrict ourselves to movements that take place in only one direction. This will make things simpler by preventing complications in the description by changing the direction in addition to the position.

3.1 Describing Motions: Velocity and Acceleration

As said in the introductory section, we restrict ourselves to motions in only one direction, also called rectilinear motions. This can happen along the horizontal, as in a car on a flat road; in the vertical, as in free fall; or along an inclined plane. At the moment, we are not interested in the cause of the movement, but simply want to give a quantitative description of the motion itself. In addition, the moving object should move as a fixed body and not rotate, so either we consider an extremely small, almost pointlike particle or an extended object, in which all parts are rigidly connected.

If we want to know where an object is at a given time, i.e., we want to locate it, we need to determine the position relative to a given reference point. This is the same as the distance measurement in the preceding chapter, where we have determined the length of a rod by measuring two positions. This reference point is often called the origin (or zero position) of an axis, which we will here call the x-axis:

The position along this axis, the *location x*, is measured in multiples of an axis section (e.g., 1 cm). The direction in which the object is moving is indicated by the sign:

If an object moves from one location x_1 to another location x_2, the change in location (the displacement) is as follows:

$$\Delta x = x_2 - x_1 \qquad\qquad (3.1)$$

As before, we indicate the change in a certain quantity by the symbol Δ in front of the quantity. Because the motion can go back and forth, the change in location does not have to be the same as the traveled path length, and the two should not be confused:

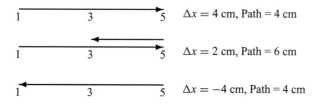

Since we are only considering motions in one direction (or dimension) here, the location is given by just one number, a coordinate. In such a case, one also says that there is one degree of freedom.

In Section 3.4, we will consider more general motions in space. Then, there are in general three numbers necessary to specify a location, the three coordinates of three-dimensional space. The change in location then is also indicated by three numbers (three coordinates or a *magnitude* and a three-dimensional *direction*, given by two angles, namely the latitude and the longitude). In three-dimensional space, the change in location also is a (three-dimensional) *vector*.

But let's return to rectilinear motion: a compact form of rectilinear motion is a graph that represents x as a function of time t. As an example, Figure 3.1 shows the height above the ground reached by the armadillo, shown in Figure 3.2 during its jump.

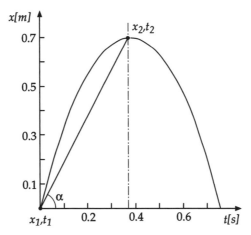

Figure 3.1 Jump height above ground x [m] as a function of time t [s] for the jumping armadillo.

Figure 3.2 Jumping armadillo by Bianca Lavies.

The diagram in Figure 3.1 describes not only the movement, but also reveals how fast the armadillo moves. Several physical variables are associated with the statement "how fast." One is the *mean velocity*:

$$\langle v \rangle = \frac{\Delta x}{\Delta t} = \frac{x_2 - x_1}{t_2 - t_1} = \frac{\text{Change in location}}{\text{Time interval}} \tag{3.2}$$

The mean velocity corresponds to the slope $\tan\alpha$ of the straight line connecting the starting point $x_1(t_1)$ with the end point $x_2(t_2)$ of the interval. The mean velocity $\langle v \rangle$ also has a magnitude and a direction. The sign of $\langle v \rangle$ is determined by that of Δx, since Δt is always positive for subsequent time points.

The mean speed

$$\langle s \rangle = \frac{\text{Path length}}{\text{Time interval}} \tag{3.3}$$

is another way of quantifying "how fast." For the path of the armadillo in Figure 3.1, the definitions of $\langle v \rangle$ and $\langle s \rangle$ are the same up until the maximum of the curve is reached.

On the way back, however, this is no longer the case. At the end, the armadillo lands where it started, and we have $\langle v \rangle = 0$ because $\Delta x = 0$. $\langle s \rangle$, on the other hand, is the same for the rise and the fall. There is yet another way of quantifying "how fast" in that one wants to describe the rate of change of the motion at one specific point in time. This is the *momentary velocity*. We can obtain this momentary velocity from the mean velocity, if we make the time interval that we consider ever smaller:

$$v = \text{(momentary) velocity} = \lim_{\Delta t \to 0} \frac{\Delta x}{\Delta t} \equiv \frac{dx}{dt} \tag{3.4}$$

The addition of "momentary" is usually omitted in physics, and when we speak of velocities it is usually implied that we are talking of the the momentary velocity. If we call $x_2 = x(t)$, $t_2 = t$ and $x_1 = x(t_0)$, and $t_1 = t_0$, we obtain the following for the velocity at time t_0:

$$v(t_0) = \lim_{t \to t_0} \frac{x(t) - x(t_0)}{t - t_0} \equiv \frac{dx}{dt}\Big|_{t=t_0} \tag{3.5}$$

Mathematically speaking, this limit is called the derivative of the function $x(t)$ with respect to the variable t at position t_0. This is the velocity at time t_0, $v(t_0)$. Graphically, we can obtain the derivative by the slope of the tangent line to the curve $x(t)$ at point t_0 (see Figure 3.3), as opposed to the slope of the straight line connecting t_1 and t_2:

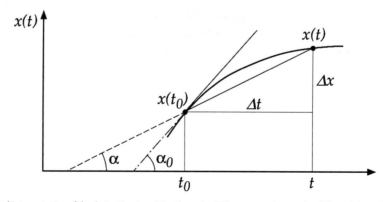

Figure 3.3 Geometrical interpretation of the derivative: transition from the difference quotient to the differential quotient corresponds to the transition from the slope of the secants to the slope of the tangents.

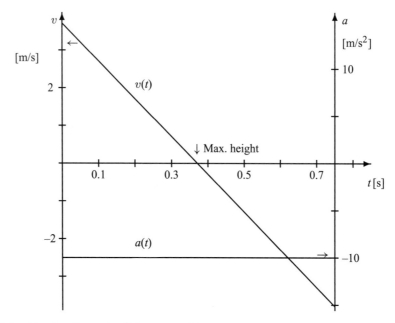

Figure 3.4 Velocity [m/s] (left scale) and acceleration [m/s²] (right scale) of the armadillo from Figures 3.1 and 3.2.

$$v(t_0) = \lim_{t \to t_0} \frac{x(t) - x(t_0)}{t - t_0} = \tan \alpha_0 \qquad (3.6)$$

For a time derivative, one can also use the following alternative notations:

$$\frac{dx}{dt}, \ \text{or} \ \dot{x} \qquad (3.7)$$

We will use the first one in the remainder of the book. The magnitude of the velocity is the speed, which can also be determined for every single time point. For our armadillo (Figure 3.1), we can determine the velocity by determining the slope of the tangent line. This gives the curve for the velocity in Figure 3.4.

When an object changes its velocity, we speak of acceleration. In everyday life, acceleration usually means a gain of velocity; in physics, the term includes both increasing and decreasing velocities, i.e., a deceleration is also called an acceleration, albeit with a negative sign. The mean acceleration $\langle a \rangle$ is defined as follows:

$$\langle a \rangle = \frac{v_2 - v_1}{t_2 - t_1} = \frac{\Delta v}{\Delta t} \qquad (3.8)$$

In analogy to the definition of the momentary velocity, we can define the momentary acceleration (and again usually leave away the characterization of "momentary" in the future):

$$a = \frac{dv}{dt} \qquad \text{or more precisely} \qquad a(t_0) = \lim_{t \to t_0} \frac{v(t) - v(t_0)}{t - t_0} = \frac{dv}{dt}\Big|_{t=t_0} \qquad (3.9)$$

The acceleration $a(t)$ is given by the slope of the tangent line for the velocity curve $v(t)$, or in other words, the time derivative of the velocity. Since the velocity itself is given by the rate of change of the position (or the time derivative of the position), the acceleration is also related to the temporal change of the position. However, now we have to consider the rate of change of the rate of change of the position, or in other words, the second derivative with respect to time:

$$a(t) = \frac{dv(t)}{dt} = \frac{d^2x(t)}{dt^2} \tag{3.10}$$

Rather than being related to the slope of the curve describing the position as a function of time, this corresponds to the curvature of this curve. The more the rate of change (i.e., the slope) itself changes, the more curved the function becomes. Also, something that is curved downward, such as the path of the jumping armadillo, has a negative curvature (and thus acceleration), whereas something curving upward has a positive acceleration.

For the armadillo (Figure 3.4) jumping up vertically, we have a one-dimensional motion in the upward direction. First it moves up, and after reaching a maximum height, it falls down again. In the velocity diagram $v(t)$, we see a straight line with a constant and negative slope: $a = -9.8$ m/s^2 (this is the acceleration due to gravity on earth). The velocity is first positive (going upward) and then negative (falling downward).

Mathematically, we can invert a differentiation by an integration. So as we can obtain $v(t)$ and $a(t)$ by differentiating $x(t)$ and $v(t)$ respectively, we can obtain $x(t)$ and $v(t)$ by integrating $v(t)$ and $a(t)$ respectively:

$$v(t) = \int_{t_0}^{t} a(t')dt' + v(t_0) \qquad x(t) = \int_{t_0}^{t} v(t')dt' + x(t_0) \tag{3.11}$$

$v(t)$ is a primitive function of $a(t)$ and $x(t)$ is a primitive function of $v(t)$. We have called the time in the integration t', in order to distinguish it from the boundary of the integral, which again is a time, t. For the case of constant acceleration, we obtain the following:

$$a(t) = a_0 \quad \Rightarrow v(t) = \int_{t_0}^{t} a_0 dt' + v(t_0) = a_0(t - t_o) + v(t_0) \tag{3.12}$$

$$t_0 = 0, v(0) = v_0 \quad \Rightarrow v(t) = a_0 t + v_0 \tag{3.13}$$

$$x(t) = \int_{t_0}^{t} v(t')dt' + x(t_0) = \int_{t_0}^{t} a_0(t' - t_o)dt' + \int_{t_0}^{t} v(t_0)dt' + x(t_0) \tag{3.14}$$

$$x(t) = \frac{a_0}{2}(t - t_0)^2 + v(t_0)(t - t_0) + x(t_0) \tag{3.15}$$

$$t_0 = 0, v(0) = v_0, x(0) = x_0 \quad \Rightarrow x(t) = \frac{a_0}{2}t^2 + v_0 t + x_0 \tag{3.16}$$

$$t_0 = 0, v(0) = 0, x(0) = 0 \quad \Rightarrow t = \sqrt{2}\sqrt{\frac{x(t)}{a_0}} \tag{3.17}$$

We had already determined the scaling law corresponding to this in Section 2.2.2 using dimensional analysis, but now we actually do know that the pre-factor is $\sqrt{2}$! If we plot the time course of $v(t)$ and $a(t)$ like we did for the armadillo in figure 3.4 we can also

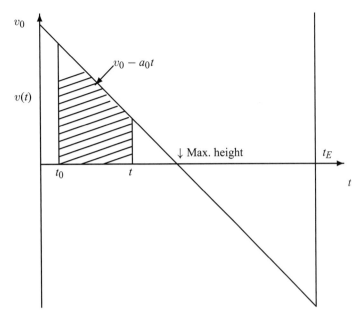

Figure 3.5 Velocity profile for a constant acceleration. The dashed area underneath the curve corresponds the the distance traveled between time t_0 and time t. t_E marks the end of the motion, which for the armadillo in Figure 3.1 is when it reaches ground again. The jumping and falling heights are the same, which implies that the areas of the triangles above and below the axis are equal.

determine the integral graphically. As Figure 3.5 shows, the area underneath the curve for the velocity corresponds to the travelled path.

3.2 Periodic Motions: Oscillations

An oscillation is a very special form of movement, namely one which repeats itself after a certain time. This time after which the motion repeats itself is called the period of the oscillation and is the most important variable that describes an oscillation. For some types of oscillation, harmonic oscillations, this period, together with the amplitude, i.e., the distance that the object moves maximally, is sufficient for a complete description of the movement.

In general, we speak of an oscillation when a physical quantity changes in time around an average, resting value. These general oscillations can also have increasing or decaying (damped) amplitudes in time; see Section 3.8.1. In that case, they are no longer strictly periodic. Oscillations can run by themselves or be forced externally, which we will discuss in depth in Section 4.2.

Further examples of oscillating systems are found in the central and inner ear, where the eardrum, the hammer, and the basilar membranes begin to vibrate according to the arriving sound waves, thus allowing the sensors in the ear to send appropriate signals to the brain. Vibrating electric charges generate electromagnetic waves (light, radiowaves, etc.), molecules, and atoms in the lattice of the solids oscillate around their rest position, their kinetic energy corresponding to the stored heat. We will see later that also more abstract quantities such as the numbers of animals or molecules can oscillate (see Section 3.5). What is striking about all of this is that the mathematics remains the same for every instance that we discuss, which is why we can illustrate the basic tenants of oscillations on a very simple example, while still getting useful information on more interesting problems. So for now, let us study the motion of an almost friction free rider on an air cushion rail (Figure 3.6).

In the first case (Figure 3.7), the rider runs along the rail without being accelerated (i.e., at a constant velocity) until it reaches the end of the rail, where its direction of motion and therefore the velocity are suddenly reversed by the collision with an elastic wall.

In the second case (Figure 3.8), the rider is attached at both ends using a spring, such that it is always drawn back to the starting point with a force that increases with the distance it departs from the rest position in the middle.

In both cases, the position of the rider can be registered with an electrical sensor at fixed time intervals. The two figures show such measurements for both cases, with and without

Figure 3.6   Air cushion rail.

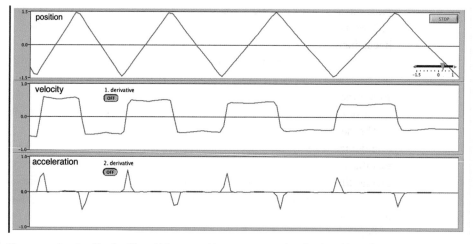

Figure 3.7   Measurement protocol for the rider, which moves with a constant speed on the air cushion rail.

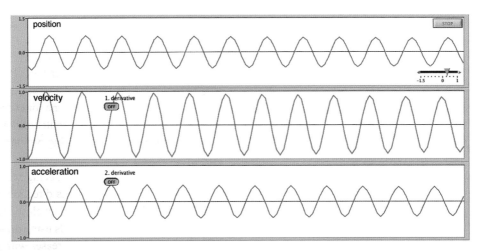

Figure 3.8 Measuring protocol for the rider on the air cushion rail, which is periodically oscillating back and forth due to the coupling with a spring.

springs. Because the rider travels between the ends of the rail regularly, we get a periodic motion that repeats with a certain period in both cases. By calculating the difference in position between each measuring point, we can also determine the velocity numerically, and again from changes in the velocity we obtain the acceleration. Notice that because we keep looking at differences, the fluctuations in the measurements are enhanced from position to velocity to acceleration, respectively. This is because, as we have seen in Section 2.1, the errors of the two differing positions are added (in squares) to obtain the error in the difference. Therefore, the relative error of the difference is much greater than the relative error in the positions.

In the first case, without acceleration, we experimentally obtain the relations that we have dealt with earlier for the special case of constant acceleration. At least we obtain this everywhere except at the ends, where the direction of motion is reversed, and we therefore must have an acceleration. Everywhere else, however, we have $a(t) = 0$, and therefore $v(t) = v_0$ on the way there and $v(t) = -v_0$ on the way back. That is, we measure a linear increase or decrease in the position. The fact that the rider repeatedly travels the same distance between the two ends makes its movement a periodic movement, which is represented here by a triangular curve. Because the position dependence is periodic, so are the velocity and acceleration. We will see in Section 3.3 in the context of Fourier expansions why this is so and how we can get from one periodic function to another.

The second measurement also shows a periodic movement that, however, appears rounder than the triangular curve of the rider without the springs. What we should notice is that the velocity and the acceleration that we have calculated from the position have a very similar time dependence to that of the position itself and look as if they are simply shifted. The acceleration actually is just the reflection of the position! Mathematically, this curve corresponds to a sine or a cosine function depending on where we put the zero point of our time axis. If the period T is the time the rider needs for the round trip and the maximum distance from the center of the rail that it reaches is the amplitude A, then we can describe the measured curve by the following function, which only has A and T as parameters, as already stated:

$$x(t) = A \cos\left(\frac{2\pi}{T}t\right) = A \cos(\omega t) \qquad \omega \equiv \frac{2\pi}{T} \qquad (3.18)$$

From this we can see that $x = A$ for times $t = 0$, $t = T$ and all further multiples of the period T. This implies that we have chosen time zero at the maximum excursion of the rider. By differentiating this we find the velocity, and with another derivative the acceleration:

$$v(t) = \frac{dx}{dt} = -\frac{2\pi}{T}A \sin\left(\frac{2\pi}{T}t\right) = -\omega A \sin(\omega t)$$

$$a(t) = \frac{dv}{dt} = -\frac{4\pi^2}{T^2}A \cos\left(\frac{2\pi}{T}t\right) = -\omega^2 A \cos(\omega t) \qquad (3.19)$$

These relations are confirmed by the measurement protocols. The transition from cosine to sine as we go from position to velocity corresponds precisely to the displacement of the curve by a quarter period. This also makes sense intuitively, because such a shift means that at the point where the position has a maximum, i.e., at the ends, the velocity becomes zero. This must be that way, since there the velocity is changing its direction, i.e., changing its sign, and hence has to be zero. Likewise, the velocity has its maximum where the position goes through the center, which also agrees with the observation, since at this point both springs are equally pulled out and thus the rider starts to be decelerated by the one which gets pulled out more strongly. We have here also introduced a new quantity ω, which is also called the angular frequency. We will see later what this angular frequency actually is; currently, we have only introduced it to make the equations a little shorter. The higher ω, the greater the amplitude of velocity and acceleration. Again, this makes sense, since a higher angular frequency means that the entire amplitude has to be traversed in less time, which is only possible if the speed is higher. What is even more interesting to note, however, is that the position $x(t)$ and the acceleration $a(t)$ have the exact same time dependence and are directly proportional to each other, i.e., $x(t) \propto a(t)$. This actually defines a harmonic oscillation, because this effectively sets out an equation that can only be solved by a sine or a cosine. We will look at this in detail in the following chapter. In terms of physics, we will see in Chapter 6 that acceleration is coupled to an acting force, such that these harmonic oscillations describe the case of a force proportional to the deflection, which in the rider was the role of the springs we have used.

In general, an oscillation occurs when there are restoring forces, which attempt to force the system back into the resting position when it is deflected. Therefore, in the rest position, the potential energy is always minimal. Again, we will return to this type of description when we cover conservation of energy in more detail (in Section 6.4).

We have seen that there are very specific cases of oscillations that are described by the fact that the acceleration is proportional to the deflection and that these correspond physically to the oscillation of a pendulum, where restoring force is proportional to the extension. This can happen in many different kinds of pendulums that we can all study with quantitative experiments. The basic equation that we have seen in order to obtain harmonic oscillations was that the acceleration, i.e., the curvature of position with time or the second derivative of position with time, had to be directly proportional to the negative position, i.e., $\frac{d^2x}{dt^2} = -\omega^2 x$. This negative sign means that physically, the force needs to be "restoring," i.e., it counteracts a deflection. This is what a spring does. When we pull out a spring, the spring pulls back with an equal force. If the force were to amplify the deflection, this deflection would grow unboundedly, which does not happen. Therefore, the negative sign makes sense physically.

So basically, we have the fundamental description for harmonic oscillations at hand in the preceding equation. The only thing that then describes the kind of system we are looking at is the angular frequency ω. To make this more explicit, let us discuss three different experimental realizations of harmonic oscillations in parallel to see that the same equations can describe quite different situations, but that if we have solved one, we have actually solved them all, which is one of the most useful things in the mathematical description of nature. The three systems are a horizontal spring connected to a frictionless sliding mass, a vertical spring on which a mass hangs, and a pendulum (see Figure 3.9).

In the first case, we want to describe the horizontal motion in the x-direction. If we apply the fact that the resulting force on the bob is given by its mass times the acceleration (see Section 6.2), we get the following:

$$m\frac{d^2x}{dt^2} = -kx$$

since the resulting force is given by a spring and the force is given by the spring constant k and the extension x. The second case is similar, but we now want to describe the position in the vertical direction, y, starting from a certain rest position y_1. In this case, if we add up the forces and equate them to the mass times the acceleration, we obtain the following:

$$m\frac{d^2y}{dt^2} = -k(y - y_1) - mg$$

where mg is the force due to gravity pulling on the vertically mounted bob. Finally, we do the same for the pendulum, where we describe the change in the angle and again we have to take into account the forces of gravity pulling on the bob, which gives the following:

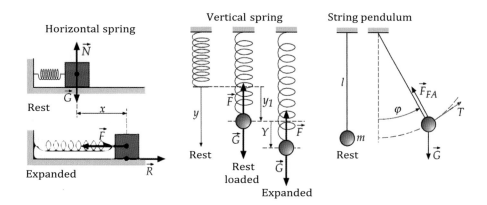

Figure 3.9   Three examples in which the force is proportional to the local change and therefore are capable of oscillation.

$$ml\frac{d^2\phi}{dt^2} = -mg\sin\phi \simeq -mg\phi$$

where we have used a Taylor expansion of the sine for small deflection angles in the approximation on the right-hand side.

While all of these equations look different and describe different situations, we can actually rewrite them and rename the variables, such that all of them look exactly the same and we have the basic differential equation for a harmonic oscillation:

$$\frac{d^2f}{dt^2} + \omega_0^2 f = 0$$

In the first case, f will be the horizontal position x and ω_0 is $\sqrt{k/m}$. Similarly, the second case has f stand for the vertical position y and ω_0 is $\sqrt{k/m}$. In this case, we have also defined y_1 as mg/k, such that the gravitational force is no longer part of the description (which makes sense, since this is constant and therefore only gives a constant shift). Finally, the third case has f stand for the deflection angle ϕ and ω_0 is $\sqrt{g/l}$.

Actually, we have seen this same equation before simply by describing the motion of a harmonic oscillator (a sine or a cosine), where the acceleration mirrored the displacement. Furthermore, we know from considering the units that ω_0 has to be a rate or angular frequency, and we will show shortly that it is in fact the angular frequency of the observed oscillation. For a general solution, we can therefore write down what the solution should be (make an ansatz), such that

$$f = f_0 \cos(\omega t + \delta),$$

This is not a large leap for us, since we actually already know that this will be the case from what we looked at before. What we will be able to do with this ansatz is see that ω_0 is in fact the angular frequency of the oscillation as well as study how the starting conditions influence the solution. The constant phase shift δ allows us to discuss solutions of the form $\cos(\omega t)$ as well as $\sin(\omega t)$ at the same time, since they correspond to phase shifts of $\delta = 0$ and $\delta = -\frac{\pi}{2}$ respectively. By taking the derivative with respect to time, we get the velocity:

$$\frac{df}{dt} = -f_0\omega \sin(\omega t + \delta)$$

which yields for the acceleration:

$$\frac{d^2f}{dt^2} = -f_0\omega^2 \cos(\omega t + \delta) \qquad = -\omega^2 f$$

Inserting this into the original equation, we obtain the following:

$$\omega_0^2 f = \omega^2 f.$$

This is always true as long as $\omega = \omega_0$. So even though f varies with time, the solution where $\omega = \omega_0$ works at any time, since the functional dependence of f drops out. Our ansatz therefore fulfills the equation if ω_0 is the angular frequency of the oscillation.

For such a harmonic oscillation, we call f_0 the *amplitude*, δ the *phase*, and ω_0 the *eigenfrequency*. The prefix *eigen*, from the German word for "own," denotes that the oscillating system will vibrate at this angular frequency once it has been started by whatever means. Actually *ansatz* and *eigen* are two of the very few instances of German words that have made it into the English language (the only others I know are *kindergarten*, *angst*, and *schadenfreude*).

We can now return to our three examples and reinsert the original variables for f. Before doing so, we want to still study how the initial conditions determine the solution, i.e., when we obtain a solution of the form $\cos \omega_0 t$ and when we obtain a solution of the form $\sin \omega_0 t$. Let's start at $t = 0$ with an amplitude f_0 and a mass at rest $df/dt = 0$. Then we find $\delta = 0$ and $f = f_0 \cos \omega_0 t$. If we start at $t = 0$ at the resting position, but with a speed v_0, we find $\delta = \frac{\pi}{2}$ and therefore $f = (v_0/\omega_0) \sin \omega_0 t$. From the initial conditions for the starting position and the starting speed, we can in general determine δ and f_0. The two preceding cases are just two reasonably common special cases.

This gives the following solutions for our three examples:

$$\text{horizontal spring}: \quad x = x_0 \cos(\omega_0 t + \delta) \quad \text{where} \quad \omega_0 = \sqrt{\frac{k}{m}}$$

$$\text{vertical spring}: \quad y = y_1 + y_0 \cos(\omega_0 t + \delta) \quad \text{where} \quad \omega_0 = \sqrt{\frac{k}{m}}$$

$$\text{mathematical pendulum}: \quad \phi = \phi_0 \cos(\omega_0 t + \delta) \quad \text{where} \quad \omega_0 = \sqrt{\frac{g}{l}}$$

The solution of the pendulum we had already found by dimensional analysis, when we looked at walking speeds, but now we know the prefactor. What we can read from this list is that while all of the systems show the same harmonic oscillation, the angular frequency ω_0 is determined by the specific physical properties of the system, i.e., $\omega_0 = \sqrt{k/m}$ for the spring and $\omega_0 = \sqrt{g/l}$ for the pendulum. This means that we can learn something about these physical properties of a system from the observation of the frequency of vibration. This is essentially the basis of all kinds of spectroscopic methods used to determine the chemical composition of substances, some of which we have discussed in the introduction and others we will see later on. Finally, the amplitude is determined by the amount of energy that is present in the oscillating system, i.e., the maximum amount of deflection.

In the case of idealized pendulums, the differential equation can also be deduced directly from the equations of motion, which we shall discuss in Section 6.2. The total energy of a harmonic oscillating system is constant, since the solutions found here describe idealized systems in which the masses swing on the springs without stopping or where the pendulum continues forever. In reality, all oscillations will stop at some point if they are not actively kept in motion. What happens in this case we will look at later in Section 3.8.1.

3.3 Describing Any Oscillation in Terms of Harmonic Ones: Fourier Series

The oscillations we initially saw in Section 3.2 were only characterized by their periodicity. However, we have seen that harmonic vibrations are easier to describe, since we need much fewer parameters for a complete description of the motion. Therefore, it would be nice if we could capture all periodic movements in the same style, namely, as a sum of harmonic (sine and cosine) functions. This is actually possible and has been mathematically formulated by Joseph Fourier. Thus mathematically, it can be said that every periodic function can be interpreted as the sum of harmonics or, more specifically and precisely, if $u(t)$ is a periodic function with a period T, i.e., $u(t + T) = u(t)$, then we find the following:

$$u(t) = \sum_{n=0}^{\infty}(A_n \cos(\omega_n t) + B_n \sin(\omega_n t)) \qquad \text{where} \quad \omega_n = \frac{2\pi n}{T}$$

If we know the Fourier coefficients A_n and B_n, of which in general there can of course be infinitely many, we directly know $u(t)$. Practically, however, only a small number of Fourier coefficients are necessary to give a reasonable description for most functions, one encounters. Conversely, if we know $u(t)$, we could directly calculate A_n and B_n from this:

$$A_0 = \frac{1}{T}\int_0^T u(t)dt, \qquad\qquad B_0 = 0$$

$$A_n = \frac{2}{T}\int_0^T u(t)\cos(\omega_n t)dt, \qquad\qquad B_n = \frac{2}{T}\int_0^T u(t)\sin(\omega_n t)dt \ (n \geq 1)$$

Thus the Fourier coefficients and the function itself are two sides of the same coin, and it depends on the application which point of view is easier to think about intuitively. The Fourier coefficients or their graphical representation, the so-called frequency spectrum, are very well suited for characterizing complex periodic functions.

As an illustration of this, let us look at the triangular oscillation we have seen earlier. If we limit the extension to 1 and want to have a triangular oscillation with a period T, our function can be described by the following two sections:

$$u(t) = 4\frac{t}{T}, \quad -T/4 \leq t \leq T/4; \quad u(t + T) = u(t)$$

$$u(t) = 2 - 4\frac{t}{T}, \quad T/4 \leq t \leq 3T/4; \quad u(t + T) = u(t)$$

With the preceding definitions, we can then calculate the corresponding Fourier coefficients:

$$A_n = \frac{2}{T}\int_{-T/4}^{T/4} 4\frac{t}{T}\cos(\omega_n t)dt + \int_{T/4}^{3T/4}\left(2 - 4\frac{t}{T}\right)\cos(\omega_n t)dt = 0$$

and

$$B_n = \frac{2}{T}\int_{-T/4}^{T/4} 4\frac{t}{T}\sin(\omega_n t)dt + \int_{T/4}^{3T/4}\left(2 - 4\frac{t}{T}\right)\sin(\omega_n t)dt = \frac{8}{\pi^2}\frac{(-1)^{(n-1)/2}}{n^2}\text{for } n \text{ odd}$$

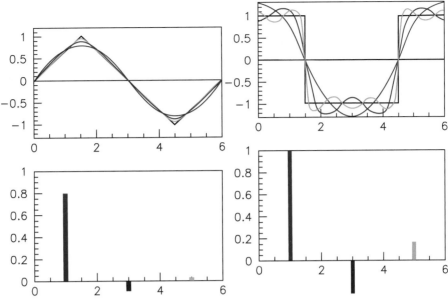

Figure 3.10 The triangular oscillation (top-left) we have observed for the position in the air cushion experiment and the square oscillation (top-right) we have observed for the velocity can be decomposed into their Fourier components. The original function is given by the black curve and successive approximations in terms of increasing Fourier components are given in red, blue, and green respectively. The bottom graphs show the frequency spectrum, i.e., the amplitudes of the corresponding terms in the Fourier series.

and we can therefore write the triangular oscillation as

$$u(t) = \frac{8}{\pi^2}\left(\sin \omega t - \frac{1}{3^2}\sin(3\omega t) + \frac{1}{5^2}\sin(5\omega t) + \dots\right)$$

This indicates a sum of harmonic oscillations, where increasing frequencies have smaller and smaller weight. This is illustrated in Figure 3.10, where successive approximations to the triangular oscillation are shown.

*For Specialists
One of the great advantages of describing osillations in terms of harmonic functions lies in the fact that we can take derivatives and integrals of harmonic functions much more easily. So, for instance, if we want to take the derivative of a function $u(t)$, we can do this to the Fourier expansion of the function:

$$\frac{du(t)}{dt} = \sum_{n=0}^{\infty}\frac{d}{dt}(A_n \cos(\omega_n t) + B_n \sin(\omega_n t)) = \sum_{n=0}^{\infty}(-A_n\omega_n \sin(\omega_n t) + B_n\omega_n \cos(\omega_n t))$$

Suppose that we compare this function to the Fourier expansion of the derivative itself:

$$\frac{du(t)}{dt} = \sum_{n=0}^{\infty}(A'_n \cos(\omega_n t) + B'_n \sin(\omega_n t)),$$

We find by comparing the coefficients in the sum that $A'_n = B_n \omega_n$ and $B'_n = -A_n \omega_n$. Analogously, we find for the coefficients of the second derivative $A''_n = -A_n \omega_n^2$ and $B''_n = -B_n \omega_n^2$. So, for instance, we can take the triangular oscillation we observed earlier and directly get out the Fourier components of the velocity, which should give a square wave oscillation, as: $A'_n = \frac{2\pi n}{T} \frac{8}{\pi^2} \frac{(-1)^{(n-1)/2}}{n^2} = \frac{16}{T\pi} \frac{(-1)^{(n-1)/2}}{n}$ for odd values of n and $B'_n = 0$. In fact, this is exactly the Fourier expansion one finds for a square-wave oscillation. Finally, the acceleration is given by $A''_n = 0$ and $B''_n = \frac{32}{T^2}(-1)^{(n+1)/2}$ for odd values of n.

We will get back to Fourier expansions and spectra for even more general types of functions when we discuss waves in Chapter 4.

3.4 Motions in Two and Three Dimensions

While we now have a good arsenal of tools to describe motions in one direction, the movements, which we initially said we wished to describe, are, however, by no means confined to simple one-dimensional cases. In Chapter 1, we have already seen the example of *Cataglyphis* ants and their paths in search for food. These take place in a flat desert and thus in two dimensions. It is no longer sufficient to only consider the time dependence of the position in a single direction. As shown in Figure 3.11, we can describe the motion using two such dependencies. That is, the location of the ant is described by a two-dimensional vector.

Also the armadillo will not actually jump directly into a single vertical direction, but will also move a little in the plane. So we actually need three directions (or coordinates) for such movements, which we will describe. In other words, our general description of a location will have to be a three-dimensional vector, corresponding to the three dimensions of the world we live in.

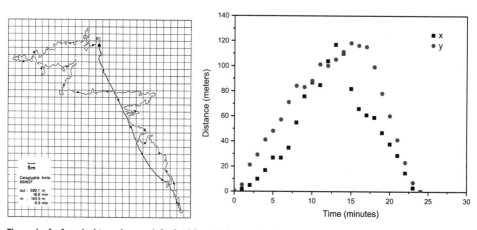

Figure 3.11   The path of a *Cataglyphis* on the search for food from Wehner and Wehner, 1990, used with permission from Taylor & Francis Ltd. Associated space-time diagram.

3.4.1 Vectors

As we have seen in the description of the movement of *Cataglyphis*, we have to divide the plane into two different coordinates. Very often, these are two directions that are perpendicular to one another. In this case, one speaks of Cartesian coordinates (named after Rene "I think therefore I am" Descartes). Since the two directions are perpendicular to one another, a movement in the one direction is never a movement into the other. This is why we can in this case describe the movement simply as the pair of movements in the two directions. This pair, summed up in one quantity, is the vector of movement. The individual directions are the components of the vector. If all the components are taken together and moved in their respective directions, the total direction of the vector and its magnitude are obtained. If I go one step each to the right and forward, I have actually moved diagonally to the forward right and covered a distance of $\sqrt{2}$ steps.

We write the vector as the sum of its components (in two dimensions, these are x, y) multiplied by the cartesian unit vectors $\vec{e}_1, \vec{e}_2$, pointing in the x and y directions respectively:

$$\vec{r} = x\,\vec{e}_1 + y\,\vec{e}_2 \tag{3.20}$$

Here the magnitude of the vector is given by its length

$$|\vec{r}| = \sqrt{x^2 + y^2} \tag{3.21}$$

and the direction is given by the following:

$$\tan\theta = \frac{y}{x} \tag{3.22}$$

In three dimensions, we accordingly need three different coordinates to specify the position in x, y, and z, where we have three different unit vectors pointing in these different directions, such that the position vector becomes the following:

$$\vec{r} = x\,\vec{e}_1 + y\,\vec{e}_2 + z\,\vec{e}_3 \tag{3.23}$$

Again, the length of the vector is given by its magnitude

$$|\vec{r}| = \sqrt{x^2 + y^2 + z^2} \tag{3.24}$$

and because we are now dealing with a three-dimensional object, the direction has to be specified not just by a single angle, but by two angles, corresponding to the latitude and the longitude, where the longitude, or azimuth angle, is given by the following:

$$\tan\phi = \frac{y}{x} \tag{3.25}$$

and the latitude or inclination is given by the following:

$$\cos\theta = \frac{z}{\sqrt{x^2 + y^2 + z^2}} \tag{3.26}$$

Thus irrespective of how we want to represent the position, we always need the same number of components as the vector has dimensions.

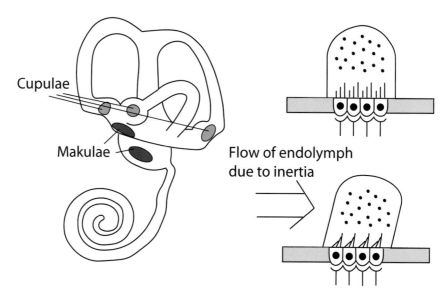

Figure 3.12 The accelerometer in the middle ear. The three intertwined tubes of the vestibular organ (labyrinth on the left) measure the three components of the acceleration via the inertia of the endolymph.

Since we are moving in a three-dimensional world, we must also measure three components of the acceleration to determine our state of motion, i.e., the components a_x, a_y, a_z. We humans do this in the inner ear (as most other animals do as well), with a combination of three circularly shaped tubes, the vestibular organ (see Figure 3.12). These three vestibules are positioned perpendicularly to each other and thus form the basis for measuring the three components of a cartesian coordinate system for the acceleration acting on our head. What happens is that when the head is accelerated, the fluid in the vestibules, the endolymph, begins to flow. We'll see where that comes from when studying mechanics. The inertia of the liquid implies that the endolymph does not strictly follow the acceleration of the head, hence a flow is induced relative to the vestibules, which are fixed to the head. This flux causes a force on the cupulae, which are fastened at the bottom in the tubes. These cupulae are connected to neurons, which pass the signal of the measurement to the brain.

3.4.2 Differentiating the Vectors of Position and Velocity

As we have just discussed, position, velocity, and acceleration are given by vectors in a three-dimensional space. We have also said that velocity and accelerations are given by derivatives of the position. This begs the question: how do I differentiate a vector?

Vectors are actually a special kind of multivalued function of several variables. Here we consider vector fields in the strict sense as a three-valued function $\vec{r}(x, y, z)$ defined at every point in three-dimensional space. The vector should also be able to change over time, so we must have the following:

$$\vec{r} = \vec{r}(x, y, z, t) \tag{3.27}$$

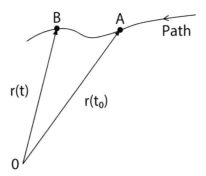

Figure 3.13 The position vector $\vec{r}$ contains all the information about the distance to the point of reference: $OA \equiv |\vec{r}| =$, i.e., length of the position vector and direction in space. $\vec{r}$ depends on time.

In the previous section, we have written a vector as a combination of its components r_x, r_y, r_z and the cartesian unit vectors $\vec{e}_1, \vec{e}_2, \vec{e}_3$:

$$\vec{r} = r_x\,\vec{e}_1 + r_y\,\vec{e}_2 + r_z\,\vec{e}_3 \qquad (3.28)$$

We can apply the sum and product rules for derivatives to this equation and since the unit vectors of the coordinate system do not change with time, we basically only have to take the time derivatives of the components of the vector and combine them with the same unit vectors:

$$\frac{d\vec{r}}{dt} = \frac{dr_x}{\partial t}\,\vec{e}_1 + \frac{dr_y}{dt}\,\vec{e}_2 + \frac{dr_z}{dt}\,\vec{e}_3 \qquad (3.29)$$

As a special example, let's look at the position vector and its derivatives.

If an object moves along an arbitrary path in three-dimensional space, we mark its position relative to a point of reference by the position vector $\vec{r}$. A two-dimensional example of this is the path of *Cataglyphis* in Figure 3.11. A short part of such a path is shown in an idealized way in Figure 3.13.

If we have a time-dependent position vector, which then is given by the time dependence of the different coordinates multiplied by the respective unit vectors, that can be summarized as follows: Given the time dependence of a position vector:

$$\vec{r}(t) = x(t)\,\vec{e}_1 + y(t)\,\vec{e}_2 + z(t)\,\vec{e}_3 \qquad (3.30)$$

The derivative of the position vector with respect to time gives the velocity $\vec{v}$, and the second derivative gives the acceleration $\vec{a}$:

$$\vec{v}(t) = \frac{d\vec{r}(t)}{dt} = \frac{dx(t)}{dt}\vec{e}_1 + \frac{dy(t)}{dt}\vec{e}_2 + \frac{dz(t)}{dt}\vec{e}_3 \qquad (3.31)$$

$$\vec{a}(t) = \frac{d^2\vec{r}(t)}{dt^2} = \frac{d^2x(t)}{dt^2}\vec{e}_1 + \frac{d^2y(t)}{dt^2}\vec{e}_2 + \frac{d^2z(t)}{dt^2}\vec{e}_3 \qquad (3.32)$$

3.5 A Circular Motion Is an Oscillation

A special case of a planar, two-dimensional oscillation that occurs frequently in nature is when an object moves on a circular path. If we look at the motion separately in the two directions, we see that the same functions are produced here as in the one-dimensional oscillation that we have considered previously. The only difference is that the two components are offset by a quarter period. Thus, if the motion in the x-direction corresponds to a cosine, the motion in the y-direction is a sine. The amplitude of these two oscillations is the radius of the circle for both cases. Finally, the circular path is defined precisely by the fact that the position is always at the distance of the radius r from the center of the circle. Mathematically, the position vector must satisfy the following condition:

$$r = |\vec{r}(t)| = \text{const} \tag{3.33}$$

But $\vec{r}(t)$ is not at all constant; it actually rotates around the circle and thus continuously changes direction. Similarly, $\vec{v}(t) = d\vec{r}/dt$ is nonzero and not constant. It also rotates around the circle and constantly changes direction. If we consider the circular motion as described by the two harmonic oscillations indicated previously, the position vector is given by the following:

$$\vec{r}(t) = r \begin{pmatrix} \cos(\omega t) \\ \sin(\omega t) \end{pmatrix} \tag{3.34}$$

Because sine squared plus cosine squared is always one no matter what the angle, i.e., $\sin^2 \alpha + \cos^2 \alpha = 1$ for every α, we also see that this position fulfills the condition for the circular orbit, i.e., we have $|\vec{r}(t)| = r = const$. If we now want to calculate the velocity of this orbit, we have to take the derivative of the two components with respect to time, as stated earlier. This means that the sine becomes a cosine and the cosine becomes a sine or written as an equation:

$$\vec{v}(t) = r \begin{pmatrix} \frac{d\cos(\omega t)}{dt} \\ \frac{d\sin(\omega t)}{dt} \end{pmatrix} = r\omega \begin{pmatrix} -\sin(\omega t) \\ \cos(\omega t) \end{pmatrix} \tag{3.35}$$

Again we see from the addition of sine and cosine in squares that the magnitude of the velocity, i.e., the speed, is constant for this circular orbit and given by $|\vec{v}(t)| = r\omega$. So even though the velocity is constantly changing due to the constant change in direction, such a harmonic circular motion corresponds to a motion with constant speed. From this we also get an intuition for the meaning of the angular frequency ω. The preceding speed increases with the radius of the circle, which makes sense, since for a given period, the path traveled per period and hence the speed goes directly with the radius. However, the other determining part of the speed is how fast the angle within the circle changes. This rate of change of the angle then is the angular frequency!

But what is the direction of the velocity with respect to position? To determine this, we can calculate the scalar product of $\vec{r}$ and $\vec{v}$. This corresponds to the projection of the two vectors onto each other and thus gives the angle between them. Thus we have to determine the following:

$$\vec{r}(t) \cdot \vec{v}(t) = r \begin{pmatrix} \cos(\omega t) \\ \sin(\omega t) \end{pmatrix} \cdot r\omega \begin{pmatrix} -\sin(\omega t) \\ \cos(\omega t) \end{pmatrix} = \qquad (3.36)$$

$$= r^2\omega(-\cos(\omega t)\sin(\omega t) + \sin(\omega t)\cos(\omega t)) = 0, \qquad (3.37)$$

which implies that the position and the velocity are perpendicular! As can be seen in the sketch, the direction of the velocity is identical to the direction of the tangent to the orbit, i.e., the circle with radius r. We actually could have known this without any calculation, since we had previously identified a derivation with the tangent and the tangent of a circle is always perpendicular to the radial direction.

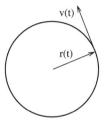

If we consider once more the previous velocity vector for the harmonic oscillation, we should notice something. The x component of the velocity has the same form as the y component of the position. There is a sine. Similarly, the y component of the velocity is given by the x component of the position. So we could also take this description of the orbit and write this as two other equations for x and y:

$$\frac{dx}{dt} = -\omega y$$
$$\frac{dy}{dt} = \omega x$$

We have now read this directly from the special form for the harmonic circular orbit. We could, however, also arrive at this system of equations, by just looking at the orbit itself. If the position describes a circle rotating with an angular frequency ω, the x component changes just as the angle ωy is added. Furthermore, we know that x decreases when we push a part of the position into the y direction. Therefore, the equation for the temporal change of x must be negative. When y is changed, on the other hand, we obtain the exact opposite; here y increases when we are at a certain x position and the size of this increase is ωx. Writing this as differential equations, i.e., identifying the rate of change with the change we have identified, we directly obtain the two equations we had obtained from considering the special property of the sines and cosines. However, the exact same equations could also be obtained from a population dynamics model if x and y are not positions of a circular orbit, but population numbers of different species that mutually influence their rate of proliferation. We will look at this in more detail, but first we will discuss the acceleration in circular motion.

What is the acceleration $\vec{a} = d\vec{v}/dt$ in circular motion? Well, we have seen in one-dimensional oscillations that the acceleration is directly given by the negative position. Mathematically speaking, we had to multiply the position with $-\omega^2$ to obtain the acceleration. This will have to be the same for circular motion, since we are dealing with two oscillations, where we can do the same operation for both components. $\vec{a}$ will thus point toward the center of the circle and have the magnitude $|\vec{a}_N(t)| = r\omega^2$.

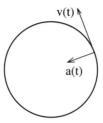

This acceleration is also called *centripetal acceleration*. The index N denotes that the direction of this centripetal acceleration is normal, i.e., perpendicular, to the orbit and thus the velocity.

Should there be a component of the acceleration parallel to the orbit, we would call this the *Tangential-component a_T*. For a circular orbit with constant speed, we do, however, have the following:

$$a_T = 0 \tag{3.38}$$

because the speed, i.e., the velocity in the direction of the orbit, does not change.

(*) For Specialists

We can show the relations that we have derived for harmonic circular motions also in a more general setting. For any circle, we have the following:

$$\vec{r} \cdot \vec{r} = r^2 = \text{const} \qquad \Rightarrow \qquad \frac{dr^2}{dt} = 0 \tag{3.39}$$

This derivative can be rewritten using the product rule, in the following form:

$$0 = \frac{dr^2}{dt} = \frac{d\vec{r}}{dt} \cdot \vec{r} + \vec{r} \cdot \frac{d\vec{r}}{dt} = 2\vec{v} \cdot \vec{r} \tag{3.40}$$

This is the result we had already obtained for the harmonic circular motion, namely $\vec{v} \perp \vec{r}$.

We can show for general circular motions with constant speed what the acceleration is. We have by definition the following:

$$v = |\vec{v}(t)| = \text{const} \tag{3.41}$$

and we use the same trick as before:

$$\vec{v} \cdot \vec{v} = v^2 = \text{const} \qquad \Rightarrow \qquad \frac{dv^2}{dt} = 0 \tag{3.42}$$

which by virtue of the product rule implies the following:

$$0 = \frac{dv^2}{dt} = \frac{d\vec{v}}{dt} \cdot \vec{v} + \vec{v} \cdot \frac{d\vec{v}}{dt} = 2\vec{a} \cdot \vec{v} \tag{3.43}$$

We therefore find that $\vec{a} \perp \vec{v}$. Since we are moving in a plane, $\vec{a}$ and $\vec{r}$ therefore have to be either parallel or antiparallel. As we have just seen, $\vec{v} \cdot \vec{r} = 0$, which implies $0 = \frac{d(\vec{v} \cdot \vec{r})}{dt}$, and again using the product rule

$$0 = \frac{d(\vec{v} \cdot \vec{r})}{dt} = \frac{d\vec{v}}{dt} \cdot \vec{r} + \vec{v} \cdot \frac{d\vec{r}}{dt} = \vec{a} \cdot \vec{r} + v^2 \tag{3.44}$$

We divide by r solve for $\vec{a}$:

$$\vec{a} \cdot \frac{\vec{r}}{r} = -\frac{v^2}{r} \tag{3.45}$$

The scalar product $\vec{a} \cdot \frac{\vec{r}}{r}$ gives the component of $\vec{a}$ in the direction of $\vec{r}$, which again is the result we have found earlier.

3.6 Circular Motion as a Population-Dynamic System

Now we have a good grasp of oscillations and circular motions, you may think that this is all fine and well, but why would I want to describe a pendulum in many different ways, which all turn out to be the same thing? Well, let's look again at the equations describing a circular orbit by describing the changes of the x and y components, respectively:

$$\frac{dx}{dt} = -\omega y$$
$$\frac{dy}{dt} = \omega x$$

We had arrived at these equations from considerations about the positions in a circular orbit. But we could also proceed quite differently. Imagine we want to describe an ecological system of two species that are interacting. Let us take foxes and rabbits, the number of rabbits being x and the number of foxes y. If there are many foxes, the rabbits are eaten, so the change in the number of rabbits (dx) is negatively affected by the number of foxes (y). The change in the number of foxes (dy) is therefore positively influenced by the number of rabbits (x). If both of these processes happen with the same rate ω, we have just described the two preceding equations!

We therefore already know what the solution to the dynamical system of rabbits and foxes is, since we know the solution of these two equations. We had already seen this earlier in the description of the circular orbit. So there will be an oscillation in both x and y, which are, however, shifted by a quarter period (a sine wave and a cosine wave) respectively. This means that also the numbers of rabbits and foxes will oscillate as a function of time, with a phase shift of a quarter period. This also makes sense when considering the ecological situation. At first, the foxes can feast on the rabbits, which makes their number increase, but then the number of rabbits decreases, which after a while (the phase shift) also decreases the number of foxes. When there few enough foxes, the population of the rabbits can recover and the cycle (same word as circle!) starts anew. What we thus see once again is that the equations basically only describe relations between things, and these things can be anything as long as they follow the relations set out in the equation. Then if we know the behavior of the equation, we know how the things we want to describe will interact. To get to that understanding of the behavior, we can use any of the things that follow the equation and thus get a simplified picture of things.

This general behavior can also be read qualitatively from the data shown in Figure 3.14 for lynxes and snow hares, but we can see that the oscillations are not nice sines and cosines. In order to describe this better, we must adapt the equations somewhat, which we will do for specialists in the next chapter.

If we did not know the solution, we should be able to deduce from the two equations that the curve actually describes a circle, and that the x and y components are oscillating. We'll look at this now. If we simply want to describe the orbit, we actually need not consider the temporal component of the changes. This means that we can eliminate the time from the system of the two equations. For this, we can look at a small time interval dt, by rearranging the equations to solve for dt. For the equation for the temporal change of x, we for instance

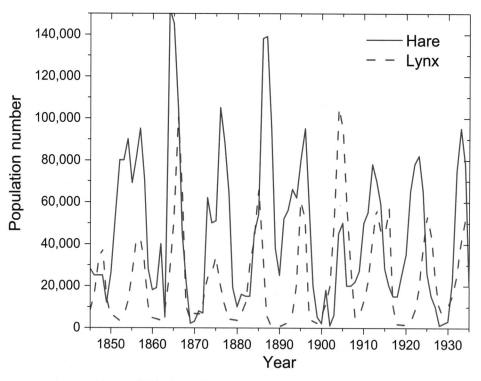

Figure 3.14 Measurement series for the population numbers of lynxes and snow hares over several years. It can be seen that the populations oscillate with a certain period, but is phase shifted by roughly a quarter period. Data from MacLulich, 1937.

obtain $dt = -\frac{dx}{\omega y}$. We can do the same for the equation describing the temporal change of y, which yields $dt = \frac{dy}{\omega x}$. Both of these time intervals have to be equal, so we can set these two results equal and obtain the following:

$$\frac{dy}{dx} = -\frac{x}{y}$$

We could also have obtained this by simply dividing the two preceding equations. What we have achieved now is that we have a description of the dependence of the y component on the x component, that is, a direct description of the orbit. If we want to actually have to orbit itself, we will now have to solve this equation. To do this, we take the x dependent terms on one side and the y dependent ones on the other to obtain:

$$ydy = -xdx$$

We can integrate this and have the following:

$$x^2 + y^2 = const$$

which is the definition of a circle! If we want to derive the oscillation from the two equations, then we must look at the change of the change, i.e., the acceleration, in one component. We must therefore once again perform mathematically a time derivative. For the equation describing the rate of change in x, we therefore obtain $\frac{d^2x}{dt^2} = -\omega\frac{dy}{dt}$. But we know that $\frac{dy}{dt}$ is directly proportional to x or more precisely that $\frac{dy}{dt} = \omega x$. We can put this into the previous equation and then have a description for only the time course of x:

$$\frac{d^2x}{dt^2} = -\omega^2 x$$

which is just the equation that describes an oscillation with an angular frequency of ω!

3.7 *Nonlinear Dynamical Systems in Ecology

We have described in Section 3.6 a population of rabbits and foxes by means of two coupled differential equations, which have given rise to oscillations. We have also seen that such oscillations occur in nature, but on closer inspection the data do not quite match up with the expectation based on these equations. A very important point is that the populations we have described can be negative, which of course cannot be the case for real population numbers. We can solve this problem quite simply by saying that the quantities x and y we have used in the equation represent the deviation from a mean population. In case these deviations are small compared to the average population, the equations can remain the same and we would get the same result. However, the data show that in the example of the lynxes and snow hares in Canada (and also more generally in other data we do not show), the deviations cannot necessarily be assumed to be small. This means that we have to change the equations that describe the system. There are three points that we need to change if we want to make the description a bit more realistic. First, the rabbits multiply (like bunnies). That is, the change of the rabbit population (dx) must also depend on the number of available rabbits (x). Secondly, the number of rabbits eaten, or the increase in the population of foxes by eating rabbits, does not only depend on the respective number of rabbits or foxes. The foxes must meet rabbits in order to eat them. The probability of this is therefore the higher the more rabbits and the more foxes there are. That is, the (negative) change of the number of rabbits (dx) and the (positive) change of the number of foxes (dy) have to depend on the product of the numbers of rabbits and foxes ($x \cdot y$). Finally, the change in the number of foxes (dy) also depends on their own number (y), but negatively, because if there are too many foxes they start to compete with each other, leading to selection pressure. We now put all of these insights into equations, by adding all of these contributions for the changes in numbers of rabbits and foxes, and obtain the following:

$$\frac{dx}{dt} = \alpha x - \beta xy$$

$$\frac{dy}{dt} = \delta xy - \gamma y$$

The constants α, β, γ and δ describe the respective rates for the processes of reproduction (α), eating (δ), and being eaten (β), as well as intraspecies competition (γ). This system of equations is known as the Lotka–Volterra equations due to the two people who first developed it. It describes a general problem of two coupled predator and prey species. These equations are now much more complicated than those that we derived earlier for the circular orbit, and it can no longer be solved analytically in all circumstances, which is why it is usually solved numerically for certain cases of interest. Such a numerical solution is shown in Figure 3.15 on the left. It can be clearly seen that there is an oscillation of the population numbers, but they do not correspond to a sines or cosines. However, the shift by about a quarter period is still there. These curves are now much more similar to the data we had seen earlier.

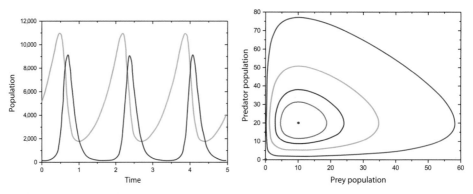

Figure 3.15 Left: Numerical solution of the Lotka–Volterra equations. Plotted in red is the population of the foxes and green is the population of the rabbits. The rates in the simulation are $\alpha = 2$/year; $\gamma = 10$/year; $\beta = 0.001$/(year*fox); $\delta = 0.002$/(year*rabbit). Right: The phase space plot of the predator–prey system. The curves correspond to different values of the constants of the integration, which effectively corresponds to different initial conditions in populations.

While there is no general solution to these equations we can discuss in detail, there are still several aspects about the solution of the system we can see without solving it completely. This gives us a more intuitive understanding of the system and therefore we will look at a few properties of this solution. On the one hand, we can look at what happens when populations actually do not change. Such a steady state will exist if, for instance, the increase in the number of rabbits by their reproduction is just compensated by the number of rabbits being eaten by foxes. However, this will surely depend strongly on the rates in the equation. The question is, however, which of the rates will be decisive for the respective populations? To find out, let's look at the two equations and ask ourselves what the populations are in equilibrium. In this case, the numbers of rabbits and foxes do not change, so dx/dt as well as dy/dt both must be equal to zero. We thus obtain the following:

$$\alpha x_0 - \beta x_0 y_0 = 0$$

$$\delta x_0 y_0 - \gamma y_0 = 0$$

Here, x_0 and y_0 denote the population numbers of rabbits and foxes in equilibrium respectively. In the first equation, we can factor out x_0, whereas in the second equation, we can factor out y_0. For both, we can then divide by the factored-out equilibrium population (it should be nonzero). This yields the following:

$$\alpha - \beta y_0 = 0$$

$$\delta x_0 - \gamma = 0$$

or:

$$\alpha = \beta y_0$$

$$\delta x_0 = \gamma$$

and hence

$$y_0 = \frac{\alpha}{\beta}$$

$$x_0 = \frac{\gamma}{\delta}$$

This means that the equilibrium population of foxes does not depend on the intrinsic competition between foxes, but only on the reproduction rate of the rabbits. Likewise, the number of rabbits does not depend on their rate of reproduction! This would not really have been expected intuitively. Historically, such a counterintuitive observation was the reason for the establishment of these equations. When looking at the number of fish caught in the Adriatic after the First World War, it was observed that the lack of fishing due to the war had not led to an increase in the fish population, but rather to an increase in the population of sharks, i.e., the predators of the fish. This is exactly the behavior expected from the Lotka–Volterra equations.

There is still more that we can say about the solutions to the equations. Just as we were able to derive the circular motion from the harmonic oscillation, we can directly look at the dependence of the number of rabbits on the number of foxes. We therefore again divide the two equations and get an equation describing the dependence of $y(x)$ on x:

$$\frac{dy}{dx} = \frac{y}{x}\frac{\delta x - \gamma}{\alpha - \beta y}$$

We can solve this equation in a similar way as we did for the harmonic oscillation. We move all terms with x and dx on one side of the equation and all terms with y and dy on the other, which gives the following:

$$\left(\frac{\alpha}{y} - \beta\right) dy = \left(\delta - \frac{\gamma}{x}\right) dx$$

This can be integrated on noth sides, and we obtain the following:

$$\alpha \ln(y) + \gamma \ln(x) - \beta y - \delta x = const$$

These curves are shown for different values of the constant in Figure 3.15 on the right. Along these curves, the system runs at an angular velocity (the angular frequency) of $\omega = \sqrt{\alpha\gamma}$, such that one can imagine the general nature of the oscillation corresponding to the numerical solution. For example, if the number of rabbits increases, the number of foxes remains constant at first, then slowly increases, but only increases as the number of rabbits decreases. The number of foxes increases even further, almost until the number of rabbits has reached their minimum when there is an sharp decrease in the number of foxes, and then the cycle starts again from scratch.

3.8 *Damped and Coupled Oscillations

3.8.1 Damped Oscillations

If there is friction in the system (i.e., the system is dissipative), the energy stored in the pendulum is no longer constant, and it decreases slowly. Correspondingly, the equation that describes the oscillation changes. The amplitude of the oscillation has to decrease with time, as is also the case of a damped oscillation. The corresponding differential equation must therefore have additional terms. In corresponding equation of motion is that of a friction force. If we start from a viscous friction, the force is proportional to the velocity with which the object moves. That is, in the general equation, there is an additional term proportional to $\frac{df}{dt}$. In addition, because the units have to match up, there has to be an additional corresponding time scale, which we will call τ_0. We thus obtain the following as the basic equation:

$$\frac{d^2f}{dt^2} + \frac{1}{\tau_0}\frac{df}{dt} + \omega_0^2 f = 0$$

In essence, τ_0 must describe the time scale on which the amplitude decays. This is why it is called the decay time. Looking at the equation, we now have two time scales that describe the motion, on the one hand the oscillation period T, and on the other hand the decay time τ_0. Depending on how these two times compare to each other, we will expect different behavior. If the decay time is greater than the period, the pendulum will oscillate. After all, there is still amplitude present after a single period of oscillation, so there will certainly be some kind of to-and-fro motion. If the decay time is shorter than the period, then there will be no oscillation at all, because the amplitude has already decayed away before the pendulum has moved back and forth. We see this intuitive solution to the problem when we consider a pendulum damped in a liquid. There we have two clearly separated regimes for the solution. On the one hand, a vibration, whose amplitude, however, decreases with time. On the other hand, if the liquid is too viscous, we have a movement that no longer oscillates, but where the movement is so strongly damped that the object returns to the resting position directly.

We want to make this description a bit more quantitative and see exactly where the border between the two regimes lies. Let us first consider the case of simple decay. The amplitude can then be described by an exponential decaying function. It must be exponential, since in the equation all terms are proportional to different derivatives of the function, and the exponential function is precisely that function whose derivatives are proportional to itself. Therefore, we "guess" the solution as an exponential function with an unknown decay time τ: $f(t) = f_0 \cdot \exp(-t/\tau)$. If we insert this solution to the equation, we can determine the decay time τ:

$$\left(\frac{1}{\tau^2} - \frac{1}{\tau \cdot \tau_0} + \omega_0^2\right) f_0 \cdot \exp(-t/\tau) = 0$$

This has to be true for all times t, which implies that we can divide by the exponential time dependence as well as the amplitude f_0 to obtain an equation for τ:

$$\left(1 - \frac{\tau}{\tau_0} + \tau^2 \omega_0^2\right) = 0$$

We solve this quadratic equation and obtain the following:

$$\tau = \frac{1/\tau_0 - \sqrt{1/\tau_0^2 - 4\omega_0^2}}{2\omega_0^2} = \frac{1}{\omega_0^2}\left(\frac{1}{2\tau_0} - \sqrt{\frac{1}{4\tau_0^2} - \omega_0^2}\right)$$

or

$$\frac{1}{\tau} = \frac{1}{2\tau_0} + \sqrt{\frac{1}{4\tau_0^2} - \omega_0^2}$$

If the decay time τ_0 is very short (that is, it has very strong damping), the preceding equation yields a time scale for the falloff of the excitation of $\tau = \tau_0$, which is what we would have expected intuitively. If the damping becomes somewhat lighter, however, the decay time is also affected by the frequency (or period). Finally, when $4\omega_0^2 > 1/\tau_0^2$ or $\omega_0 > 1/(2\tau_0)$, then we suddenly have a square root of a negative value in the solution for τ, which cannot be. Therefore, the simple decay of the amplitude will no longer occur for such long decay times. Specialists can read later how one can still make sense of this via imaginary or complex numbers and also how this solution describes the oscillating solution.

But in fact, we expected this, since in this case the decay time is longer than the period of the fundamental oscillation, and then there can be no purely exponential decay of the amplitude. As already indicated, in our intuitive expectations, in this case, we have to find an oscillation, albeit with a decreasing amplitude. The essential point is, however, that we now have an exact, quantitative limit for the occurrence of this behavior, where we would not have expected the factor 2 intuitively.

Now let us also describe the decaying oscillation. This can be described as a function $f(t) = f_0 \cdot \exp(-t/\tau) \cdot \cos(\omega t)$, where we have chosen the time such that $t = 0$ corresponds to the maximum amplitude, f_0, which then decreases exponentially. If we insert this into the equation, we must work a little more since we now have the product of two time-dependent functions. This means that we must use the product rule. Then we get the following:

$$\left(\frac{1}{\tau^2} - \frac{1}{\tau \cdot \tau_0} - \omega^2 + \omega_0^2 \right) f_0 \cdot \exp(-t/\tau) \cdot \cos(\omega t) + \left(\frac{2\omega}{\tau} - \frac{\omega}{\tau_0} \right) f_0 \cdot \exp(-t/\tau) \cdot \sin(\omega t) = 0$$

Now we can no longer simply ignore the time-dependent function and thus forget how the exponential case still went. For the exponential part, this also applies here, but then we still have terms with sine and cosine functions. We are lucky, however, because the sine and the cosine are independent functions, which means that they are never zero at the same time. Therefore, the prefactors of all the sines must be zero, as well as the prefactors of all the cosines. Only then is the equation satisfied at all times. That is, we have two equations that determine this condition:

$$\frac{1}{\tau^2} - \frac{1}{\tau \cdot \tau_0} - \omega^2 + \omega_0^2 = 0$$

and

$$\frac{2\omega}{\tau} - \frac{\omega}{\tau_0} = 0$$

This is very convenient because we have two unknown variables that we really want to determine: the frequency of the oscillation ω and the decay of the oscillation τ. The second of these equations gives us a direct relationship between τ and τ_0, namely $\tau = 2 \cdot \tau_0$. We can use this in the first equation and obtain the following:

$$-\frac{1}{4\tau_0^2} - \omega^2 + \omega_0^2 = 0$$

This results in the angular frequency of the damped oscillation to be $\omega = \sqrt{\omega_0^2 - \frac{1}{4\tau_0^2}}$.

Here, too, there is a limit to the description where the expression underneath the square root becomes negative, namely, $1/\tau_0^2 > 4\omega_0^2$, which is exactly the opposite of the previous case. Thus, these two solutions describe all possible cases. It should be noted that the decay of the amplitude is the fastest on the boundary between these two cases, that is, if $\omega_0 = 1/(2\tau_0)$, and occurs within the half period of an oscillation. Thus we see where the factor of two, which we had found previously and which we would not have expected intuitively, comes from. This damping, which causes the oscillation to come to a standstill the fastest, one

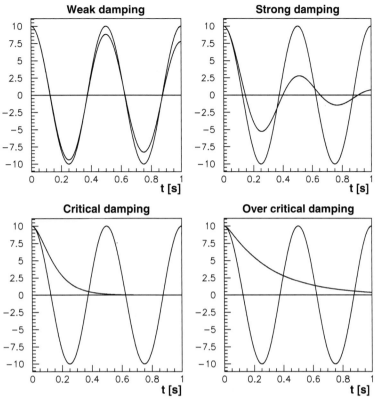

Figure 3.16 Various cases of a damped oscillation. With very weak damping, an almost harmonic oscillation results: when the damping gets stronger, we obtain an exponentially decaying oscillation. In the overcritically damped case, an exponential decay results, which is the fastest when the damping is exactly critical.

also calls critical damping. The various cases of damped oscillations that we've discussed here are graphically illustrated in Figure 3.16.

(*) For Specialists

A uniform description of the process is obtained by using complex numbers. Then one can use the direct approach via the exponential function for all cases, the cosine and sine functions being described by imaginary exponentials: $\exp(ix) = \cos(x) + i\sin(x)$. The imaginary part of the solution thus describes the angular frequency of the oscillation, while the real part describes the decay rate. The solution ansatz is then always $f(t) = f_0 \exp(i\gamma t)$, such that $\gamma = \omega + \frac{i}{\tau}$. If we put this into the equation, we obtain the following:

$$-\gamma^2 + \frac{i\gamma}{\tau_0} + \omega_0^2 = 0$$

Solving for γ gives us the following:

$$\gamma = -\frac{i}{2\tau_0} + \sqrt{-\frac{1}{2\tau_0^2} + \omega_0^2}$$

This describes both cases. If the term underneath the square root is negative, γ is purely imaginary, and we have an exponential decay, while if the term under the square root is positive, then this square root is the real part, that is, the angular frequency, and $2\tau_0$ is the decay time.

3.8.2 Coupled Oscillations

If several oscillating systems interact with one another, that is, are coupled together, new eigenfrequencies belonging to certain collective oscillation states occur, in addition to the natural frequencies of the individual oscillators. These natural oscillations of the coupled system can be selectively stimulated by the choice of suitable initial conditions. The more systems interact, the greater the number of characteristic vibrational states and the more diverse are the phenomena. We restrict ourselves here to the case of two harmonic oscillators connected to each other by a coupling spring. This coupling is important in Chapter 4 to make the jump from vibrations to waves.

Whether we are dealing with two pendulums connected by a spring, or with two balls connected elastically to the walls and connected to each other, the fundamental vibrations of the two systems are the same, namely, a vibration with the center of gravity at rest and harmonic variation of the relative distance or a vibration with an unchanged distance and a harmonic oscillation of the center of gravity. The corresponding eigenfrequencies are determined from the equations of motion. Two identical, undamped linear oscillators have the same eigenfrequencies in the uncoupled case:

$$\omega_0 = \sqrt{\frac{k}{m}} \qquad \frac{d^2 x_{1,2}}{dt^2} + \omega_0^2 x_{1,2} = 0$$

If the connecting spring has the spring constant k', we must take into account the corresponding spring force in our description. Then we get two coupled equations describing the motion of the two "pendulums":

$$m\frac{d^2 x_1}{dt^2} = -kx_1 + k'(x_2 - x_1)$$

$$m\frac{d^2 x_2}{dt^2} = -kx_2 - k'(x_2 - x_1)$$

With the substitution $\frac{1}{2}(x_2 + x_1) \equiv x_S$ (position of the center of mass) and $x_1 - x_2 = x_R$ (relative distance of the two masses), we obtain two new equations for x_R and x_S by adding and subtracting the two equations respectively:

$$\frac{d^2x_S}{dt^2} = -\frac{k}{m}x_S = -\omega_S^2 x_S \qquad \frac{d^2x_R}{dt^2} = -\frac{k+2k'a}{m}x_R = -\omega_R^2 x_R$$

The two eigenfrequencies that appear in the two decoupled equations for x_S and x_R are called *normal frequencies* of the system of coupled oscillators:

$$\omega_S = \sqrt{\frac{k}{m}} = \omega_0 \qquad \omega_R = \sqrt{\frac{k+2k'}{m}} = \omega_0\sqrt{1+2\frac{k'}{k}}$$

If the coupling is weak ($k' \ll k$), we can approximate the term underneath the square root using Taylor expansion as $\sqrt{1+\epsilon} \approx 1 + \epsilon/2$ or otherwise put the following:

$$\omega_R \approx \omega_0\left(1+\frac{k'}{k}\right) = \omega_0 + \Delta\omega \qquad \Delta\omega \ll \omega_0.$$

For the first normal mode (center of mass (a)) the oscillations are in phase, whereas they are phase shifted by 180 degrees for the second normal mode (relative distance (b)). If the initial conditions are chosen correctly, these normal modes can be excited, which will then correspond to the following oscillations:

$$x_S = x_{S0}\sin(\omega_0 t - \delta_S) \qquad\qquad x_R = x_{R0}\sin((\omega_0 + \Delta\omega)t - \delta_R)$$

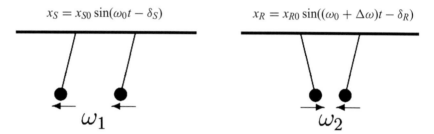

In general, one excites a mode of oscillation that is a superposition of the two normal modes. For example, if mass 2 is at the resting position for time zero and mass 1 is at its maximum amplitude x_0, then the following occurs:

$$t = 0: \quad x_1 = x_0, \ x_2 = 0 \quad \Rightarrow x_R = x_1 - x_2 = x_0 \quad x_S = \frac{1}{2}(x_1 + x_2) = \frac{x_0}{2}$$

$$x_R = x_0 \cos\omega_0 t \qquad x_S = \frac{x_0}{2}\cos((\omega_0 + \Delta\omega)t)$$

$$\Rightarrow \quad x_1(t) = x_0\cos(\omega_0 t)\cos\frac{\Delta\omega}{2}t \qquad x_2(t) = x_0\sin(\omega_0 t)\sin\frac{\Delta\omega}{2}t$$

The amplitude of the main oscillating cos or sin is slowly modulated with a frequency $\Delta\omega/2$. This is called a *beating*. Figure 3.17 shows the time dependence of the motion of the two oscillators in case $\Delta\omega/\omega_0 = 0.2$. When x_1 has a node, x_2 has an antinode and vice versa. Thus the coupling transfers energy from one oscillator to the other and vice versa. The time required for complete transmission corresponds to a quarter of the period of slow amplitude modulation:

$$\tau = \frac{1}{4}\frac{2\pi}{T} = \frac{\pi}{(\omega_R - \omega_S)} = \frac{\pi}{\Delta\omega}$$

τ increases the weaker the coupling is.

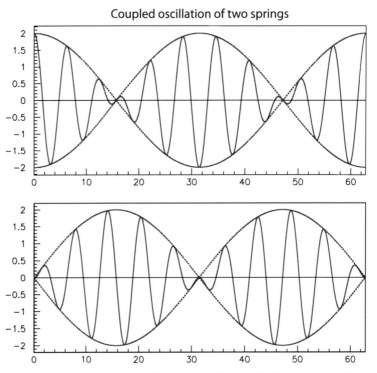

Coupled oscillation of two springs

Figure 3.17 Movement of a system of two identical coupled oscillators. Top: oscillator 1, which has been deflected for $t = 0$; bottom: oscillator 2, which for $t = 0$ is in the rest position. Spring constants: k' (coupling) $= 0.2k$. The dashed curves show the slow variation of the amplitude with the frequency $\Delta\omega/2$.

Such a beating we can also view as an example of a discrete Fourier series. We have two oscillations $u_1(t) = u_0 \cos((\omega - \Delta\omega)t)$ and $u_2(t) = u_0 \cos((\omega + \Delta\omega)t)$. The total motion is given by the sum of these two osciallations: $u(t) = u_1(t) + u_2(t) = u_0(\cos((\omega - \Delta\omega)t) + \cos((\omega + \Delta\omega)t))$. Using the sum rule for the cosine, $\cos(\alpha \pm \beta) = \cos(\alpha)\cos(\beta) \pm \sin(\alpha)\sin(\beta)$, we obtain the following

$$u(t) = u_0(\cos(\omega t)\cos(\Delta\omega t) + \sin(\omega t)\sin(\Delta\omega t) + \cos(\omega t)\cos(\Delta\omega t) - \sin(\omega t)\sin(\Delta\omega t))$$

The two terms containing a sine cancel, and we obtain for the total oscillation similar to the preceding result:

$$u(t) = 2u_0 \cos(\Delta\omega t)\cos(\omega t)$$

This means that the beating consists of an oscillation with a mean frequency ω, which is modulated on a time scale $1/(2\Delta\omega)$. This means that after a time $\pi/(2\Delta\omega)$, both oscillations cancel only to enhance each other at a time $\pi/(\Delta\omega)$. When we will look at the Fourier series of this oscillation, we already know from the setting of the problem that we will just have two Fourier components at the two frequencies $\omega - \Delta\omega$ and $\omega + \Delta\omega$.

The essential point is that the coupling allows the first pendulum to transmit its excitation to the second. Thus, excitation can propagate over a whole chain of coupled pendulums. This will be described as a wave in Chapter 4.

Exercises

3.1   Kinematics

An oscillating motion can be described by a harmonic function (i.e., a sine or a cosine), where the period of the oscillation is related to the angular frequency ω: $T = 2\pi/\omega$. If the motion starts at x_0 and zero speed at time $t = 0$, we have $x(t) = x_0 cos(\omega t)$.

(a) Determine the velocity of the oscillating object at any time t. What is the maximum velocity the object can obtain?

(b) Determine the acceleration of the object for any time t. What is the maximum acceleration? Can you express the time dependence in terms of a dependence on $x(t)$?

(c) Back to velocity. Can you also express the speed as a function of $x(t)$ in a similar fashion as in (b)? For this purpose, use that $cos^2(\alpha) + sin^2(\alpha) = 1$ for any value of α.

3.2   Circular motion

Two vectors are perpendicular if their scalar product is zero.

(a) On a circular orbit, the distance from the center is constant, i.e., $\frac{d}{dt}|\vec{r}| = 0$. Where does the velocity point in this case?

(b) If there is a constant speed, i.e., $\frac{d}{dt}|\vec{v}| = 0$, where does the acceleration point? Which direction is this in a circular orbit with constant speed?

3.3   Damped oscillation

Consider a pendulum, where a ball with radius $r = 1.0(1)$ cm and mass $m = 0.10(1)$ kg hangs from the ceiling on a string of length $\ell = 1.00(5)$ m and swings in oil of viscosity $\eta = 0.20(2)$ Pas.

(a) What is the frequency of the pendulum?

(b) What is the uncertainty of this frequency?

(c) What is the time after which the pendulum has lost half its energy? The potential energy of the pendulum depends on the square of its amplitude.

(d) What is the uncertainty of this time?

Quiz Questions

3.1   Kinematics

A train moves as a function of time as shown in Figure 3.18. What can you say about this motion?

A   The train is always speeding up.

B   The train is always slowing down.

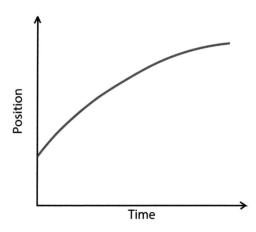

Figure 3.18 A train's motion as a function of time.

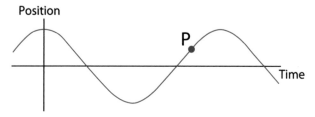

Figure 3.19 A mass is oscillating on a spring in the x-direction.

    C The trains speeds up and slows down alternatingly.
    D The speed is constant.

3.2 Kinematics 2

What is true for a constant (positive) acceleration?

    A The velocity increases as t^2.
    B The velocity increases as t.
    C The velocity decreases as $1/t$.
    D The velocity decreases as $-t$.
    E The speed remains constant.
    F The position remains constant.

3.3 Oscillation

A mass is oscillating on a spring in the x-direction as shown in Figure 3.19. What can you say about the velocity and acceleration in point P?

    A The velocity is positive and the acceleration is positive.
    B The velocity is positive and the acceleration is negative.
    C The velocity is negative and the acceleration is positive.
    D The velocity is negative and the acceleration is negative.
    E The velocity is positive and the acceleration is zero.

F The velocity is zero and the acceleration is positive.

G The velocity is negative and the acceleration is zero.

H The velocity is zero and the acceleration is negative.

3.4 Oscillation 2

Which of these conditions needs to be fulfilled in order to observe an oscillation?

A There must be a stable equilibrium state.

B There is little or no friction.

C There was at some point a disturbance in the system.

D There must be a spring somewhere in the system.

E There must be a circular motion somewhere in the system.

F There must be gravity in the system.

4   Resonances and Waves

4.1 How Resonances and Waves Determine How We Interact with the Environment

Our senses rely heavily on the interaction with waves from the outside world. Whether it is our ears that hear sound waves or our eyes that see light waves, most of the interactions that we have with the outside world takes place via waves that are transmitted from the objects we are interested in to our senses. Therefore, in order to properly understand our senses, we have to know how waves are created and transmitted and also how they can be sensed.

When a wave hits our senses, vibrations are excited in parts of our senses. In our ears, sound waves excite vibrations in the basilar membrane, which changes the conformation of hair cells that lead to the firing of neurons. This is done in a frequency-specific way and only those parts of the basilar membrane in the cochlea that are fit for this frequency are resonantly excited. Similarly, in our eyes, electrons in the receptor molecules in our retina are excited to vibrate, which again results in the firing of nerves. All of these processes of excitation of our senses by different kinds of waves are due to the parts being resonant with certain frequencies of the waves. Thus in order to get a grasp of how this works, we will have to see how vibrating systems act when forced externally as well as when and how resonance occurs. This will be done in Section 4.2.

But also in the field of instrumental methods, there are vibrations and waves that determine how a great variety of these methods work. In the case of nuclear resonance, it is again electromagnetic (radio) waves that stimulate an oscillation of nuclear spins; in seismology, sound waves are used in the earth in order to learn something about the structure of rocks (and where to find oil). In diffraction experiments, e.g., crystallography using X-rays, it is the superposition of many different partial waves after the interaction with the crystal that gives us the diffraction pattern and thus the crystal structure.

However, while waves and vibrations are intimately coupled, they are not the same thing. In order to understand this difference also in terms of a mathematical description, we will have to be very clear about this distinction before we can treat how waves, vibrations, and resonances go together.

A wave is a *disturbance* that propagates in some medium (this can be a gas, a liquid, a solid, or something more abstract like the electromagnetic field). In a water wave on a lake or in the ocean, the disturbance is a local shift of liquid layers that spread across the surface. In the case of a wave in a rope, which we produce, for example, by a hitting a cable under tension, the disturbance consists in a localized deflection of cable elements from their normal position. Such a disturbance travels along the medium, i.e., along the rope. A vibration, on the other hand, consists of the periodic oscillation of an object around its rest position. A vibration can therefore *not* propagate, in contrast to a wave. The excitation of vibrations by a wave needs the wave itself to be periodic, such that the disturbance passes through the point we want to excite periodically. These then typically are harmonic waves, which we will look at in Section 4.3.

This basically comes from the fact that, in general, we can consider a wave to be a superposition of many vibrational systems that are coupled to each other. In the suspended chain of oscillators that we have discussed, the disturbance consists in a deflection of the individual oscillators from its rest position, and this is transmitted to its nearest neighbor by the coupling of the pendulums. In a very long chain of these, we thus get a traveling wave. If the disturbance consists of a vibration rather than a single pulse, the wave will also be periodic and can be described by harmonic waves.

We can imagine a solid (see Section 7.1) to be a large number of atomic oscillators (small balls) that are connected by springs to a resting position in a three-dimensional lattice. If we produce a perturbation at one point, for example by pulling out one of the balls from its resting position, then this perturbation propagates through the different oscillators in the crystal by means of their coupling. This is the mechanism of sound wave propagation in solids. The disturbance associated with a sound wave consists of a shift of atoms, groups of atoms, or even macroscopic layers. This also applies to sound waves in gases. The displacement of gas layers is accompanied by a migrating pressure wave, a local increase or decrease of the pressure. Different couplings and/or restoring forces (shearing, compression) lead to correspondingly different sound waves in gases, liquids, and solids. In the case of electrical waves in cables, light, and other electromagnetic waves, the disturbance occurs in temporally and locally variable electric and magnetic fields.

4.2 Forced Oscillations and Resonance

4.2.1 The Forced Harmonic Oscillator

We have described vibratory systems and oscillations in detail in Section 3.2 when the oscillation is left to evolve on its own. However, if we want to describe how an external excitation can drive an oscillating system, we have to understand the motion of a driven or forced oscillation. For this, we apply an oscillating, external force $F_{ext} = F_0 \cos(\Omega t)$ onto an oscillatory system. Thus, the equation describing the motion for the oscillation (without damping) is given by the following:

$$m\frac{d^2x}{dt^2} = -kx + F_{ext} = -kx + F_0 \cdot \cos(\Omega t)$$

If, as in the preceding, we convert this equation into a generic equation with the eigenfrequency $\omega_0 = \sqrt{k/m}$, we get the following:

$$\frac{d^2x(t)}{dt^2} + \omega_0^2 x(t) = F_0/m \cos(\Omega t)$$

Because the pendulum is excited with a certain fixed frequency, we would expect that the pendulum will also oscillate with this externally prescribed frequency (Ω), at least after a long time. This is because we start with the system at rest and then nothing actually happens in the absence of an excitation, therefore the excitation will decide how the system behaves. However, we also know that on its own, the oscillation will have a natural frequency (or eigenfrequency) ω_0. Therefore, we might expect that something special happens when these two frequencies are the same, i.e., when $\Omega = \omega_0$. Since ω_0 is the frequency at which the oscillation runs by itself, it will be particularly easy to excite the pendulum at this frequency. Thus, when the pendulum oscillates at the excitation frequency after a long time, the amplitude of this oscillation will be very large if the excitation frequency corresponds to the eigenfrequency. We can make this more quantitative by assuming that after a long time, a vibration with the excitation frequency solves the preceding equation, i.e., that the solution is of the form: $x(t) = x_0 \cos(\Omega t)$. We insert this ansatz (or guess for a solution) into the equation as we did earlier to obtain an equation describing the amplitude x_0:

$$(-\Omega^2 + \omega_0^2)x_0 \cos(\Omega t) = F_0/m \cos(\Omega t)$$

This means that the acceleration changes periodically with the frequency Ω, as is the case for the excitation. But the maximum excursion of the oscillation, i.e., the amplitude, is constant in time, which we can see if we divide by the cosines on both sides. However, this amplitude does depend on the excitation frequency, or more specifically on the difference of the excitation frequency and the eigenfrequency. For the full quantitative dependence, we obtain the following:

$$x_0(\omega_0^2 - \Omega^2) = F_0/m$$

If the frequency with which we drive the oscillation (Ω) is the same as the eigenfrequency (ω_0), then the amplitude is multiplied by something very small (actually zero if they are absolutely equal) on the left-hand side of the equation, while the right-hand side corresponds to the externally applied excitation, i.e., something that is nonzero. In this case, the amplitude becomes arbitrarily large. Physically, this means that a vibration can be excited to a very large amplitude, provided that the driving happens at the eigenfrequency. This large excitation using a small stimulus is called *resonance*. We can make this still somewhat more quantitative by solving the preceding equation for x_0, to obtain the following:

$$x_0 = \frac{F_0}{m(\omega_0^2 - \Omega^2)}$$

Here we can consider three limiting cases:

We have just discussed the resonant excitation. If the excitation frequency corresponds to the eigenfrequency, the denominator is zero and thus we obtain an arbitrarily large (infinite) amplitude of the excited oscillation. In a real system, the amplitude is not, of course, infinitely large, since other effects come to bear. For example, a spring would be overdrawn if the amplitude of oscillation becomes too large and hence would no longer act as a spring.

The second limiting case is that of a very small excitation frequency. In this case, we can neglect Ω in the preceding equation and get an amplitude of $x_0 = F_0/(m\omega_0^2) = F_0/k$. We could have obtained this directly, since a small excitation frequency means that we stimulate very slowly. In the extreme case, we do not actually excite any oscillations during the time we observe. That is, we expect an extension of the spring that corresponds simply to the applied force divided by the spring constant.

The final limiting case is that of a very fast, high-frequency excitation. Then Ω is much bigger than ω_0 and we can in principle neglect ω_0 in the preceding equation. Physically, this means that within a natural oscillation of the pendulum we have very many periods of excitation. Therefore, it will be impossible to excite a sizable amplitude. In the limiting case, the amplitude $x_0 = -F_0/(m\Omega^2)$, thus the amplitude tends to go to zero for very high frequencies. In addition, we see that the amplitude now has a negative sign, which means that the oscillation of the pendulum is exactly the opposite of the excitation. The cosine function is thus just shifted by half a period (or 180 degrees). This phase shift will be important in the case of exciting a damped oscillation.

We thus see that we can excite vibrations even by weak external forces, as long as we hit the right frequency. This may occur with physical structures such as bridges, but also with electrons in atoms or nuclear spins. If the excitation corresponds to the frequency of the natural oscillation, a large oscillation can result, or a large part of the excitation is absorbed by the oscillation. Such processes of resonance absorption form the basis of all spectroscopic methods used in physics, chemistry, and biology. The excitation allows the eigenfrequency to be determined very precisely, which in turn depends on the physical (or chemical) properties of the substance, such as, for example, the binding energy. These properties also depend on the environment, which for example is used in nuclear magnetic resonance (NMR) spectroscopy to study the molecular structure of complex molecules.

4.2.2 *The Damped, Driven Oscillation

Some of the problems in the preceding description (such as infinite amplitudes or sudden changes in phase) no longer appear if we take into account that in any real system there will be some form of friction. Thus if we want to get rid of these problems, we have to consider the case where the oscillation is damped. In this case, the equation describing the motion changes according to the following:

$$m\frac{d^2x}{dt^2} = -kx - f\frac{dx}{dt} + F_{ext} = -kx - f\frac{dx}{dt} + F_0 \cdot \cos(\Omega t)$$

This can also be described in the form of a generic equation:

$$\frac{d^2x(t)}{dt^2} + 1/\tau_0 \frac{dx(t)}{dt} + \omega_0^2 x(t) = F_0/m \cos(\Omega t)$$

Here $1/\tau_0 = f/m$ is the natural time scale for the damping that we have already discussed in the treatment of the freely oscillating, damped pendulum in Section 3.8.1.

We could solve this problem in the same way as we did for the undamped case, but this proper quantitative treatment is only interesting for specialists. What we are really interested in is how the damping changes the ideal behavior of an undamped resonance discussed earlier. As such, this is a good example of how one can solve equations without properly solving them directly, but by using the intuition, or commonsense knowledge of the system, to get a semiquantitative description. Qualitatively, we can base ourselves on the findings that we have obtained, on the one hand, for the freely oscillating, damped vibration, and on the other hand, over for the driven oscillation without damping. First, the damping must lead to the fact that the excited amplitude no longer diverges on resonance. This is because, even if the attenuation is small, it will eventually start to kick in and oppose the excitation, after which there will be no further excitation possible, thus leading to a maximum in the excited amplitude. So we have to find something that cancels this division by zero. In addition, we know from the freely oscillating system that the damping introduces an additional time scale, namely, the time at which the pendulum ceases to oscillate by itself. Again, we will expect qualitative differences depending on whether this time scale is larger or smaller than the period. If the damping time is larger than the period, the pendulum will actually oscillate and we will have to see something similar to the case without damping, so there should be a maximum excitation at a frequency approximately equal to the eigenfrequency. Remember, the oscillation frequency for the damped case is not exactly the eigenfrequency. Since the amplitude is not infinite, the excited amplitudes will spread to more frequencies, so the resonance curve will have to become wider.

If the damping time is less than the period, we will not be able to stimulate proper oscillations. This means that there will be no maximum in the resonance curve, and the excited amplitudes will be relatively small, since all the excitations will be dampened quickly. But if we are exciting very slowly, we should obtain the same result as without damping, since in that case we have effectively no speed and thus also no damping. These qualitative results are presented in Figure 4.1.

With these qualitative considerations, we can now also try to find the mathematical solution that has these physical properties. To do this, we begin with the resonance curve without damping:

$$x_0 = \frac{F_0}{m(\omega_0^2 - \Omega^2)}$$

If, as we have said earlier, we wish to obtain an amplitude that is always finite, we can simply add a constant to the denominator. The only physical quantity obtained by introducing damping into the situation is τ_0. In order to have the correct units in the constant that we add, we find that if we add $1/\tau_0^2$ to the difference between the two frequencies. This, however, does not fully solve the problem, because if we were to go to even higher excitation frequencies there would be a point at which the denominator would be again by zero (because of the negative sign of Ω^2). So what we actually have to look at is the absolute value of the amplitude of the oscillation. This would be as follows:

$$x_0 = \frac{F_0}{m(\sqrt{(\omega_0^2 - \Omega^2)^2} + 1/\tau_0^2)}$$

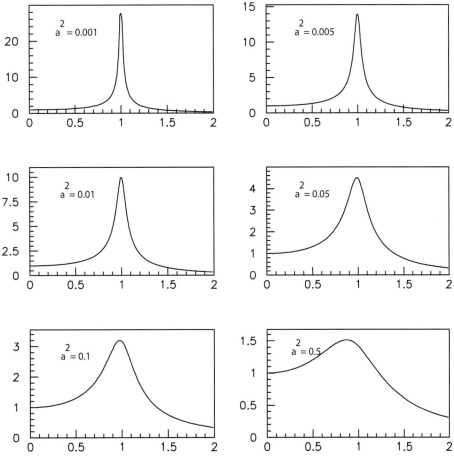

Figure 4.1 Frequency dependence of the amplitude for a resonant system near resonance $\Omega/\omega_0 = 1$. From top-left to bottom-right, the six images correspond to six different values of damping $a \equiv 2/(\tau_0 \cdot \omega_0)$, namely $a^2 = 0.001, 0.005, 0.01, 0.05, 0.1, 0.5$.

But if we consider the limiting case where the damping is very strong, i.e., takes place in a very short time or in other words that τ_0 almost zero, then the amplitude would depend on τ_0 in this case, even if the excitation were very slow ($\Omega \simeq 0$). This, however, should not happen, as we have said before, since slow excitations should yield the same result as the undamped case. Therefore, the correction factor also has to include the excitation frequency to some positive power, such that it disappears for slow excitations. Again, we have to find the proper units to determine what this power should be, and the simplest solution is the following:

$$x_0 = \frac{F_0}{m(\sqrt{(\omega_0^2 - \Omega^2)^2} + \Omega/\tau_0)}$$

This function has the essential characteristics that we have demanded, but the sum of a square root and a direct term looks somewhat strange when this equation is to come from a uniform description. If we want to treat the part with τ_0 and the part with $(\omega_0^2 - \Omega^2)$ on equal footing, while keeping all of the other properties that we have just discussed the same, we will have to put the part with τ_0 underneath the square root. In order to have the proper units, however, have to take the square of the term first. This yields the following:

$$x_0 = \frac{F_0}{m\sqrt{(\omega_0^2 - \Omega^2)^2 + (\Omega/\tau_0)^2}}$$

Interestingly, this is exactly the solution to the proper derivation that we have spared us.

For Comparison, the Exact Derivation for Specialists

Here, too, the pendulum will have to oscillate at the same frequency with which it is excited. That is, the amplitude is finally described by an oscillation: $x(t) = x_0 \cos(\Omega t + \delta)$. Here, however, we have now taken the view that the attenuation can cause a phase shift between the oscillation and the excitation. Without damping, we have already seen the fact that for high frequencies the oscillation opposes the excitation, i.e., there is a phase shift of 180 degrees. However, this phase shift can have all possible values in the case of a damped oscillation. This means that in the description, we will not only obtain cosine terms, but also sines. We could describe this with the help of complex numbers, but we can also explicitly take the sine and cosine terms. If we use the above amplitude in the vibration equation, we obtain the following:

$$\left((-\Omega^2 + \omega_0^2)\cos(\Omega t + \delta) - \Omega/\tau_0 \sin(\Omega t + \delta)\right) x_0 = F_0/m \cos(\Omega t)$$

To separate the various influences of the sine and cosine oscillations, we must write the terms with the phase shift as pure sine or cosine functions. For this, we need the two trigonometric relations that determine the sine and the cosine of the sum of two angles:

$$\cos(\alpha + \beta) = \cos(\alpha)\cos(\beta) - \sin(\alpha)\sin(\beta)$$

and

$$\sin(\alpha + \beta) = \sin(\alpha)\cos(\beta) + \cos(\alpha)\sin(\beta)$$

This yields a description of the excitation of a damped pendulum:

$$(-\Omega^2 + \omega_0^2)(\cos(\Omega t)\cos(\delta) - \sin(\Omega t)\sin(\delta)) - \Omega/\tau_0(\sin(\Omega t)\cos(\delta)$$

$$+ \cos(\Omega t)\sin(\delta)) = \frac{F_0}{x_0 m}\sin(\Omega t)$$

Here we can separate the terms containing only cosines and the terms containing only sines, thus obtaining the following:

$$\left[x_0\left((-\Omega^2 + \omega_0^2)\cos(\delta) - \Omega/\tau_0 \sin(\delta)\right) - F_0/m\right]\cos(\Omega t)$$

$$= x_0\left((-\Omega^2 + \omega_0^2)\sin(\delta) + \Omega/\tau_0 \cos(\delta)\right)\sin(\Omega t)$$

Since sines and cosines are always shifted by 90 degrees, this equation can only be true if the prefactors in front of sines and cosines are both zero respectively. That is, the preceding equation effectively gives us two equations that we can use to determine both x_0 and δ. We obtain the following:

$$x_0 \left((-\Omega^2 + \omega_0^2) \cos(\delta) + \Omega/\tau_0 \sin(\delta) \right) = F_0/m$$

$$(-\Omega^2 + \omega_0^2) \sin(\delta) = \Omega/\tau_0 \cos(\delta)$$

The second of these equations directly yields the phase shift given by the following:

$$\tan(\delta) = -\frac{\Omega}{\tau_0(\omega_0^2 - \Omega^2)}$$

In the high- and low-frequency limits, a phase shift of zero and 180 degrees is obtained respectively, just as with the pendulum without damping. For resonant excitation, i.e., $\Omega = \omega_0$, the tangent diverges, which corresponds to a phase shift of 90 degrees. We can also explain this from the physical point of view, because when we excite a pendulum, e.g., a swing, with the resonant frequency, we give the most excitation when the swing is in full swing. At that point, the damping is largest, because of the highest speed, and we must use the most excitation to maintain the movement.

The amplitude of the excited oscillation is obtained from the first of the preceding equations when we use the relation for the phase shift. To do this, we need the relationship between the tangent and the sine or the cosine of an angle: $\sin(\delta) = \tan(\delta)/\sqrt{1 + \tan^2(\delta)}$ and $\cos(\delta) = 1/\sqrt{1 + \tan^2(\delta)}$. Using the equation for the tangent of the phase shift, this yields the following:

$$\sin(\delta) = -\frac{\Omega}{\tau_0\sqrt{(\omega_0^2 - \Omega^2)^2 + (\Omega/\tau_0)^2}}$$

$$\cos(\delta) = \frac{(\omega_0^2 - \Omega^2)}{\sqrt{(\omega_0^2 - \Omega^2)^2 + (\Omega/\tau_0)^2}}$$

These we insert in the equation containing x_0 and obtain the following:

$$x_0 \left(\frac{(\omega_0^2 - \Omega^2)^2}{\sqrt{(\omega_0^2 - \Omega^2)^2 + (\Omega/\tau_0)^2}} + \frac{\Omega^2}{\tau_0^2 \sqrt{(\omega_0^2 - \Omega^2)^2 + (\Omega/\tau_0)^2}} \right) = F_0/m$$

or

$$x_0 \left(\sqrt{(\omega_0^2 - \Omega^2)^2 + (\Omega/\tau_0)^2} \right) = F_0/m$$

Finally, we obtain the absolute value of the amplitude as follows:

$$x_0 = \frac{F_0}{m\sqrt{(\omega_0^2 - \Omega^2)^2 + (\Omega/\tau_0)^2}}$$

This means that when the oscillation is damped, no resonance catastrophe occurs; the amplitude of the excited oscillation will not diverge even with resonant excitation. In addition, the frequency at which the maximum excitation is reached is shifted to somewhat smaller frequencies. By taking the derivative of the amplitude and setting this to zero, we obtain a maximum excitation at the frequency $\Omega_{max}^2 = \omega_0^2 - \frac{1}{2\tau_0^2}$. Thus, an overdamped system no longer has a maximum in the amplitude as a function of excitation frequency. If the attenuation is very small (τ_0 very large), then the curve we have already discussed results in the limit value. The various resonance excitations are shown graphically in Figure 4.1. From this illustration, it is also apparent that, with increasing attenuation, more and more frequencies can lead to an excitation, but the excited amplitude is also becoming smaller and smaller. This is exactly the same as in our qualitative treatment.

4.3 One-Dimensional and Harmonic Waves

4.3.1 One-Dimensional Waves

As already described, a collection of oscillating systems can form the medium for the propagation of a wave. Important terms for the description of such waves are the *propagation speed*, the magnitude (amplitude) of the disturbance (also called *excitation*), and the direction of the disturbance relative to the propagation speed. If the speed of propagation and the direction of the disturbance are perpendicular to one another, one speaks of transversal waves' if the two are parallel, the waves are longitudinal. Examples of purely transverse waves are the waves of the electromagnetic spectrum, light, X-rays, etc. Sound waves in gases are purely longitudinal, and sound waves in solids can be both longitudinal and transversal. We begin with the case of a one-dimensional wave, i.e., a wave that propagates only in one single direction. As an illustration on which to develop an understanding of the different concepts and the mathematical description, we choose a long, elastic rope under tension, which is locally deformed, for instance by plucking it. This deflection propagates along the rope and in this case presents the excitation of the wave we wish to describe in the end.

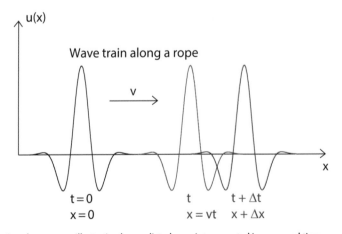

Figure 4.2 A wave train moving along a rope, illustrating how a disturbance is transported in space and time.

The disturbance that travels along the rope and is generated at time $t = 0$ and location $x = 0$ is described by a function $u(x, t)$. Here, x is the coordinate along the rope. The excitation u, in this case, is the displacement of the rope perpendicular to its direction and propagates with the velocity v.

We assume that the disturbance (i.e., the wave) is not dampened, that is, it does not change its amplitude and also retains its shape. Such waves are called *dispersion-free*. The fidelity in shape of the disturbance implies for the function of space and time describing the disturbance that

$$u(x, t) = u(x + \Delta x, t + \Delta t)$$

This means we have the same functional form at a later time $t + \Delta t$, but in another place, $x + \Delta x$. If we know that the disturbance propagates with a speed v, then we know that Δx and Δt are directly related to each other via

$$v = \frac{\Delta x}{\Delta t}$$

Thus, we can show that the separate space and time dependencies of wave $u(x, t)$ can be described by a function $\tilde{u}$ of only a single variable of the type

$$u(x, t) = \tilde{u}(x - vt)$$

If we insert this into the equation we obtained from the fidelity of shape, we get the following:

$$u(x + \Delta x, t + \Delta t) = \tilde{u}(x + \Delta x - v(t + \Delta t)) = \tilde{u}(x - vt + (\Delta x - v\Delta t)) = \tilde{u}(x - vt) = u(x, t)$$

The variables x and t are therefore no longer independent, but are linked to one another by the propagation velocity v.

While the preceding wave is propagating in the positive x direction, a wave that propagates in the negative x direction is described by the function $u(x, t) = u(x + vt)$. The sign can be easily explained. If a wave propagates in the positive x−direction, the front part of the wave, the *wavefront* (with a positive x relative to the center of the wave) arrives earlier (i.e., at a shorter time) at a fixed position x. In the opposite direction, this part arrives later.

4.3.2 The Wave Equation for a Rope

Let us be more quantitative and describe the disturbance of a rope under tension in terms of the spatial and temporal change of the disturbance as a function of this disturbance itself. The rope is under tension in the x-direction and is deflected in y-direction, perpendicular to x. The wave is thus a transverse wave and we want to describe the x and t dependencies of the disturbance $y(x, t)$. For a small piece of the rope of the length dx at position x, we have given the equation of motion in the y-direction by the following:

$$m\frac{\partial^2 y}{\partial t^2} = F_y(x) - F_y(x + dx)$$

Here F_y is the projection of the tensional force onto the y-direction. The tensional force will always point along the rope. This means that its projection onto the y-direction is directly given by the total tensional force F and the slope of the rope at position x, i.e., $\alpha(x)$: $F_y(x) = F\alpha(x)$. The slope of the rope at x is nothing but the derivative of the function $y(x, t)$ as a function of space x, i.e., $\alpha(x) = \frac{\partial y}{\partial x}(x)$. Inserting this for the tensional forces, we obtain the following for the equation of motion:

$$m\frac{\partial^2 y}{\partial t^2} = F\left(\frac{\partial y}{\partial x}(x) - \frac{\partial y}{\partial x}(x + dx)\right) = Fdx\frac{\partial^2 y}{\partial x^2}$$

Finally, we want to have a description independent of the small length dx. For this purpose, we have to describe the mass m of the small volume as a function of this length as well: $m = \rho A dx$, where ρ is the density of the material and A is the cross-sectional area of the rope. This then gives the relation between the temporal change of the disturbance with the spatial dependence of the disturbance we have been looking for:

$$\frac{\partial^2 y}{\partial t^2} = \frac{\sigma}{\rho}\frac{\partial^2 y}{\partial x^2}$$

Here, $\sigma = F/A$ is the tensional stress of the rope. The prefactor on the right-hand side of the equation has the units of a speed squared, such that we can rewrite the equation in the following form:

$$\frac{\partial^2 y}{\partial t^2} = v^2\frac{\partial^2 y}{\partial x^2}$$

where v is the propagation speed of the disturbance. This *wave equation* is generally valid for all types of waves and not just those in a rope under tension.

4.3.3 Harmonic Waves

An important class of undamped waves are the harmonic waves. This is because any disturbance can be interpreted as a superimposition of harmonic waves (also the form of a pulse described previously). This is discussed in more detail in sections on Fourier expansions (3.3) and transforms (4.6). For a harmonic wave, the excitation is a harmonic function (i.e., a sine or a cosine) both in space and time:

$$u(x, t) = u(x \pm vt) = u_0 \sin k(x \pm vt)$$

The argument of the sine, the so-called *phase* of the wave, must be dimensionless. Therefore, we had to introduce the additional factor k, the so-called *wave number*, with the dimension of an inverse length. (Note that the k introduced here has no relation whatsoever to the spring constant k. Because of the limited number of letters in the alphabet, it can sometimes happen that different physical properties are traditionally denoted by the same letter.) The variable u_0 is called *amplitude*.

Graphically, we can reproduce a harmonic wave in two different types of pictures. In the location image, u is plotted as a function of x at a certain time t, whereas in the temporal image, u is displayed as a function of t at a fixed location x. Both are given by sine curves for harmonic waves, but their physical significance is quite different.

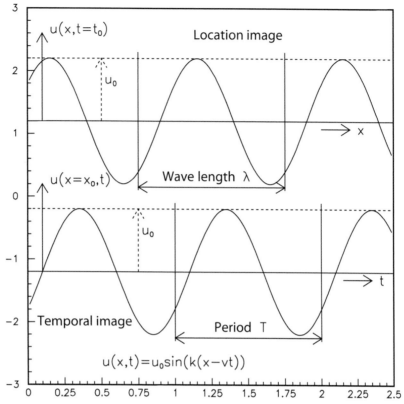

Figure 4.3 The spatial and temporal images of a harmonic wave, illustrating the wavelength, the period, and the amplitude.

In the location image, the spatial period is called *wavelength* and is denoted by the letter λ. With $u(x, t) = u(x + \lambda, t)$ we have the following:

$$u_0 \sin(kx + k\lambda - kvt) = u_0 \sin(kx - kvt)$$

$$\Rightarrow k\lambda = 2\pi, \quad \lambda = \frac{2\pi}{k}$$

λ has the units of [Length].

In the temporal image, the period of oscillation, also called *duration of vibration*, appears. With $u(x, t) = u(x, t + T)$, we find the following:

$$u_0 \sin(kx - kvt - kvT) = u_0 \sin(kx - kvt)$$

$$\Rightarrow kvT = 2\pi, \quad kv = \frac{2\pi}{T} \equiv \omega$$

The angular frequency ω has units of [Time^{-1}].

Instead of the angular frequency ω, we can also use the *frequency* ν:

$$\omega \equiv 2\pi\nu, \quad \nu = \frac{1}{T} \quad \Rightarrow \quad kv = 2\pi\nu$$

The wave speed (also called the *phase velocity*) v is given by the following:

$$v = \frac{\omega}{k} = \lambda \nu = \frac{\lambda}{T}$$

The propagation speed of a wave is usually determined by the medium, that is, it does not change for a particular type of wave and a fixed medium. If you change the frequency ν, therefore, λ will have to change accordingly. Smaller wavelengths correspond to higher frequencies and vice versa.

With these different relations, one obtains the following possibilities to write down the relationship defining a harmonic wave (the most commonly used one is the second, indicated by (*)):

$$
\begin{aligned}
u &= u_0 \sin(kx - kvt) \\
&= u_0 \sin(kx - \omega t) \ (*) \\
&= u_0 \sin(2\pi(x/\lambda - \nu t)) \\
&= u_0 \sin(2\pi(x/\lambda - t/T))
\end{aligned}
$$

The similarity of the mathematical description of harmonic waves and vibrations often creates a certain amount of confusion. So let us state the distinction between the two once again: A *wave* is described by a function $u(x, t) = u_0 \sin(kx - \omega t)$ dependent on both location *and* time. An *oscillation* of a physical quantity f is described by a function $f(t) = f_0 \sin(\omega t - \delta)$ of time *only*.

Mathematically, we can show that a harmonic wave satisfies the wave equation. We obtain this by taking the derivative of the disturbance with respect to time, at a fixed position twice, as well as with the derivative with respect to position at a fixed time, also twice:

$$\left(\frac{du}{dx}\right)_{t=\text{const.}} = u_0 k \cos(kx - \omega t) \qquad \left(\frac{d^2u}{dx^2}\right)_{t=\text{const.}} = -u_0 k^2 \sin(kx - \omega t)$$

$$\left(\frac{du}{dt}\right)_{x=\text{const.}} = -u_0 \omega \cos(kx - \omega t) \qquad \left(\frac{d^2u}{dt^2}\right)_{x=\text{const.}} = -u_0 \omega^2 \sin(kx - \omega t)$$

$$\Rightarrow \text{ Wave} - \text{equation} : \qquad \frac{\partial^2 u}{\partial t^2} = v^2 \frac{\partial^2 u}{\partial x^2}$$

However, the wave equation also holds for every other function $u(x, t) = u(x - vt)$, that is, for any wave that shows fidelity in shape, i.e., a wave without damping. Again, this can be shown by inserting the preceding ansatz into the wave equation by taking the second spatial and temporal derivatives.

The physical content of the wave equation, as we have seen, is obtained by examining a concrete situation in which waves appear. In the theory of electricity (electrodynamics), the fundamental equations that connect electric fields and magnetic fields are Maxwell's equations (discussed later). For electromagnetic waves ($u = E, B$), the preceding wave equation follows from Maxwell's equations, which can also be extended to three spatial dimensions. The propagation speed is then given by the speed of light. Many mechanical quantities can occur as excitation or disturbance in a wave, e.g., pressure p or displacement $\vec{r}$. The corresponding wave equation with velocity v then follows from the fundamental equations of mechanics. We will look more closely at these basic equations later, but in essence they will have the same form as the equations discussed earlier.

If we find such a wave equation in the treatment of a system, we know that waves can occur and we know their speed of propagation.

4.4 Waves Are Transporting Energy

With a continuous wave of any type, energy is transported. The energy flux of a wave depends on the kinetic energy of the oscillating elements, as well as the propagation velocity. As an example, consider again the wave traveling along a rope transversally.

The energy of the moving rope element dx with the mass $dm = A\rho dx$ (ρ = density, A = cross-section) is as follows:

$$dE = \frac{dm}{2}v_{max}^2 = \frac{A\rho dx}{2}u_0^2\omega^2$$

Here, we have used the fact that the maximum speed of the rope element, v_{max}, is given by the maximum of the time derivative of the disturbance $u = u_0 \sin(kx - \omega t)$:

$$v_{max} = \left(\frac{\partial u}{\partial t}\right)_{max} = u_0\omega$$

Thus, the energy dE that traverses the rope cross-section during the time dt is given by the following:

$$\frac{dE}{dt} = \frac{A\rho dx}{2dt}u_0^2\omega^2 = \frac{\rho A}{2}u_0^2\omega^2 v$$

This energy that flows through the cross-section A per unit of time is called the *intensity* of the wave I:

$$I = \frac{dE}{dt} \cdot \frac{1}{A} = \frac{\rho v}{2}u_0^2\omega^2$$

This definition of the intensity is valid for all types of waves. Therefore, we find that

$$\begin{aligned} \text{intensity} \quad &\propto \quad u_0^2 \quad &&\text{amplitude of the wave squared} \\ &\propto \quad \omega^2 \quad &&\text{frequency of the wave squared} \\ &\propto \quad v \quad &&\text{propagation speed of the wave} \end{aligned}$$

Because $v = \lambda \cdot \nu$, we also find that the smaller the wavelength, the larger the transported energy.

The unit of intensity is $[I] = W/m^2$. We will see later that in the case of sound, another unit is often used to indicate intensities based on its logarithm. This is the decibel (dB).

The definition of the intensity of a wave can also be used to determine the distance dependence of the intensity of a given source. For an undamped wave, the total energy transported is constant. This means that

$$\frac{dE}{dt} = IA = const$$

Otherwise put, there is an equation of continuity (see Section 7.6.1) for the intensity of the wave. Thus, if we want to determine the intensity at two different points 1 and 2, we know that we must have $I_1 A_1 = I_2 A_2$, where the areas are those that are traversed from the source at the respective points. For a point source, which emits the wave uniformly in all three spatial directions, the corresponding surface therefore is that of a sphere with a radius given by the distance at which we determine the wave's intensity, i.e., $I_1 4\pi r_1^2 = I_2 4\pi r_2^2$. Therefore, we obtain the following for the distance dependence of the intensity of a point source:

$$\frac{I_1}{I_2} = \frac{r_2^2}{r_1^2}$$

4.5 The Physiology and Physics of Hearing

In the introduction, we alluded to how the understanding of resonances and waves can help us in understand our senses, such as hearing. Now that we have laid the foundations in terms of the description of waves and resonance, we can discuss some physiological aspects of hearing, that is, how sound waves are recorded and processed in the ear, whose structure is shown in Figure 4.4. As we have seen, the proper physical unit of (sound) intensity would be W/m^2. But the physiologist, who is interested in how noisy a lecture hall is, would not be helped much by such a measure. Noise only disturbs when you can actually hear it. Ultrasound makes no noise to humans, but bats use it for navigation. But also in the region where the ear is sensitive, not all frequencies are equally captured. The ears' highest sensitivity is in the frequency range around 3 kilohertz (kHz) – it is not without reason that babies cry on this frequency. Here the parent already reacts at an intensity of 10^{-13} W/m^2. Even at a frequency of 1 kHz, the sound wave requires a tenfold intensity, as Figure 4.5 shows. This frequency range around 3 kHz also corresponds to the high notes of musical instruments. Therefore, the sensitivity of the ear leads to the fact that different instruments in an orchestra need more or less personnel depending on the notes typically played by the instruments. Often a single piccolo flute is juxtaposed to six basses. At even higher frequencies, the sensitivity decreases again and frequencies above 20 kHz are usually inaudible. This sensitivity decreases with age, and older people typically already have difficulty hearing frequencies above 10 kHz. This basically comes from the fact that the basilar membrane loses some of its flexibility with age, thus changing its resonance properties.

Not only the frequency dependence but also the fact that most of our sensory organs (in this case, the ear) are not actually measuring instruments with a linearly calibrated scale

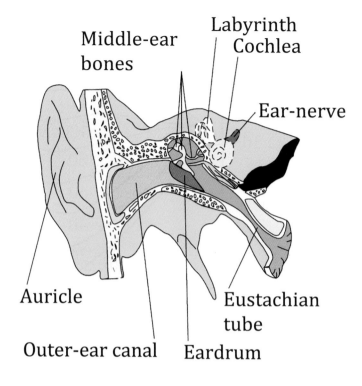

Labyrinth
Cochlea
Middle-ear
bones
Ear-nerve
Auricle
Eustachian
tube
Outer-ear canal   **Eardrum**

Figure 4.4 Structure of the human ear: auditory canal, eardrum, and central ear with auditory ossicles and eustachian tube; inner ear with the three rings of the equilibrium organ (labyrinth); and the snail, on which the auditory nerve begins.

makes the physical intensity useless as a measure from a physiological point of view. If the ear actually did perceive linearly, it would never be able to process the intensity range it does, covering many decades and occurring in daily life. Consequently, the ear (as well as the eye) reacts logarithmically to external stimuli. Therefore, when a sound intensity is specified, we should use a logarithmic scale, which is defined as follows:

$$N\,[\text{Decibel (dB)}] = 10\log_{10}\frac{I}{I_0}$$

The dimensionless quantity N is a measure of sound intensity relative to a reference intensity I_0. This reference intensity determines the zero point of the scale and for sound has been chosen to approximate the limit of perception of the human ear at $\nu = 1,000$ hertz (Hz), i.e., $I_0 = 10^{-12}$ W/m². At this frequency, therefore, an example intensity of $I = I_0 \equiv$ 0 dB, $I = 10I_0 \equiv 10$ dB, and $I = 100I_0 \equiv 20$ dB. This means that, for example, $I = 1$ W/m² corresponds to $\log_{10}(I/I_0) = \log_{10}(1/10^{-12}) = 12$ Bel or 120 dB.

The frequency dependence of the perceived loudness is taken into account in the dBA scale (Figure 4.6). It describes at which loudness a large number of test persons perceive a sound of a certain frequency as approximately equal to a normal tone of a certain dB value. For the standard frequency $\nu = 1000$ Hz, the dBA and dB scales are the same. Because of its frequency independence, the dBA scale can also be used for noise, containing multiple

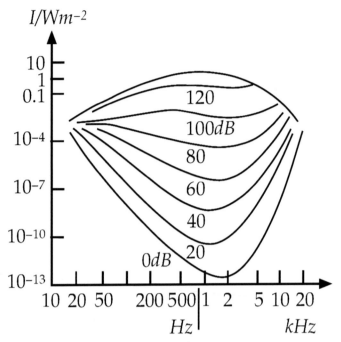

Figure 4.5 Frequency dependence of hearing. The curves indicate the sound intensities and frequencies of the human ear, which correspond to a certain impression of loudness. The uppermost curve corresponds approximately to the pain threshold, the lowest to the hearing threshold.

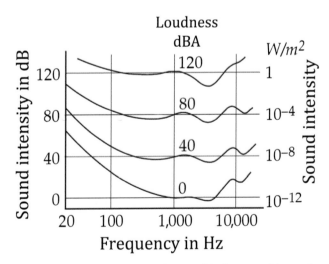

Figure 4.6 dBA values on the decibel scale of the sound intensity as a function of the frequency of the sound wave.

Table 4.1 Typical intensity levels of sounds measured in dBA	
Type of sound	dBA value
Rushing of leaves/quiet room	10/20
Conversation/street noise	60/70
Loud music	80
Passing train	Up to 100
Jet plane engine at a distance of 4 m	130
Damage to the ear depending on time	Starting at 80

frequencies. Examples of typical dBA values are given in Table 4.1. The hearing range of human beings starts at 20 Hz, and the upper limit is between 10 and 20 kHz. It is higher for younger people than for older people.

Sound waves coming from the air into our ears stimulate the eardrum to vibrate. In order to transmit the incident energy to the entrance membrane of the inner ear (oval window) as far as possible without reflections, the ankle bones, anvil, and stirrups lying between them are used. We will look at the basics behind the reflection of waves in Section 4.7.1. From the oval window, the waves travel along the so-called basilar membrane in the cochlea, which is shaped such that different frequencies can produce large amplitudes by resonance at different positions along the membrane (see Figure 4.7). The geometrical arrangement of the basilar membrane thus ensures that we convert the complicated temporal change in pressure into a collection of tones of different frequencies. Only where the basilar membrane is resonantly excited, the nerve cells, which are associated with the hair cells underneath them, are excited to fire. The frequency information of the sound is thus converted into location information on the basilar membrane, or a sequence of nerve signals, which is then sent to the brain.

A harmonic wave, which excites only a definite part of the basilar membrane, is perceived as a pure tone. What constitutes such a pure tone, and how we can perceive different sounds as the sum of pure sounds, corresponds to the Fourier expansions we have seen in Section 3.3, but we will look at this again more generally in Section 4.6. If two harmonic waves of the same frequency occur, their effect depends on their relative phase. If the two waves are in phase, the total amplitude is the sum of the individual amplitudes. If they are out of phase by 180 degrees or π, the difference of the amplitudes results. These so-called interference effects will play an important role in the further description of how waves behave, and we will treat them in detail in Section 4.7.4. As far as sound goes, they are for instance of great importance in the design of the acoustics of rooms, such as concert halls.

4.6 Fourier Transforms

We have so far often only looked at waves, which are harmonic functions of space and time. For example, this was the case at the very beginning of this chapter with location

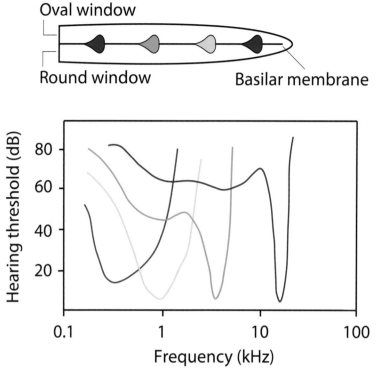

Figure 4.7 Schematic representation of the basilar membrane and its resonance frequencies at different positions, as indicated by the hearing threshold as a function of frequency. This shows how the ear transforms a temporal intensity signal into a frequency signal, effectively performing a Fourier transform. Data from Kiang, 1980.

and temporal pictures for harmonic waves. We have, however, also seen in the treatment of oscillations that we can describe any periodic function as a sum of sine and cosine oscillations. This is something we want to generalize now.

We have already seen that the ear divides a sound wave into its Fourier components. The magnitude of the excitation of the basilar membrane at a certain place corresponds precisely to the excitation amplitude at the resonance frequency of this position. The nerve signals, which the ear then sends out after this spatial separation of the signal, corresponds exactly to a frequency spectrum of the sound heard. However, the Fourier decomposition is also found in technical instruments, which we will learn about more in later chapters, such as nuclear magnetic resonance, mass spectrometers, electrocardiograms, and electroencephalograms. So far, in particular in hearing, we have considered functions of time, which we have described as sums of temporal harmonics. However, the same decomposition could also be carried out in space, where a wave also oscillates within a wavelength. In that case, instead of the angular frequency ω, which corresponds to the period T, we have to use the wave number k, corresponding to the wavelength λ, and we can perform a Fourier decomposition in space as follows:

$$u(x) = u(x + \lambda) = \sum_{n=0}^{\infty}(A_n \cos k_n x + B_n \sin k_n x) \qquad \text{mit } k_n = \frac{2\pi n}{\lambda}$$

$$A_0 = \frac{1}{\lambda} \int_0^{\lambda} u(x)dx, \qquad\qquad B_0 = 0$$

$$A_n = \frac{2}{\lambda} \int_0^{\lambda} u(x) \cos(k_n x)dx, \qquad\qquad B_n = \frac{2}{\lambda} \int_0^{\lambda} u(x) \sin(k_n x)dx \quad (n \geq 1)$$

Here, too, there are a variety of applications, the most prominent being the scattering methods, e.g., crystallography, where the spatially periodic crystal produces a scattering image corresponding to the Fourier decomposition of this structure. In the case of crystals, the scattered image is again a periodic structure. We will deal with this in more detail in Section 5.3.5.

Also non-periodic functions can be decomposed into Fourier components. However, since there is no longer a basic frequency (we can describe a non-periodic function as a periodic function with an infinite period, $(T \to \infty)$), there are no longer Fourier coefficients A_n and B_n belonging to fixed values of $\omega = \omega_n$. Rather, we describe the frequency spectrum using functions $A(\omega)$ and $B(\omega)$, which represent a continuous frequency distribution and the Fourier sum becomes a Fourier integral. This can be made plausible by the fact that the higher terms in the Fourier series always correspond to shorter times. The first Fourier component is then also exactly at the frequency, which corresponds to the original period of the function. So if we want to come to longer and longer periods of description, eventually having an infinitely long period, we must always look at smaller and smaller frequencies. The differences of the frequencies to be considered are then arbitrarily small, which corresponds to a continuous distribution of the Fourier components and infinitely small steps in the sum, which is otherwise known as an integral.

Mathematically speaking, this means that we can represent any function $u(t)$ as the integral of harmonic functions:

$$u(t) = \int (A(\omega) \cos(\omega t) + B(\omega) \sin(\omega t))d\omega$$

Here, the functions $A(\omega)$ and $B(\omega)$, are the Fourier components or Fourier transforms that completely describe the given function. They are calculated exactly the same as the discrete Fourier components via an integral of the function multiplied by an oscillation with the frequency ω. Since ω in this case can have all the values and not only multiples of the fundamental frequency, the Fourier transform becomes a continuous function rather than a collection of discrete Fourier components. As we have said earlier, in a given system, which is excited by different harmonic waves, the resonance behavior should directly reflect the Fourier components. This is, for example, given for the basilar membrane, where the resonance properties directly give out a frequency spectrum of the heard sound.

In general the difference in the two ways of looking at a function, either in its time dependence or in its frequency spectrum, lies in the fact that the two variables are inversely proportional to each other. This means that Fourier components corresponding to small frequencies correspond to long times and vice versa. This can be illustrated in the Fourier

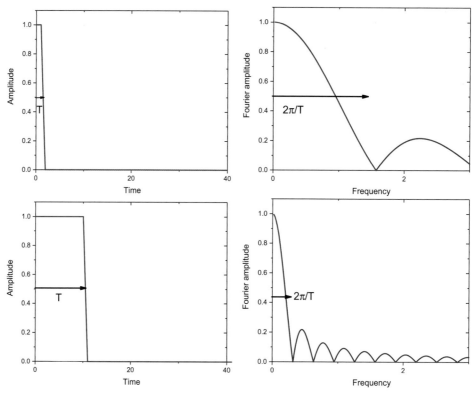

Schematic representation of the Fourier transform of a sharp excitation. In the case of a short excitation, many frequencies are required in the Fourier transform, which results in a broad curve. For a longer excitation, the Fourier transform becomes narrower.

decomposition of a box function of different widths, shown in Figure 4.8. In Section 4.8, we will see that this is a different formulation of the uncertainty principle of quantum mechanics (where we describe matter waves).

As a practical example, we can see this behavior in the fact that the width of the Fourier transform of a resonant excitation must correspond to a decay time of the corresponding oscillation. We can make this more plausible by remembering the sound of a beat. The farther apart the two frequencies were (i.e., the broader the frequency spectrum or Fourier decomposition is), the shorter the time between the beating of the signal. The beat corresponds to a discrete Fourier decomposition, so if we go to a continuous Fourier transform, the end result is no longer periodic because we have summed many closely located frequencies, but the oscillation is still zero in the same time scale – the individual oscillations are still the same. That is, we have a decay time equal to the inverse width of the Fourier transform. To do this, we consider the previous beating more quantitatively. The Fourier spectrum had two maxima at $\omega + \Delta\omega$ and $\omega - \Delta\omega$. We can also conceive this as a maximum at the frequency ω with a width of $2\Delta\omega$, that is, an oscillation with the frequency ω falling off at the time $1/(2\Delta\omega)$. But since it is a discrete Fourier decomposition, this will continue periodically in the case of the beat. Because all frequencies in the interval occur in the continuous case, the signals cancel out for longer periods and the signal has disappeared after a time scale of $1/(2\Delta\omega)$.

Likewise, we can make it plausible that a maximum in the Fourier transform corresponds to the frequency of an oscillation. Depending on how high that maximum is, the frequency will be more clearly resolved. In the extreme case, the maximum is a single Fourier component that corresponds directly to a harmonic oscillation with a single frequency.

4.7 The Principle of Superposition

The basic physical principle behind the Fourier transform is the *principle of superposition*. If several wave trains travel in the same medium, they can simply be added up to give the total excitation in the medium at a given point – just as we have seen in the Fourier decomposition. This means that depending on the different relative amplitudes, waves can amplify or reduce each other at different positions. Figure 4.9 illustrates such a situation using the example of two wave trains that run along a rope in opposite directions. At the point (and during the time) they overlap, both increased and reduced excitation can be observed depending on the exact time and place.

The superposition principle applies to any kind of excitation and to any type of (linear) wave, which makes it extremely useful, since we can therefore separate a complicated wave field into numerous simple single waves that we can follow and understand.

4.7.1 Reflection and Transmission of Waves

One instance where this is useful is when a wave hits a boundary and a part of it is reflected. The ensuing wave form can be quite complicated, but in fact consists of two simple waves traveling in opposite directions. Such boundaries appear at many places, such as in a sound wave propagating in air on a wall; a sound wave in a rod reaching its end; an electromagnetic wave (light) entering the glass of a pair of spectacles from the air; the depth of water in a lake drastically changing near a shallows and the lake's surface waves changing their speed; etc. In all cases, what is important is that the wave speed is different on the two sides of the boundary. In this case, the waves are changing. If they impinge on the boundary at an angle, they change direction (are refracted); we will deal with this more closely in Section 5.2. But even if the wave impinges perpendicularly onto the boundary, something happens. One part of the wave continues to penetrate into the new medium behind the boundary, while another part is reflected and runs backward in the original medium. The first part is called *transmission*, while the second is called *reflection*.

Figure 4.10 illustrates how a sound wave splits into two parts when it passes from medium 1 to a new medium 2 at an interface. As we have seen, the wave speed is determined by the materials properties of the density ρ and the elastic modulus E in the form $v = \sqrt{E/\rho}$, such that different materials will have different speeds of sound, or in other words, the wave speed changes at the interface.

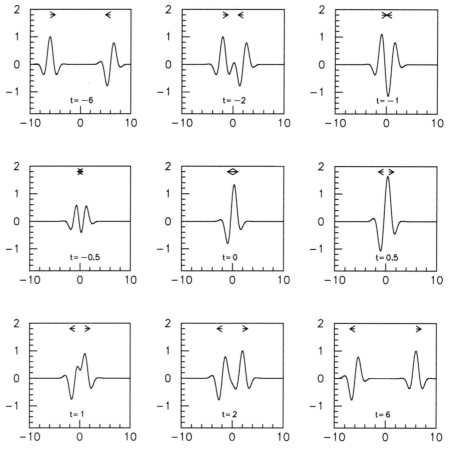

Figure 4.9 | Superposition of two wave trains in a rope under tension. Shown are nine spatial images at different times t. The waves entering from the left and from the right overlap for a short time interval. The total excitation can be stronger or weaker than the single wave itself. The centers of the wave trains are marked with arrows.

In order to quantitatively determine what the reflected and the transmitted parts are for a given situation, we concentrate on sound waves, and further restrict ourselves to harmonic disturbances. The latter is, however, only to keep the mathematical discussion as simple as possible; the results apply to every wave form. Near the interface, the different parts of the wave add up according to the superposition principle. In medium 1, these are the incoming wave and the reflected wave, which move in opposite directions in the same medium. If we make a list of all the parts of the wave we have to keep track of, we obtain the following:

Medium 1	**Medium 2**
	$x = 0$

Incoming wave: $u_i = A\cos(k_1 x - \omega_1 t)$ Outgoing wave: $u_t = C\cos(k_2 x - \omega_2 t)$
$(\to +x)$ $(\to +x)$

Outgoing wave: $u_r = B\cos(k_1 x + \omega_1 t)$
$(\to -x)$

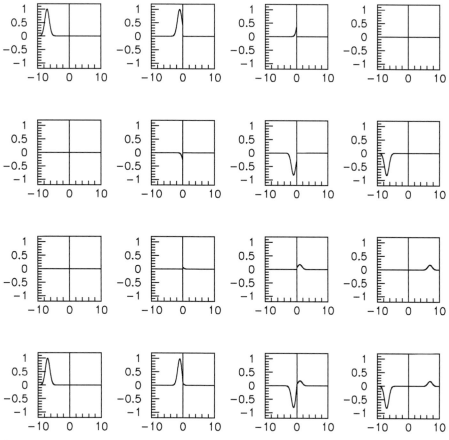

Figure 4.10 Reflection of a sound wave. Each line shows four spatial images of a sound wave that are registered at different times and which encounters a boundary between two media at $x = 0$ (medium 1: $x < 0$, medium 2: $x > 0$). The top line shows only the incoming wave form; the second only the reflected; the third the transmitted; and the last line the total sound wave, that is, the sum of all three parts. ($r = (\rho_1 v_1)/(\rho_2 v_2) = 0.1$, $v_1 = v_2$.)

If we place the x coordinate such that its origin lies at the boundary, we can directly determine the time dependence of the different parts of the excitation at the boundary for all times by inserting $x = 0$ in the preceding equations. Since we know that the excitation of the interface has to be the same from both sides (otherwise, there would be gaps in the material), we can obtain a condition for the amplitudes of the reflected and transmitted parts:

$$u_i(0) + u_r(0) = u_t(0) \qquad \Rightarrow \quad A\cos(-\omega_1 t) + B\cos(\omega_1 t) = C\cos(-\omega_2 t)$$

Furthermore, we know that the frequencies of the wave have to be the same on both sides, $\omega_1 = \omega_2$, because otherwise again there would be gas in the material at different times, which cannot be. Therefore, the cosines on both sides are the same and drop out, to give us the first condition for the different amplitudes, as follows:

$$A + B = C$$

However, we have two unknown amplitudes (that of the transmitted wave and that of the reflected wave), so we need another condition to find out the full picture. Apart from the fact that the total amplitude needs to be the same at both sides of the boundary, the force on the boundary element also needs to be the same from both sides, as otherwise the material would rupture at the boundary. This force is given by the slope of the excitation times the tension (or elastic modulus), as we have seen previously. Mathematically speaking, this means that the derivative with respect to space multiplied by the elastic modulus of the material on the respective side needs to be the same on both sides at $x = 0$, as follows:

$$E_1 \left(\frac{\partial u_i}{\partial x}(0) + \frac{\partial u_r}{\partial x}(0) \right) = E_2 \frac{\partial u_t}{\partial x}(0)$$

Alternatively, again inserting the expressions for the different parts of the preceding wave into this equation, we obtain the following:

$$E_1(k_1 A \sin(-\omega_1 t) + k_1 B \sin(\omega_1 t)) = C k_2 \sin(-\omega_2 t) E_2$$

Again, the frequencies must be the same on both sides, $\omega_1 = \omega_2$, such that the sines drop out of the equation and we are left with the following:

$$E_1 k_1 (-A + B) = -C k_2 E_2$$

or

$$(A - B) = C \frac{k_2}{k_1} \frac{E_2}{E_1}$$

By making use of the fact that $v = \lambda \cdot \nu = \omega/k$ and the fact that the frequencies are the same in both media, we see that $\frac{k_2}{k_1} = \frac{v_1}{v_2}$ and we can rewrite this further by replacing the ratio of wave numbers by the inverse ratio of wave speeds to obtain the following:

$$(A - B) = C \frac{v_1}{v_2} \frac{E_2}{E_1}$$

As a final step, we use the relation for the propagation velocity as a function of the characteristic constants of the material $v = \sqrt{E/\rho}$ to define the *sound impedance* $Z = v\rho = \sqrt{E\rho} = E/v$, such that we obtain the following condition:

$$(A - B) = C \frac{Z_2}{Z_1}$$

The two relations for $A + B$ and $A - B$ can now be solved for the two unknowns B and C, which will have to depend on the impedances of the two materials. For this, we add the following two relations:

$$A + B + (A - B) = 2A = C + C \frac{Z_2}{Z_1} = C \frac{Z_1 + Z_2}{Z_1}$$

Otherwise, we can add the following:

$$\frac{C}{A} = \frac{2Z_1}{Z_1 + Z_2}$$

Inserting this into $A + B = C = A\frac{2Z_1}{Z_1+Z_2}$ directly gives the following:

$$\frac{B}{A} = \frac{Z_1 - Z_2}{Z_1 + Z_2}$$

Basically, we now know everything about the reflection and transmission of the incoming wave. However, usually the intensity of a wave is a very important property as well, which is why we also want to determine the ratios for the intensities for the different parts of the wave. As previously described, the intensity is given by the following:

$$I = \frac{\rho}{2} u_0^2 \omega^2 v$$

Given that the two parts of the wave are in the same medium, the coefficient of reflection (R), which is the ratio of the reflected intensity to the incident intensity, is given by the following:

$$R = \frac{|B|^2}{|A|^2} = \left(\frac{r-1}{r+1}\right)^2 \qquad r \equiv \frac{Z_1}{Z_2}$$

The transmission coefficient (i.e., the ratio of the transmitted intensity to the incident intensity) needs to take into account the different propagation speeds and densities, yielding the following:

$$T = \frac{|C|^2}{|A|^2}\frac{Z_2}{Z_1} = \frac{4r}{(r+1)^2}$$

If we add these two relations, we find that $R + T = 1$, which means that the incoming intensity is completely split into a transmitted and a reflected part and the process of passing from one medium to another does not lead to the gain or loss of any intensity. In fact, this is something that we would have expected based on conservation of energy (see Section 6.4), and we could also have used this as a condition to determine the distribution into transmitted and reflected parts, which would of course have given us the same result.

From the preceding, we see that it is the difference in the sound impedance Z of the two media that determines how much of the wave is reflected at an interface. Figure 4.11 illustrates different situations from a hard to a soft material and vice versa.

We can look at some extreme cases to simplify the relations and get an intuitive feel for what reflection and transmission entails.

First, if the two materials have the same impedance, there is effectively no interface, since the important properties of the material are the same. Hence, we have $Z_1 = Z_2$ and we obtain $B = 0, \ C = A$. There is no reflection and all the intensity is transmitted (unchanged): $R = 0, T = 1$.

If the material for the incoming wave is much harder than the one for the transmitted wave, i.e., $Z_1 \gg Z_2$, and we obtain $B = A, \ C = 2A$, which implies that everything is reflected: $R = 1, T = 0$. Everything is reflected: $R = 1, T = 0$. The total amplitude at point $x = 0$ is $A + B = 2A$, which would correspond to a *free end* in a rope.

Finally, if the material for the incoming wave is much softer than the one for the transmitted wave, i.e., $Z_1 \ll Z_2 \ \Rightarrow \ B = -A, \ C = 0$, again everything is reflected: $R = 1, T = 0$. However, the amplitude of the incoming wave is opposite to the amplitude of the reflected wave. This is also called a phase jump, and this will be important in optics

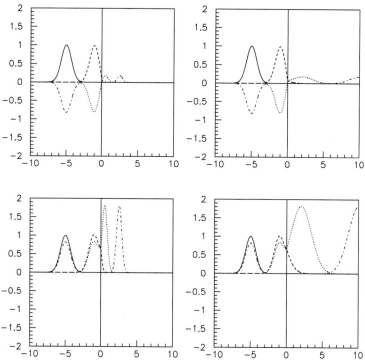

Figure 4.11 Reflection and transmission of sound waves at the boundary surface of two materials. Upper-left: $r = 0.1$, $v_1 = 2v_2$, i.e., transition from soft to hard material with reduction of the propagation speed; upper-right: $r = 0.1$, $v_2 = 2v_1$, i.e., transition from soft to hard material with increasing the propagation speed; lower-left: $r = 10$, $v_1 = 2v_2$, i.e., transition from hard to soft material with reduction of the propagation speed; lower-right: $r = 10$, $v_2 = 2v_1$, that is, transition from hard to soft material with increase of the propagation speed. The four different curves show spatial images of the total excitation at four different times: (1) continuous: before the impingement on the boundary surface, only an incoming part recognizable; (2) long dashed: the beginning of the wave train has already reached the boundary; (3) short dashed: the end of the wave train reaches the boundary, reflected, and continuous parts are clearly recognizable; and (4) short-long dashed: only reflected and transmitted parts are visible.

in Section 5.3.3. The amplitude at point $x = 0$ is $A + B = 0$, which corresponds to a *fixed end* in a rope.

4.7.2 Sonar

As an example of how waves behave at boundaries, let us consider how a sound wave in air can penetrate a body filled with water. This actually happens in our ear, where the sound waves in air have to excite the basilar membrane, which is inside a liquid. This transition happens in the middle ear. Air has a density of about $\rho_1 = 1.2$ kg/m^3 and a sound velocity of $v_1 = 330$ m/s. Water, however, has a density of $\rho_2 = 10^3$ kg/m^3 and a sound velocity of about $v_2 = 1,600$ m/s. Thus we get $\rho_1 v_1 = 400$ kg/(m^{2}s) and $\rho_2 v_2 = 1.6 \times 10^6$ kg/(m^{2}s), meaning that we have to pass sound waves into a very hard material, corresponding to

Figure 4.12 A bat can locate a flying moth in total darkness and also detect its speed (Doppler effect; see Section 4.9) by emitting sound waves (ultrasound: $\nu \approx 80$ kHz, not audible by humans) and registers the signals reflected by the moth.

the third case in the previous section. If we make this a little more precise and take into account a Taylor expansion up to the second term, we get $T \simeq \frac{4Z_1}{Z_2}$. With the numbers from the preceding, we get a transmitted intensity of $T = 0.001$, so only about one-thousandth of the intensity of the sound that comes from the air is transmitted into the water (and vice versa). This can be easily experienced when you try to talk under water, where you have to scream very hard in order to be just about heard. However, our ear is highly sensitive and can make out extremely small sound intensities. This means that the ear has found a way to bypass this small coefficient of transmission in some way. This is the reason for the construction in the middle ear encompassing the anvil, hammer, and stirrups, which greatly amplify the vibrations of the air by means of a lever, and then apply these directly to the auditory canal, where a reduction in diameter also results in an additional reinforcement. In fact, all of this results in a net *gain* of the sound intensity in the middle ear! If a wave is sent into an unknown medium, then the change of the sound impedance ρv relative to the incoming medium can be deduced from the reflected part and the spatial distance to this or other boundaries from the runtime of the reflected signal. This method is used in geology using seismic waves to find oil deposits or investigate the structure of the mantle of the earth. In ships and submarines, this sonar is used to measure the water depth, and similarly, bats, whales, and dolphins emit ultrasound signals, whereby they communicate with each other, locate prey, or orient themselves in the dark (see Figure 4.12). Ultrasound is also used in prenatal diagnostics in medicine (see Figure 4.13). Here, too, it must be ensured that the sound waves from the ultrasound device can penetrate the body. Therefore, the contact is made by means of a gel that has similar elastic properties as the human body, but which can make direct contact with the ultrasonic source.

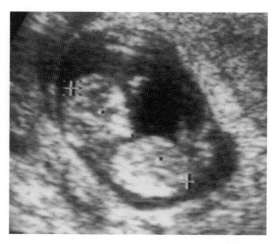

Figure 4.13 An ultrasound scan of a fetus looking for his thumb to suck. The frequency of the sound waves used is about 5 MHz.

4.7.3 Standing Waves

As we have seen previously, if a wave comes from a soft material into a hard one (or vice versa), all of the intensity is reflected such that we have two sets of waves counterpropagating in the medium. Actually, if we take a medium with two ends (thus limiting the propagation of the wave spatially), we always have two counterpropagating parts that under the right conditions can form standing waves. These are excitations that show *nodes*, where there is no excitation at all, and *antinode* points (or lines or surfaces) of maximum deflection. In principle, standing waves can occur in one-, two- and three-dimensional systems, where the nodes are points, lines, and surfaces respectively. As you can imagine, in a three-dimensional systems with surfaces of no excitation, the description can become quite complicated. Therefore, we want to illustrate the concepts on the basis of a one-dimensional system, a vibrating string, and will only describe higher-dimensional systems qualitatively.

So consider a string, which is firmly clamped at both ends, such as in a violin or in a piano. If we pluck this string, the wave pulse thus produced is reflected at the clamped end and superimposes with the incoming one. If we repeat the plucking at suitable intervals, two excitations can always meet such that the amplitude is amplified (or reduced). This corresponds to the so-called fundamental and higher harmonics of the string, which are shown in Figure 4.14. The lowest driving frequency at which a standing wave occurs is the one where there are nodes only at the two ends of the string, and this is called the fundamental vibration. Here, the wavelength is twice the length of the string. In the case of integer multiples of the fundamental frequency, higher harmonics occur, the wavelengths of which correspond to integral fractions of twice the string length.

We can now make this more quantitative. Figure 4.15 illustrates the situation. Let's call the transverse deflection of the string $u(x, t)$ and have it clamped at $x = 0$ and $x = L$, which means that $u(0, t) = u(L, t) = 0$. The wave coming from the right ($x > 0$) to the clamping point at $x = 0$ is $u(x, t) = u_0 \cos(kx + \omega t)$, and it superimposes with the reflected wave

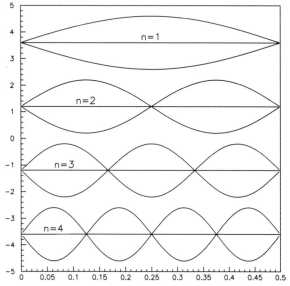

Figure 4.14 Standing waves in a string of length $L = 0.5$ m. In the fundamental vibration, all the string elements move up and down in phase. For the wavelength of this, we find $\lambda = 2L = 1$ m. For the higher vibrational modes, or harmonics (offset in the vertical axis), one finds $\lambda_n = 2L/n$, where n is an integer. A node always appears on the clamped ends.

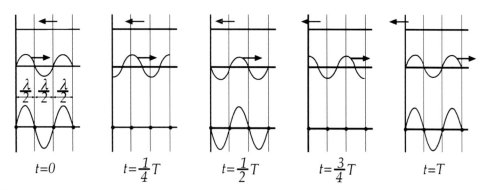

Figure 4.15 The five spatial images of the incident and reflected harmonic waves, which are recorded at different times, show how the superposition of the two waves can lead to a standing wave in which the nodes and bellies are stationary.

having inverted amplitude and moving to the right given by $u(x,t) = -u_0 \cos(kx - \omega t)$. We have just discussed the reflection at a fixed end having an inverted amplitude. The total displacement of the string then is the sum of these two waves, as follows:

$$u(x,t) = u_0 \cos(kx + \omega t) - u_0 \cos(kx - \omega t) = -2u_0 \sin kx \sin \omega t$$

Because we have clamped the rope at the two ends (at $x = 0$ and $x = L$), the displacement must disappear at the ends. For $x = 0$, this is already included in the approach, but for

$x = L$, this implies a condition for the standing wave and thus tells us something more about the system:

$$u(L, t) = -2u_0 \sin(kL) \sin(\omega t) = 0$$

Because this has to be valid for all times, we know that the prefactor has to fulfill the following:

$$\sin kL = 0,$$

$\sin(\omega t)$ changes with time and u_0 is an amplitude and therefore should be nonzero. This sine is zero exactly when its argument is a multiple of 180 degrees, i.e., π, which is why we obtain the following condition:

$$kL = n\pi, \quad n = \text{integer}$$

If we call the wave number k, which corresponds to the integer n by k_n, and indicate the corresponding wavelength by λ_n, we find the following

$$k_n = \frac{2\pi}{\lambda_n} = \frac{n\pi}{L} \Rightarrow \lambda_n = \frac{2L}{n}$$

The wave number and the frequency are related via $\omega_n = k_n v$ such that we obtain the fundamental and harmonic frequencies:

$$\omega_n = \frac{n\pi}{L} v$$

In all of the preceding, which concerned traveling waves, we have said that everything was described by the wave speed, which in turn was given by the properties of the wave-carrying medium. Now for standing waves, there are additional requirements determined by the spatial limitation of where the waves can occur, specifying the wavelengths or frequencies, which can only occur in discrete values (which are *quantized*). The value $n = 1$ corresponds to the fundamental (or basic) frequency, as follows:

$$\omega_1 = 2\pi\nu_1 = \frac{\pi}{L} v, \quad \lambda_1 = 2L$$

The higher harmonics correspond to the numbers $n = 2, 3 \ldots$, that is, multiples of the fundamental frequency, as we have already noted. The first harmonic differs from the fundamental vibration in frequency by a factor of two. In music, this step from the fundamental frequency to the first harmonic corresponds to increasing the tone by one octave.

In the case of nonresonant excitation, as occurs, for example, when drawing, striking, or plucking a string instrument, not only standing waves of a defined frequency are excited, but a whole spectrum thereof. The general oscillation state is then a superposition of harmonic oscillations,

$$u(x, t) = \sum_n A_n \cos(\omega_n t + \delta_n) \sin k_n x$$

This is something we have just seen recently when treating Fourier transforms.

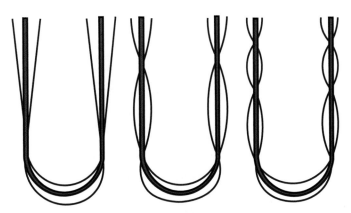

Figure 4.16 Basic and higher vibrational modes of a tuning fork.

Membranes, Hollow Bodies

As already mentioned, standing waves also occur in two- and higher-dimensional systems. Figure 4.16 shows the basic and higher vibrational modes of a tuning fork. The tuning fork is often connected to a wooden hollow body, the dimensions of which are adjusted in such a way that the fundamental frequencies of this sound body and the tuning fork are the matched (see also Section 4.7.3). The hollow body thus acts as a resonator, which transforms the sound wave produced in the tuning fork into a sound wave in a larger space and therefore is perceptible in all directions. A tuning fork without a resonating body is very difficult to hear if it is not right next to the ear. Similarly, the bodies of string instruments are constructed to allow for a clearer and louder sound by resonantly amplifying the vibrations of the strings.

Since the tones in instruments are always produced via standing waves (usually of strings or pipes), the wavelengths of the fundamental oscillations and thus the frequencies of the tones are always determined by the linear dimensions. The deeper the tones (i.e., the lower the frequencies), the larger wavelengths and thus the sound bodies and strings have to be (see Figure 4.17).

In higher-dimensional systems, because waves traveling in different directions have to be added up, the wavelengths of higher vibrational modes do not have to be integer multiples of the fundamental mode. Therefore, the vibrational states of sound bodies, oscillating glasses, or other more complex systems are very difficult to calculate and are usually obtained experimentally (or numerically on the computer). This is why instrument maker is a job that needs years of experience. One way of doing this experimentally for membranes is to spread sand on these membranes and excite a certain vibration. The sand then remains in heaps only at the nodes, thus providing a nice visual for the vibrational state (see Figure 4.18). The corresponding figures are called Chladni figures, after their discoverer.

Figure 4.17 The frequency range of the drawn string instruments and saxophones indicated by the horizontal bars is related to the size of the instrument. The frequency scale is indicated by the keys, and the frequencies increase from left to right.

Pipes and Organs

Standing waves can also be generated with sound waves in air. In the simplest case, a flute or organ pipe consists of a tube open at both ends. As explained earlier, a sound wave can be described by the displacement of the molecules $u = u_0 \sin(kx - \omega t)$ (displacement wave) or the sound pressure $p = p_0 \cos(kx - \omega t)$. The two waves are phase-shifted by a quarter period.

At the open ends of the tube, the sound pressure is negligibly small, thus there are nodes of the sound pressure wave $p(x, t)$ at the ends. If the tube has a closed end, there is a maximum in pressure there, i.e., an antinode and therefore a node in the center of the tube, as ilustrated in Figure 4.19. On the other hand, the molecules can no longer move on the wall of the closure, and thus the displacement wave needs to have a node there. The different pressures of the standing wave in a pipe are illustrated by the flames in the Ruben's tube shown in Figure 4.20.

The sound pressure wave is shifted by $\pi/2$ relative to the displacement wave. This is particularly noticeable with pipes in which standing waves are excited.

Figure 4.20 shows the basic vibration mode of a pipe open at both ends, above the displacement, below the sound pressure at different times. Similar to the string under tension, we find the following:

$$\text{(i) closed on both sides :} \quad \lambda_n = \frac{2L}{n}, \quad \omega_n = \frac{n\pi}{L} v$$

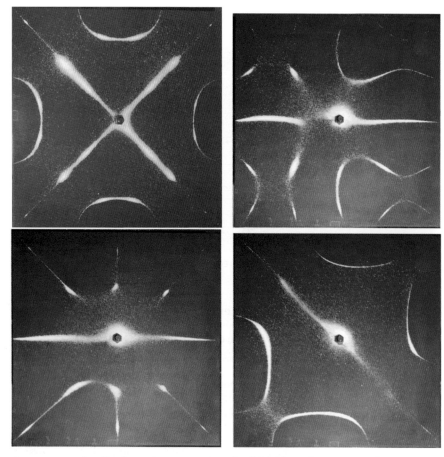

Figure 4.18 Standing waves on a plate. The observed patterns correspond to Chladni figures for a square membrane.

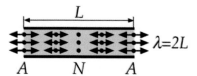

Figure 4.19 Pressure distribution in a pipe closed at both end in the basic mode of excitation. There is a node in the middle and two antinodes at either end.

Figure 4.20 A tone of a frequency is applied to a pipe, which is filled with a flammable gas. Holes in the top of the pipe let gas exit at different speeds depending on the pressure inside. When there is a standing wave excited in the tube, the flames of the exiting gas directly indicate the amplitude of the standing wave.

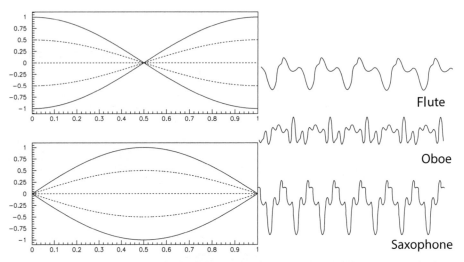

Figure 4.21 Left: Basic mode of a pipe open at both ends. Top: displacement u; bottom: pressure p. Right: wave trains of different wind instruments.

$$\text{(ii) open on both sides :} \quad \lambda_n = \frac{2L}{n} \quad \omega_n = \frac{n\pi}{L}v$$

For a pipe that is closed at one end, but open at the other, the wavelength of the basic mode is doubled:

$$\text{(iii) open on one end;} \quad \lambda_n = \frac{4L}{2n-1}, \quad \omega_n = \frac{(2n-1)\pi}{2L}v$$

These properties of standing waves in pipes lead to the following effects that you can test yourself at home:

- Blowing into different organ pipes: the longest pipes produce the lowest notes.
- Closing one end of an open pipe while blowing it: the basic frequency is halved, and the tone decreases.
- Using a light (helium)or heavy (CO_2) gas: $v(\text{He}) > v(\text{CO}_2) \quad \Rightarrow \quad \nu(\text{He}) > \nu(\text{CO}_2)$.
- Using hot gas: this leads to a higher tone because $v = \sqrt{\kappa RT/M}$.

As we have seen, in many instruments, a sound wave is first produced in a nearly one-dimensional medium, such as a thin rod, a string, or a pipe. For the best possible transmission of these waves to the surrounding space, additional resonators, e.g., the sound bodies of string instruments, are needed that are excited by the string or pipe. The waves emitted by these resonators then actually propagate in three dimensions and actually create the timbre and quality of the sound produced. The vibration of sound body causes the oscillation of the string or the air to excite harmonics of the actually excited fundamental vibration. These harmonics correspond to multiples of the fundamental frequency and can thus be seen as Fourier components of a discrete Fourier decomposition. The relative proportions of the harmonics correspond exactly to the Fourier coefficients. They determine the timbre of the instrument. Because of all this, the specific timbre of a given musical instrument is actually very different, even if the same note is played; see Figure 4.21, showing wave trains of different instruments for the same note.

4.7.4 Interference

In reflection and transmission as well as in standing waves, we have looked at very special instances of how two different types of waves overlap, where we have always used the superposition principle in order to simply add the contributions of the two different waves. We want to make this more general now and add up two plane waves of equal frequency, wavelength, and amplitude that both run in the $+x$-direction. Given that frequency, wavelength, and amplitude are the same, the only property in which the two waves differ is the phase of the oscillation, which we denote by δ. This basically describes the initial conditions of the excitation producing the wave. So we have two waves described by the following:

$$\text{Wave1} : \; u_1(x, t) = A \sin(kx - \omega t), \qquad \text{Intensity1} : \; I_0 \sim A^2$$
$$\text{Wave2} : \; u_2(x, t) = A \sin(kx - \omega t + \delta), \quad \text{Intensity2} : \; I_0 \sim A^2 \,.$$

Their superposition gives the resulting amplitude:

$$u(x, t) = u_1 + u_2 = A(\sin(kx - \omega t) + \sin(kx - \omega t + \delta)) \,.$$

With the help of the trigonometry relation

$$\sin \alpha + \sin \beta = 2 \sin \left(\frac{\alpha + \beta}{2} \right) \cos \left(\frac{\alpha - \beta}{2} \right)$$

we can write this as

$$u(x, t) = 2A \cos \frac{\delta}{2} \sin \left(kx - \omega t + \frac{\delta}{2} \right) = u_0 \sin \left(kx - \omega t + \frac{\delta}{2} \right)$$

Here, we have rewritten $2A \cos \frac{\delta}{2} = u_0$ in the second step. This formulation indicates that the resulting wave is another plane wave, with the same frequency and wavelength as the two original waves (and a phase shift halfway between the two original waves). Its amplitude u_0 and therefore also its intensity, however, depends on the relative phase and varies between the values of 0 and $2A$:

$$\text{Amplitude} : \; u_0 = 2A \cos \frac{\delta}{2}, \quad \text{Intensity} : \; I \sim 4A^2 \cos \frac{\delta}{2} \sim 4I_0 \cos^2 \frac{\delta}{2} \,.$$

For the limiting cases, in which the two waves maximally amplify (*constructive interference*) or extinguish (*destructive interference*) each other (see also Figure 5.2), one finds the following

$$u_0 = u_{0max} = 2A, \quad I_{max} \sim 4A^2 \sim 4I_0 \; \text{ for } \; \delta = 0, \pm 2\pi, \pm 4\pi \,;$$
$$u_0 = u_{0min} = 0, \quad I_{min} = 0 \; \text{ for } \; \delta = \pm \pi, \pm 3\pi, \dots \,.$$

Often, the phase shift δ of the two partial waves is caused by the fact that they have traveled different paths from a common source to the observation point where they superimpose. There is a relationship between this path-length difference Δ and the phase shift δ via the wavelength λ:

$$\delta = \frac{\Delta}{\lambda} 2\pi = k\Delta$$

This means that the phase shift is characterized precisely by the ratio of the path-length difference to the wavelength. We thus obtain constructive interference for

$$\Delta = \frac{\lambda \delta}{2\pi} = m\lambda \, ,$$

and destructive interference for

$$\Delta = (m + 1/2)\lambda \quad (m \text{ integer}) \, .$$

One of the basic interference experiments is the *double-slit experiment* (see Figure 4.22). Here, a wave hits two slits. The incident plane wave creates two spherical waves at the two slits, which then superimpose at a position (far) behind the slits. The experiment can be performed with any wave, e.g., sound waves, water waves (see Figure 4.23), microwaves, or (monochromatic) light (see Section 5.3.2). The only important point is that the distance between the slits is comparable to the wavelength.

The two slits have a distance d between them, and each originates spherical waves, which are observed on a screen at a distance D, far away from the slits. The path difference Δ to reach point $P(x)$ for the two waves originating at Q_1 and Q_2 is then given by $(x, d \ll D)$:

$$\Delta = r_2 - r_1 = \sqrt{D^2 + \left(x + \frac{d}{2}\right)^2} - \sqrt{D^2 + \left(x - \frac{d}{2}\right)^2}$$

$$\Rightarrow \Delta \simeq D\left(1 + \frac{1}{2}\left(\frac{x^2}{D^2} + \frac{xd}{2D^2} + \frac{d^2}{4D^2}\right)\right) - D\left(1 + \frac{1}{2}\left(\frac{x^2}{D^2} - \frac{xd}{2D^2} + \frac{d^2}{4D^2}\right)\right) = \frac{dx}{D}$$

Thus the phase difference is approximately

$$\delta = k\Delta = \frac{2\pi d}{\lambda D}x$$

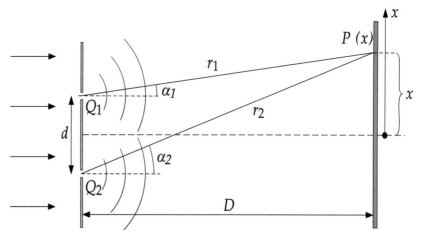

Figure 4.22 Young's double-slit interference experiment. The incident plane wave creates coherent spherical waves at the slits, which overlap at the screen. We observe the intensity distribution in points P at a distance x from the center of a screen with a fixed distance D to the slits. The intensity distribution shows a stripe pattern with alternating maximum and minimum intensity.

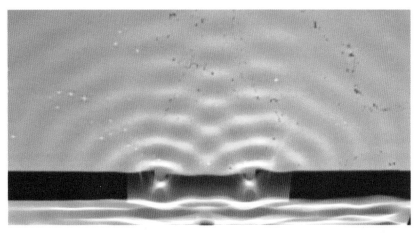

Figure 4.23 A double-slit interference experiment with water waves. The incident plane wave creates coherent spherical waves at the slits, which are superimposed with space-dependent phase shift behind the double slit. At certain angles, the amplitude is canceled out by destructive interference and at others it is enhanced by constructive interference.

which leads to an intensity distribution behind the double slit given by the following:

$$I(x) = 4I_0 \cos^2 \frac{k\Delta}{2} = 4I_0 \cos^2 \frac{kxd}{2D} = 4I_0 \cos^2 \frac{\pi xd}{\lambda D}$$

There will be maxima for the following:

$$\frac{\pi dx}{\lambda D} = m\pi \ (m \text{ integer}) \ \Rightarrow \ x = mD\frac{\lambda}{d}, \ \alpha = \frac{1}{2}(\alpha_1 + \alpha_2) \simeq \frac{x}{D} = m\frac{\lambda}{d}$$

If we carry out the double-slit experiment with sound waves, using two loudspeakers at a given distance, which are driven by a sinusoidal generator, we observe equidistant bands of loud and soft sounds, whose distance increases with decreasing distance between the sources d and increasing wavelength of the sound λ. We can also change this interference pattern by changing the phase relationship between the two sources. If the sources are in opposite phase, i.e., we add a total phase shift of π to one wave, the maxima suddenly produce minima and vice versa, since the phase shift then is $\delta = k\Delta + \pi$, which results in the preceding conditions for constructive and destructive interference being interchanged.

4.8 Wave Mechanics and Heisenberg's Uncertainty Principle

4.8.1 Particles Are Also Waves

In the course of the beginning of the twentieth century, it has been shown that there are more types of waves than the elastic ones, which we have just treated, and the electromagnetic ones that we are about to treat in Chapter 5 on optics. These other types of waves are those that describe matter itself. Waves that describe elementary particles. This is the basis of

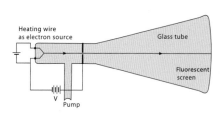

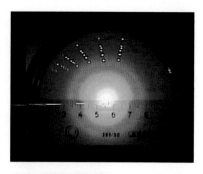

Figure 4.24 Diffraction pattern of electrons passing through a hole. Because electrons behave like waves, such a diffraction pattern is observed. The wavelength corresponding to the electron wave depends on the speed of the particles.

quantum mechanics, and the most fundamental properties of waves, i.e., interference and diffraction, can also be shown for elementary particles, such as electrons or atoms. When we send a wave through a slit, we get an intensity pattern that changes periodically or decreases at a certain angle. This angle, as we have seen, is inversely proportional to the size of the slit.

We thus also get an interference pattern when we send electrons (or atoms) through a double slit or a diffraction pattern when we send electrons through a hole (see Figure 4.24). This means that electrons behave like waves of a certain wavelength. It has been found that this wavelength is directly related to the momentum of the particle, i.e., the product of mass and velocity, $p = m \cdot v$ (see Section 6.3), as follows:

$$\lambda = \frac{h}{p}$$

Here, $h = 6.3 \times 10^{-34} Js$ is Planck's constant. Actually, often also $\hbar = \frac{h}{2\pi} = 10^{-34} Js$ is called Planck's constant, and one has to be careful if one only hears talk of Planck's constant. Typically, a physicist will directly say "h" or "h-bar" instead of Planck's constant, which indicates the exact version he or she means. These wavelike properties of elementary particles imply that if we consider the diffraction properties of the electron wave, there is a connection between the *position*, i.e., the size of the slit and the momentum or the *direction* of the particle. These two quantities, like the scattering beams in light, are connected by means of a Fourier transform, which implies the product of the width (or uncertainty) in space and the uncertainty in the momentum must be constant. This is Heisenberg's uncertainty principle, which states the following:

$$\Delta x \Delta p > h$$

Planck's constant appears here because of the relation between momentum and wavelength. For all kinds of waves, there is a similar relation between the wavelength and position $\Delta x \Delta \lambda > 1$. These basic properties of particles allow a large number of phenomena to be described coherently, and we will now look at a few examples of this and return to describe the nature of chemical bonds and the periodic table on this basis in Section 9.3.3, once we know more about electric interactions between particles.

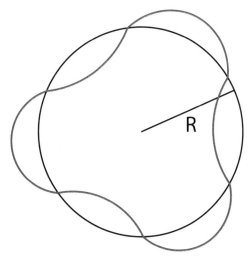

Figure 4.25 The electron in an atom performs a circular orbit around the atomic nucleus in such a way that it forms a standing wave. This state is stable.

4.8.2 The Bohr Model of the Atom

If electrons behave like waves, they can also form standing waves if they interfere with themselves. The wave of an electron moving about an atomic nucleus will have no amplitude on average, after many consecutive turns around the nucleus. This is because the starting phases of the wave are somewhat different in each of the turns, and thus if we add up many of these oscillations with slightly different phases we actually get zero. We have already seen this in the standing waves in organ pipes. There is, however, a special case where we do obtain a large amplitude. This occurs when the circumference of the orbit is just a multiple of the wavelength, as follows:

$$2\pi r = n\lambda = \frac{nh}{m \cdot v}$$

In this case, every orbit around the nucleus interferes constructively with the others, and a standing wave forms. This is illustrated in Figure 4.25 and forms the basis of the Bohr model, which gives a good description of electron binding in the atom.

If we transform the previous relation between the orbit radius and the momentum, we obtain an orbit radius, which is determined as follows:

$$r = \frac{n\hbar}{m \cdot v}$$

From the forces that act in the binding, this radius can then be calculated, which we do not want to do here. The standing waves of the electrons that result from such calculations correspond to the orbitals in chemistry, and from the structure of atoms from different electrons, the various chemical properties of the atoms of the periodic system can be explained; see Section 9.3.3.

4.8.3 Hydrogen Bonds

A somewhat different form of bond can result when a hydrogen atom is bound to oxygen or nitrogen. If there is another oxygen or nitrogen nearby, the proton of the hydrogen effectively binds to those two different binding partners. That is, the proton wave is smeared out between the binding partners, resulting in a corresponding blur in the momentum of the proton. That is, it must have an uncertainty in momentum of about $m_p v = h/x$, where x is the distance between the binding partners. The corresponding kinetic energy must then correspond to the binding energy in order to give a bound state, i.e., $E_B = m_p v^2/2 = h^2/(2m_p x^2)$. If the bond distance of these hydrogen bonds is roughly $x \simeq 1\mathring{A}$, the size of a hydrogen atom, and we insert the proton mass, we obtain a strength of these hydrogen bonds of $E_B \simeq 10^{-20}$ J. This means the hydrogen bonds lead to a reasonably strong bond compared to the thermal energy at room temperature (see Section 8.2.3), but not an extremely strong bond that can still be broken by thermal fluctuations. Therefore, the hydrogen bond that holds the various base pairs of the DNA together can also be separated relatively easily enzymatically, which allows an effective copying of the gene material encapsulated in DNA.

4.9 *The Doppler Effect

When a police car drives by us on the highway, we hear a distinct change in the frequency (i.e., the pitch) of its siren as it passes us. This applies to all waves encountered by a moving observer or emanated by a moving source. Long before this could be heard daily on highways, Christian Doppler postulated such a change in frequency for light in order to explain the different colors of stars. While we now know that the coloration of stars is not due to this effect but rather due to their different temperatures and hence their different efficiencies in burning nuclear fuel, Doppler was still correct in his description of the effect named after him, and we hear it mentioned almost every day. It is actually valid for all kinds of waves, including light (albeit slightly differently due to effects of special relativity, which do not concern us here). In addition, the effect is used as an addition to sonar by bats in pursuit of prey for estimating their speed, as well as in hospitals in prenatal diagnostics using ultrasound studying blood flow, or in astrophysics in the investigation of the early history of the universe (as it turns out, there is a Doppler shift in the coloration of galaxies rather than stars due to their movement, which allows us to study the expansion of the universe). Thus, the effect has numerous applications, so it is worth trying to understand its underlying causes.

A loudspeaker emits a wave of wavelength λ and frequency ν. If the loudspeaker moves relative to the observer (for instance, because it is placed on a truck) at speed v_S, the latter observes a wave with a shortened wavelength λ'. The shortening is just v_S/ν, as one can convince oneself by looking at the distance between the wave crests in Figure 4.26. Putting

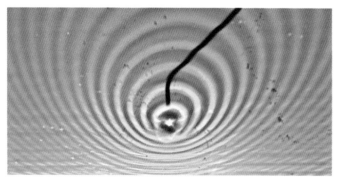

Figure 4.26 The change in wavelength for a moving source for surface waves.

this into an equation for λ' gives the following:

$$\lambda' = \lambda - \frac{v_S}{\nu} = \frac{v}{\nu} - \frac{v_S}{\nu} = \frac{v - v_S}{\nu}$$

The observed frequency therefore increases as follows:

$$\nu' = \frac{v}{\lambda'} = \frac{\nu}{1 - \frac{v_S}{v}}$$

If the source moves away from the observer, the sign of v_S changes and the observed frequency correspondingly decreases.

If the source velocity v_S is larger than the propagation speed of the wave itself, the radiated intensity is collected in a cone, the Mach cone, which moves along with the source at its tip. The observer hears a loud bang when the cone passes by, since all of the intensity collected in the cone arrives at once. This is the supersonic boom that a military jet can produce whenever it is traveling at speeds faster than sound. The Mach cone can also be observed very nicely in ships or swimming ducks, as they move at a velocity greater than that of the surface wave on a lake.

If the observer (or detector) moves rather than the source, the situation is slightly different. According to its movement, the latter sees the waves in a faster or slower sequence. Then the new frequency is as follows:

$$\nu' = \nu \pm \frac{v_D}{\lambda} = \nu \left(1 \pm \frac{v_D}{v}\right)$$

Here, the positive sign in the velocity means that the observer moves toward the source, and the negative one that the observer moves away from the source. Thus we again have a higher frequency when we move toward the source and a lower one when we move away from it.

If we observe something by means of echolocation, both of the aforementioned effects occur. This can be for ultrasound diagnostics, where we want to examine the blood flow, for example, thus the moving particle is an erythrocyte, or the case of a bat that wants to catch a flying mosquito. On the one hand, the moving particle experiences a frequency shift since it can be regarded as a moving detector. Then it reflects sound of this frequency and gives it out as a moving source. The frequency shift experienced by the ultrasound device or the bat is then given by the combination of the effects, and we measure the frequency ν'':

$$\nu'' = \nu\frac{1 + v_O/v}{1 - v_O/v}$$

where v_O is the velocity of the moving object relative to the source (i.e., the bat or the ultrasound device). Since flies typically move at speeds that are small compared to the speed of sound, we can make the approximation $v_O \ll v$ and get a frequency change of approximately the following:

$$\Delta\nu = \nu'' - \nu = \nu\left(\frac{1 + v_O/v}{1 - v_O/v} - 1\right) = \nu\frac{1 + v_O/v - (1 - v_O/v)}{1 - v_O/v} = \nu\frac{2v_O/v}{1 - v_O/v} \simeq 2\nu\frac{v_O}{v}$$

Exercises

4.1 Resonance and absorption

A mass of 0.50(1) kg is suspended on a spring having a spring constant of 5.0(1) N/m. Friction in the system leads to a damping coefficient of $f = 0.90(9)$kg/s. Now we excite this pendulum with an oscillating external force given by $F_{ext} = F_0\cos(\Omega t)$. The amplitude of this force is $F_0 = 2.0(2)$N and the period of the excitation $(2\pi/\Omega)$ is 1.00(1) s.

(a) What is the natural oscillation frequency (also called the eigenfrequency) of the swinging mass?
(b) What is the uncertainty of this frequency?
(c) What is the amplitude of the oscillation of the mass after a long time?
(d) What is the uncertainty of this amplitude?
(e) Does this amplitude decay with time due to the friction in the system? Justify your answer.

4.2 Elastic waves

(a) A string made of steel (radius $r = 0.20(2)$ mm, density $\rho = 8 \times 10^3$kg/m^3) that is 30.0(1) cm long should be drawn such that is gives out an A (i.e., a frequency of 440(22) Hz) in its basic mode when plucked. What is the tension force F with which you have to draw the string?
(b) What is the uncertainty of this force?
(c) Suppose this string has a yield tension of 0.7 GPa. What is the highest possible frequency of the basic mode that this string can support?

4.3 Elastic waves 2

A rope of length L hangs freely from the ceiling. If you briefly move the top of the rope, there is going to be a pulse of a transverse wave moving down the rope. This downward motion is due to the fact that the tension in the rope due to gravity pulls back on the excitation. What is the speed of this wave moving downward? How long does it take until the pulse arrives at the end of the rope? Compare this to the time it takes for a stone to fall the same distance.

4.4 Sound intensity

The noise level (i.e., sound intensity) next to a straight highway is 70 dB at a distance of 50 m to the highway. At what distance would the noise level be 60, 50, and 40 dB respectively? Ignore for your calculations the fact that sound is also absorbed in air.

4.5 Hearing threshold

The ear canal and thus the basilar membrane of humans is roughly 2.5 cm long. From this, we can estimate the frequency at which human hearing is most sensitive. In other words, at what frequency of oscillation will the amplitude of the response of the cochlea be largest? Consider the basilar membrane fixed at one end.

4.6 Hearing threshold 2

How sensitive would the ear have to be that we could hear thermal motion of the eardrum? Consider for this that air molecules hitting the eardrum have an energy due to thermal motion of roughly $k_B T = 4 \times 10^{-21}$ J, where $T = 300$ K is the temperature at normal conditions. Also use the fact that the ear is most sensitive at frequencies of about 1.5 kHz (see Exercise 4.5) and that the eardrum has an area of about 0.25 cm^2. Compare this to the actual hearing threshold of roughly 10^{-12}W/m^2.

4.7 Sound waves

Why is the speed of sound in liquids typically larger than the speed of sound in gases?

4.8 Sound intensity 2

(a) A point source (i.e., a small loudspeaker) emits a spherical sound wave whose intensity is damped in time as follows: $P = P_0 \exp(-t/\tau)$, where $P_0 = 125$W and $\tau = 10$s. At what distance r is the maximum sound intensity 120 dB?

(b) At this point, how long does it take until the intensity has decreased to a level of 65 dB?

4.9 Fourier decomposition

We have considered the basilar membrane in the cochlea as a frequency-measuring device, where different parts of the membrane can be excited at different resonance frequencies due to the geometry of the basilar membrane. So let us simplify this even further and treat the basilar membrane as a series of resonators each having an eigenfrequency ω_n and a damping coefficient f.

(a) What is the reverberation time of such a resonator? In other words, what is the time after which its energy has decreased by a factor of $1/e$ after its excitation has stopped?

(b) What is the frequency resolution of such a resonator? Consider the width of the resonance curve for this purpose.

(c) How are reverberation time and frequency resolution coupled? Can you use this to draw conclusions about repetition rates at which specific tones can no longer be separated? As a numerical example, consider that the relative frequency resolution is about 5% at a frequency of 100 Hz for humans. What conclusions do you draw for the typical arias of sopranos versus sonorous voices?

4.10 Fourier transform

What is the Fourier transform of a damped, harmonic oscillation? The amplitude is given by $u(t) = u_0 \exp(-\gamma t) \cos(\Omega t)$, where $\Omega^2 = \omega_0^2 - \gamma^2$. Determine the cosine component of the Fourier transform, as follows:

$$A(\omega) = \sqrt{\frac{2}{\pi}} \int_0^\infty u(t) \cos(\omega t) dt$$

To simplify the calculation, you can use complex numbers to rewrite the cosine as follows: $\cos(\phi) = \frac{1}{2}(\exp(i\phi) + \exp(-i\phi))$.

4.11 Fourier transform 2

Determine the Fourier transforms $A(\omega)$, $B(\omega)$ of a short pulse $z(t)$ of duration τ, i.e., $z(t) = 1$ for $0 < t < \tau$ and $z(t) = 0$ for all other times. Remember: $A(\omega) = \frac{1}{\sqrt{2\pi}} \int_{-\infty}^{\infty} z(t) \cos(\omega t) dt$ and $B(\omega) = \frac{1}{\sqrt{2\pi}} \int_{-\infty}^{\infty} z(t) \sin(\omega t) dt$.

4.12 Standing waves

(a) An organ pipe is filled with air (speed of sound 340(1) m/s), closed on top and has a length of 1.70(1) m. What is the basic frequency of the pipe?

(b) What is the uncertainty of this frequency?

4.13 Standing waves 2

(a) For the Ruben's tube shown in the text, a basic tone of A (corresponding to a frequency of $\nu = 440(2)$Hz) has shown five nodes. How long is the tube? For this, you need to know that the tube is filled with methane (density $0.60(1)$kg/m^3) and open at one end.

(b) What is the uncertainty of this length?

4.14 Diffraction

(a) In an experiment with microwaves, a source has been passed through a slit width a width of 10(1) cm. A receiver antenna has recorded a (first) minimum intensity at an angle of $\phi = 30(3)°$. What is the wavelength of the microwaves used? What is their frequency?

(b) What is the uncertainty of this wavelength and this frequency?

4.15 Interference

Waves in the sea have a wavelength of roughly $\lambda = 10$ m close to a harbor. If the harbor has two entrances, both 5 m wide and 50 m apart, where do you have to moor your ship such that it lies as quietly as possible? In other words, where do the waves interfere destructively, if you consider the entrances to the harbor as point sources?

4.16 Sonar and Doppler

(a) Bats emit a sound with a frequency of about 60 kHz to orient themselves. With a speed of sound of 340 m/s in air, what is the smallest insect that a bat can still resolve?

(b) If the bat is flying with a speed of 10.0(1) m/s, what is the frequency of the sonar signal reflected off an insect flying with a speed of 3.0(1) m/s toward the bat?

(c) What is the uncertainty of the observed frequency change?

4.17 Sonar and Doppler 2

(a) In ultrasound diagnostics in hospitals, typically sound waves with a frequency of 1.50000(1) MHz are used. When considering that the compressional modulus of tissue is about $K = 2.0(1) \cdot 10^9$ Pa, and the tissue's density is close to that of water, i.e., $\rho = 1.0(1) \cdot 10^3$ kg/m^3, what is the smallest object such an ultrasound can still resolve?

(b) Consider that you now intend to determine the speed of the blood flow using the same ultrasound. What is the frequency shift that you have to be able to determine? Assume that the blood flow you want to measure is close to the aorta and is about 10(1) cm/s away from the source of the ultrasound.

(c) What is the uncertainty of the observed frequency change?

4.18 Atomic Physics

(a) How would the size of atoms be changed if not electrons but muons would make up matter. Muons are known elementary particles that are the same as electrons in virtually all aspects, but weigh about 200 times more.

(b) How does the answer to (a) change if antiprotons instead of muons were orbiting the proton? Antiprotons are the same as protons, except that they are oppositely charged, i.e., they are attracted to protons in the same way as electrons, but are about 2,000 times more massive than electrons.

Quiz Questions

4.1 Resonance
Which of the following statements describes the phenomenon of resonance?

A Without damping, the maximum excitation takes place at the natural frequency.

B At its maximum, the resonance curve obtains a phase shift of 180 degrees.

C The amplitude of excitation decreases with frequency as $1/\Omega^2$ for large excitation frequencies.

D With damping, the maximum excitation is shifted to higher frequencies.

E At very small frequencies, the amplitude of excitation depends only on the amplitude of the applied force and the spring constant of the pendulum.

4.2 Properties of waves

Harmonic waves are described by different properties: frequency, angular frequency, period, amplitude, and wavelength. Which of these are directly proportional?

A Period and frequency
B Amplitude and wavelength
C Period and wavelength
D Angular frequency and period
E Angular frequency and frequency
F Period and amplitude

4.3 Properties of waves 2

Harmonic waves are described by different properties: frequency, angular frequency, period, amplitude, and wavelength. Which of these are inversely proportional?

A Period and frequency
B Amplitude and wavelength
C Period and wavelength
D Angular frequency and period
E Angular frequency and frequency
F Period and amplitude

4.4 Waves

A transverse wave on a rope is described by $y = y_0 \sin(kx + \omega t)$, where $y_0 = 0.1$m, $k = 2\pi\,\text{m}^{-1}$, and $\omega = 10\pi\,\text{s}^{-1}$. What is the frequency ν of this wave?

A 10 Hz
B 5 Hz
C 2 Hz
D 1 Hz

4.5 Waves 2

In which direction does it travel?

A +y
B −y
C +x
D −x

4.6 Waves 3

What is the speed of the wave?

A 10 m/s
B 5 m/s
C 2 m/s
D 1 m/s

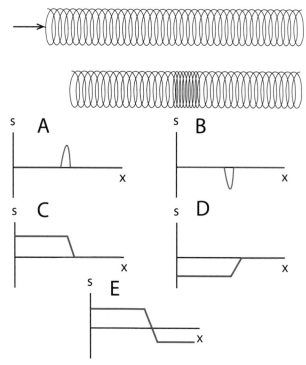

Figure 4.27   A wave pulse sent through a long spring.

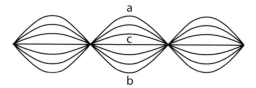

Figure 4.28   A rope attached at both ends with an excited standing wave.

4.7   Wave transport

A wave pulse is sent through a long spring (a slinky). Which of the curves in Figure 4.27 describes the displacement (s) of the spring as a function of position (x). Displacements to the right are defined as being positive.

4.8   Standing waves

A rope is attached at both ends and a standing wave is excited, as shown in Figure 4.28 (the second harmonic). The position of the rope thus oscillates between the curves marked a and b. What can you say about the vertical speed of the rope when its position corresponds to curve b? Positive velocities are upward.

A It is zero everywhere.
B It is positive everywhere.
C It is negative everywhere.
D It depends on position.

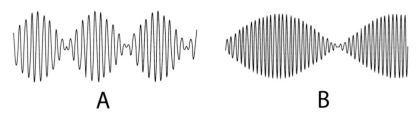

A   B

Figure 4.29 Two sources emitting sound with a small difference in frequency between them.

4.9 Sound waves

Which kind of sound waves only exist in solids?

A Longitudinal shearing waves
B Transversal compressional waves
C Transversal shearing waves
D Surface waves
E Radiowaves
F Longitudinal compressional waves

4.10 Beating

Two sources are emitting sound (or any wave) with a small difference in frequency between the sources. This leads to an on and off in the sound, illustrated by the curves in Figure 4.29. Which of the two pairs of sources contain the sources with the highest frequency? Both graphs are plotted on the same temporal scale.

A Pair A
B Pair B
C Both are the same

4.11 Standing wave excitation

The hearing canal and thus the basilar membrane of humans is about 2.5 cm long. From this, you can conclude that the ear is most sensitive for which wavelength?

A 2.5 cm
B 5 cm
C 7.5 cm
D 10 cm

4.12 Fourier transforms

Which of the sketches in Figure 4.30 correspond to pairs of functions related via a Fourier transform?

4.13 Fourier transforms 2

Which of the sketches in Figure 4.31 correspond to pairs of functions related via a Fourier-transform?

4.14 Fourier transforms 3

Which of the sketches in Figure 4.32 correspond to pairs of functions related via a Fourier transform?

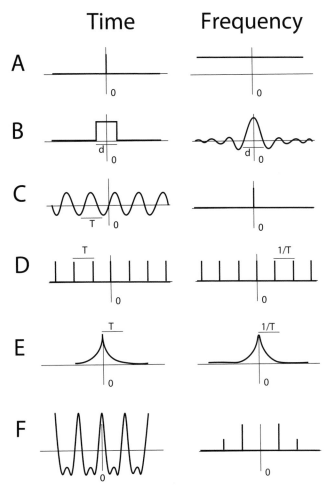

Figure 4.30 Pairs of functions related via Fourier transform.

4.15 Interference

Two tuning forks show a beating with a frequency of 5 Hz. One of the tuning forks is tuned to a frequency of 245 Hz. What frequency is the other one tuned to?

A 240 Hz

B 242.5 Hz

C 245 Hz

D 247.5 Hz

E 250 Hz

F None of the above

4.16 Interference 2

Which of the following conditions need to be fulfilled for two harmonic waves to form a standing wave?

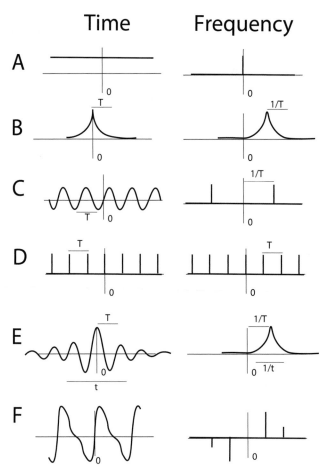

Figure 4.31 Pairs of functions related via Fourier transform.

A The waves have the same amplitude.
B The waves are polarized.
C The waves are transverse waves.
D The waves have the same frequency.
E The waves are counterpropagating.
F The phase difference is a multiple of π.

4.17 Interference 3

Which of the following conditions need to be fulfilled for two harmonic waves to show complete destructive interference?

A The phase difference can have any constant.
B The waves have the same amplitude.
C The waves are polarized.

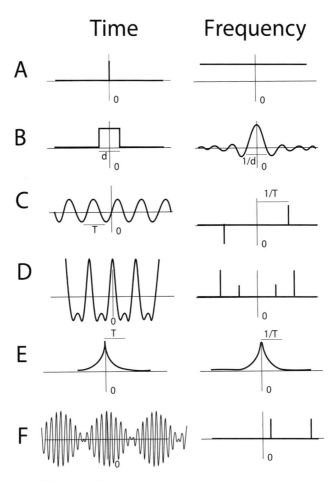

Figure 4.32 Pairs of functions related via Fourier transform.

D The path difference between the two waves is $\Delta x = \lambda(m + 1/2)$, where m is an integer.

4.18 Transition between different media

Two strings of the same material are attached in a series. On top there is a thinner one, on the bottom a thicker one. If a harmonic wave is released at the top, which properties of this wave change at the boundary between the two strings?

A Frequency
B Period
C Wavelength
D Propagation speed
E Amplitude
F Wave number

4.19 Doppler effect

A bus line has buses depart every 10 minutes in both directions. Someone tells you they had seen buses of this line in a given direction every 5 minutes. Do you believe this?

A No, this is impossible.

B Yes, they were at one of the terminals.

C Yes, they were in a bus traveling in the given direction.

D Yes, they were in a bus traveling opposite to the given direction.

Optics, Light, and Colors

5.1 How Light Interacts with Matter and What We Can Learn from This

The waves we have looked at so far were waves where the disturbance took place in a mechanical medium where we could directly see what interacted with what. It is therefore somewhat easier to introduce the concepts of waves, such as interference and diffraction in the context of these mechanical waves. However, one of our most important senses, vision, and the corresponding experimental techniques relies on entirely different waves, namely electromagnetic waves, where the disturbance is not propagating within some mechanical aspect of a medium, but in the electromagnetic field. At the end of the book (Chapters 9 and 10), we will deal with electrical and magnetic fields, but let it suffice to say here that the magnetic fields align a compass needle and the electric fields maintain an electrical current and that these fields are actually responsible for all phenomena of optics. In optics, we are dealing with optical imaging systems, such as microscopes, with how (color) vision works, and how different objects obtain different optical properties, such as colors. In order to understand these processes, however, we must briefly look at the atomic structure and use light as an electromagnetic wave. The electric fields will stimulate the electrons in the atomic shell, and this interaction is decisive for the optical properties of matter, such as fluorescence. However, coloration, as we have seen in Section 1.1.1, can also directly stem from the interference effects on an ordered lattice, such that these wave phenomena are important in optics as well. Similarly, interference and diffraction form the basis of X-ray crystallography, which is of great importance in structural biology and chemistry. Apart from using crystallography to elucidate the structures of proteins and thus understanding their functions, the importance of X-ray diffraction is illustrated by the classic case of how the determination of the structure of DNA has directly shown how the genetic code is stored and copied on DNA molecules.

Finally, optical instruments, such as microscopy, are among the most widely used tools in biology, which is why it is important that you know their basic principles. Here, the refractive index is the most important parameter, which can describe very many properties of optical systems within the framework of geometrical optics. This is of particular importance in microscopy, since the mode of operation of a microscope can thus be represented. We will, however, also have to deal with diffraction and interference phenomena in order to explain the resolution limit of microscopes. Nowadays, however, it is possible to circumvent these limitations up to a certain degree – we will briefly discuss these modern methods in connection with absorption processes and fluorescence.

As in the case of acoustics, we will deal in detail with wave transport, behavior at boundaries, resolution, interference, and diffraction, so some of the concepts will get a repeat introduction here. While sound waves correspond to mechanical deformations in liquids and solid substances or pressure fluctuations in gases, the excitation in an electromagnetic wave consists of temporally and locally changing electric and magnetic fields. This will actually influence how light is absorbed by electrically charged fundamental particles, as will be explained later in this chapter. But first, let us look again at what happens when a wave hits the boundary between two different materials, but this time for light waves.

5.2 Refraction and Reflection

When a wave reaches the boundary between two media, as we have already seen in the previous chapter, the intensity of this wave can either be transmitted or reflected. The transmitted part of the wave is transported farther in the new medium, and the reflected part goes back into the medium the wave has come from. We have looked at the case of what happens when the wave is incident perpendicular to the interface and seen that, for reasons of energy conservation (conservation of intensity) and conservation of momentum (conservation of the deflection at the boundary surface), we can determine how much of the wave is reflected and transmitted. This has depended on the impedance of the material, which depends on the speed of the wave in the medium and the density, i.e., how easily the medium is deflected. In the case of an electromagnetic wave, the amplitude (i.e., the deflection) is given by the electric field. The intensity will thus depend on the square of the electric field – more precisely, $I = c_n \epsilon \epsilon_0 E^2 / 2$. Here, ϵ is a materials constant (called the dielectric constant), which describes how easily the electrons can be displaced in the atomic shells, and we shall get to know it more closely in Chapter 9 regarding electricity. This parameter is also directly related to the refractive index via $n = \sqrt{\epsilon}$, and we can therefore see that the refractive index n will take the role of the impedance when we consider electromagnetic waves at interfaces. We can therefore simply put in the refractive index for the impedance in the relationships derived for reflection and transmission of waves in Section 4.7.1 and obtain the following:

$$R = \left(\frac{1-n}{1+n} \right)^2$$

$$T = 1 - R = \left(\frac{4n}{1+n} \right)^2$$

Here, we have considered that medium 1 has a refractive index of 1 while medium 2 has a refractive index of n. In the general case, we would have to replace n by the ratio of the refractive indices, n_2/n_1.

As is often the case, we can consider some aspects of these relationships qualitatively without going through lengthy calculations. Since we consider intensities, the relationship must contain a square, because the physically relevant interacting property is the amplitude.

Furthermore, we know that the index of refraction is the decisive quantity that describes the material whose optical properties we consider. We will exploit this in detail when we look at geometrical optics. If the index of refraction were on, or otherwise put, equal to that of the surrounding material, then we could just as well have no boundary between the two materials as the materials do not differ in the only relevant parameter. In that case, the reflectivity must be zero. This means that the reflectivity has to somehow go with $(1 - n)^2$. The point that nothing should happen with the same index of refraction is something that can also be used in imaging by looking at the refractive index of a surrounding material so that disturbing things can be made "invisible." Also, the reflectivity cannot exceed 1, there cannot be more light reflected than we put in. Therefore, the reflectivity coefficient has to be normalized by something containing the index of refraction n. For indices of refraction smaller than one (something that never happens in our universe because of the theory of relativity), this would still have to hold, which is why the normalization factor has to be $(1 + n)^2$.

Let us take a numerical example of the reflectivity and consider the incidence of light from air to glass, with a refractive index of $n = 1.5$. The reflection coefficient then is $R = ((1 - 1.5)/(1 + 1.5))^2 = ((-0.5)/(2.5))^2 = ((-1)/5^2) = 1/25 = 0.04$ or 4%. That is, when we shine light on glass (with a perpendicular incidence), 4% of its intensity is reflected. This is also the reason that windows seem to act as mirrors at night and not during the day. At night, when no light falls through the window from the outside, we only see the 4% reflected light. During the day, this reflection is also there, but it is lost underneath the background of the light that falls in from outside.

If the light does not impinge perpendicularly to the boundary, we get a refraction of the transmitted beam in addition to the reflection. This is true not only for light waves, but for any waves as we had alluded to earlier without describing the effect in detail. We want to amend this here and describe how the propagation direction of the beam changes inside the medium relative to the outside. This is clearly shown by Figure 5.1, which could be for surface waves on the ocean just as well as for light. In the two media, the waves have different wavelengths due to the different wave speed. The frequency in both media is the same because of energy conservation, but the wavelength changes according to the speed. So, we have a wavelength $\lambda_n = \lambda/n$ inside the medium. Here λ is the wavelength in vacuum and n is the refractive index as previously, which describes the reduction of the speed of the wave: $c_n = c/n$. The distance between the respective wave crests and troughs is therefore smaller inside the medium. If, at the boundary, the electric fields are to connect continuously into one another, the distance between the wave crests and troughs must be the same outside and inside the medium along the boundary. This means that the direction of the transport of the waves must differ, or more specifically that we must have $\lambda/\cos(\pi/2 - \alpha) = \lambda_n/\cos(\pi/2 - \beta)$. Finally, we know that $\cos(\pi/2 - \alpha) = \sin(\alpha)$, such that we obtain the following:

$$n = \sin(\alpha)/\sin(\beta)$$

If a light beam is incident on a medium at an angle from the vacuum, the angle of the phase planes is changed because the part of the planes that impinges on the medium first propagates more slowly than the rest, which is still in a vacuum. Therefore, the light beam

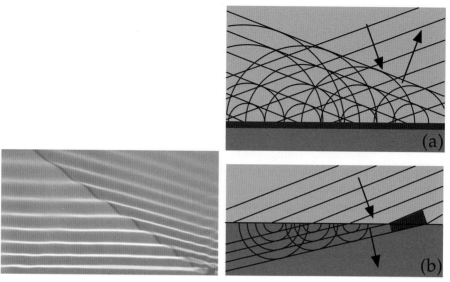

Figure 5.1 At a boundary between two media, the wavelengths of light changes, quantified by the refractive index. A wave incident at an angle to the boundary will therefore be refracted, such that the projection of the different wavelengths onto the boundary is the same in both media. This leads to the formulation of Snell's law.

is refracted. This is Snell's law of refraction that we will use in geometrical optics as the basis for the description of all types of optical instruments. But before we restrict ourselves to describing the propagation of the light only in the form of light beams, let us look at some phenomena directly related to the wave nature of light, most of which are related to the ability of light to interfere.

5.3 Interference and Diffraction

Many of the applications of optics in biology are based on the property of interference of light. In Chapter 1, we saw that the coloration of butterflies and many other animals is caused by nanometer-sized structures on the surface of their skin or scales, which allow only a reflection of a particular color by interference. Likewise, crystallography and other methods of structure determination are directly dependent on the interference of light (whether visible or X-ray). In order for light to interfere, certain properties are necessary, which we now discuss. For this, the concept of coherence is of particular importance.

5.3.1 Coherence

When we treated interference in general waves (in Section 4.7.4), we assumed that the relative phase δ between two interfering waves is independent of time. Such waves are coherent. Light waves originating from different sources are usually *not* coherent. While the emission duration of the emitting atoms is very short ($\tau \leq 10^{-8}$ s), the period of the

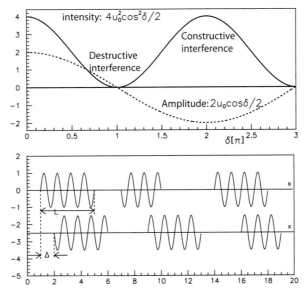

Figure 5.2 Coherence and interference: the top image shows the intensity and amplitude of a plane wave resulting from the superposition of two waves with the same wavelength and fixed phase difference δ as a function of δ. The bottom picture shows typical sequences of two independent light sources that emit waves of the same wavelength during a short time of $\tau = L/c$. The phase differences between the individual wave trains are randomly distributed.

light is much shorter ($T \simeq 10^{-15}$ s). That is, light waves consist of many individual wave trains with a length of $L \simeq c\tau \leq 3$ m, which are randomly distributed over time and can therefore easily show position differences in the range of half a wavelength (or roughly 300 nm) corresponding to phase differences in the range of π. These many contributions are averaged over during the observation time, as illustrated in Figure 5.2. Therefore, there is also a phase difference $\delta = \delta(t)$, which changes with time quickly and randomly. Our eye, and in general also optical devices, cannot follow these rapid changes, but register temporal averages. For the intensity of two superimposed waves, we therefore obtain the sum of the individual intensities:

$$\bar{I} \sim \overline{4A^2\cos^2\frac{\delta(t)}{2}} = 4A^2\frac{1}{2} = 2A^2 \sim 2I_0$$

Incoherent waves thus cannot interfere, since the intensities rather than the amplitudes are added. If we want to observe light interference, the wave of a single source has to be split into partial waves, and these must then be superimposed. As long as the path difference Δ is small compared to the so-called *coherence length* $L \simeq 3$ m, the partial waves are coherent. A *laser* (*L*ight *A*mplification by *S*timulated *E*mission of *R*adiation) emits waves whose coherence length L can be up to 1,000 km. For two wave trains not to get out of phase over this spatial distance, their wavelengths must be equal to a great level of accuracy. Lasers are thus very monochromatic light sources. In lasers that emit very short pulses, lasting only slightly longer than the period of the light, this is no longer the case. The coherence length of these lasers is therefore also a lot shorter. However, the different wavelengths are still in phase and therefore coherent. Such lasers are used, for example, in multiphoton microscopes.

5.3.2 Double-Slit Interference

One of the fundamental experiments showing the wave nature of light is *Young's double-slit experiment*. In this experiment, a single light source illuminates two slits. The incident plane wave creates two spherical waves originating at the slits, so the coherence condition can be guaranteed. We have actually already seen this experiment (see Figure 4.21) in Section 4.7.4 for sound waves and water waves. However, electromagnetic waves such as microwaves or monochromatic laser light give the same result. We again want describe the result, albeit in a different way (but as always coming to the same result). As we have seen in Section 4.7.4, one important factor for interference to be observable is that the gap between the slits is comparable to the wavelength.

The spherical waves emanating from the two slits, being a distance d apart, are observed on a screen. We can obtain the path difference Δ of the wave originating at Q_1 relative to that originating at Q_2 up to point P using geometry. Consider the situation as drawn in Figure 5.3. If we form an equilateral triangle with one side going from Q_2 to P and where the extension of the remaining side goes through Q_1, we are missing a triangle that has side lengths d and Δ from the total area covering Q_1, Q_2, and P. Then we know the following from the definition of the sine:

$$\Delta = d \sin(\beta)/ \sin(\theta)$$

We want to describe the path difference as the function of the angles of observation (i.e., α_1 and α_2), which means that we must express the angles θ and β as a function of these two

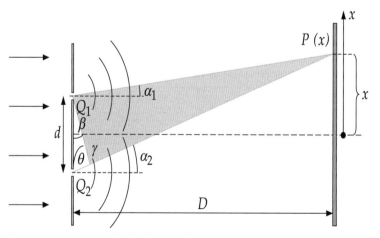

Figure 5.3   Young's interference experiment with a double slit. The incident plane wave generates coherent spherical waves at the slits, which can be superimposed. One observes the intensity distribution in points P at a the distance x from the center of a screen with a fixed distance D to the double slits. The intensity distribution shows a striped pattern with alternating maximum and minimum intensity.

angles. To do this, we first consider the angle at the base of the shaded, isosceles triangle, (γ). Because of the angular sum in a triangle being π, we obtain $\pi = 2\gamma + \alpha_1 - \alpha_2$ or otherwise put the following:

$$\gamma = \pi/2 - (\alpha_1 - \alpha_2)/2$$

With this, we obtain the two angles:

$$\theta = \pi - \gamma = \pi/2 + (\alpha_1 - \alpha_2)/2$$

and

$$\beta = \pi/2 - (\gamma - \alpha_2) = (\alpha_1 + \alpha_2)/2$$

and thus the path difference:

$$\Delta = d\sin(\alpha)/\cos(\delta\alpha)$$

where we have used $\alpha_1 = \alpha + \delta\alpha$ and $\alpha_2 = \alpha - \delta\alpha$. This means that the average angle of observation is α and the difference between the two angles is $2\delta\alpha$. If the observation screen is far away ($D \gg d$), the difference between the two angles can be neglected, such that $\cos(\delta\alpha) \simeq 1$, and we obtain the following:

$$\Delta = d\sin(\alpha)$$

If in addition the angle of observation is small, we obtain $\Delta/d = \sin(\alpha) \simeq \alpha \simeq x/D$. In other words, we obtain a phase shift of

$$\delta = k\Delta = \frac{2\pi d}{\lambda}\sin(\alpha) \simeq \frac{2\pi d}{\lambda D}x$$

as we have obtained in Section 4.7.4 using the Pythagorean theorem. This phase difference then gives the following intensity distribution on the screen of observation:

$$I(x) \sim 4A^2 \cos^2\frac{k\Delta}{2} = 4A^2 \cos^2\frac{kxd}{2D} = 4A^2 \cos^2\frac{\pi xd}{\lambda D}.$$

This is valid as long as $x \ll D$ due to the preceding assumptions. We will thus obtain maxima for the following:

$$\frac{\pi dx}{\lambda D} = m\pi \ (m \text{ integer}) \Rightarrow x = mD\frac{\lambda}{d}, \ \alpha \simeq \frac{x}{D} = m\frac{\lambda}{d}$$

We observe equidistant bands of light and dark whose distance increases with decreasing the source distance d and increasing the wavelength λ as illustrated in Figure 5.4.

If we consider how the phase shift of the two sources in the preceding example comes about, we can extend the description to more general scattering processes, as will be needed in crystallography, diffraction, or the interference leading to the coloration of butterflies. As we have found, $\delta = k\Delta = kd\sin\alpha$ is the phase shift of the two sources. If we take into account that the incident light and the light on the screen have a different direction (given by the angle α), we see that we get the same phase shift when we consider the distance of the sources multiplied by the difference of the two wave vectors $\vec{k}_i$ and $\vec{k}_a$. These are vectors that show in the propagation direction of the light and have a magnitude, which corresponds to the wave number, that is, $|\vec{k}_i| = |\vec{k}_a| = |\vec{k}| = 2\pi/\lambda$. This difference turns out to have the

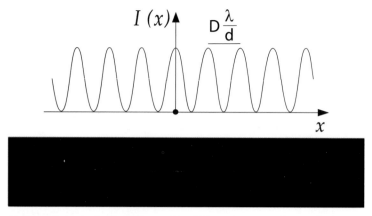

Figure 5.4 The intensity distribution behind a double slit shows a sinusoidally varying pattern that can also be seen in an optical double-slit experiment in a series of light and dark spots.

magnitude $|\vec{q}| = |\vec{k}_i - \vec{k}_a| = |\vec{k}| \sin \alpha$. Otherwise put, the phase difference corresponds to $\delta = |\vec{q}| d$. We can thus transform distributions of sources with certain distances d directly in distributions of intensities at certain angles α by means of a Fourier transform.

To see why this is, we first have to have a closer look at what kind of properties are connected via a Fourier transform. So far, we have considered Fourier transforms as a way of connecting frequencies and times. Oscillations in time correspond to a certain frequency, and these frequencies could be used to describe a time dependence using a sum of harmonic oscillations. This could, however, only be done because in an oscillation in time, the time taken times the (angular) frequency always describes a certain angle giving the phase of the oscillation. This intimate coupling between time and frequency we thus now see between distance of sources (i.e., position) and wave-vector difference. Since the phase is described by the scattering angle and the distance between scatterers, we have a pair of properties in this interference process (the scattering vector $\vec{q}$ and the distance between scatterers d), which are connected by a Fourier transform. What we can look at experimentally is the distribution of intensities on the screen (as a function of the angle, or $\vec{q}$), as we have discussed previously. If we want to determine the structure of a molecule, we are really interested in the distribution of the sources. However, we have just seen that we can obtain this distribution of the sources directly via a Fourier transform of the intensity pattern (more precisely, a Fourier transform of the amplitudes). Thus, the amplitude of the double slit as we have seen has been a modulated oscillation. We have already seen such a Fourier transform when we discussed beats, i.e., two sources with a constant frequency difference. This frequency picture corresponds directly to a double slit when we replace axis header from frequency to scatterer distance. In the Fourier transforms, we consequently have to exchange the axis headers from time to scattering angle, such that the on–off sound of a beat corresponds to the light and dark stripes on the screen. If we can only use the intensity, as is the case in crystallography, things are a little more complicated, since we only obtain the distance of the two slits, but not their positions from the light and dark pattern. This "phase problem" is of great importance in the determination of protein structures from X-ray data and is very complicated to solve.

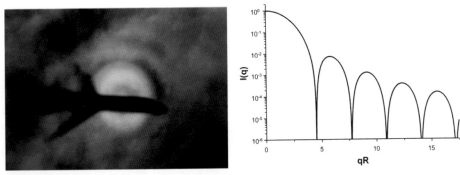

Figure 5.5 Diffraction of white light on water droplets in a mist or cloud can lead to beautifully colored patterns, such as the glory shown left. This is due to the diffraction maxima being wavelength dependent and therefore at different angles for the different colors.

An example of the interrelationship between the distribution of scatterers and the scattered intensity can be observed in an airplane when it flies over a closed cloud (or even better, mist) cover. If you can see the shadow of the aircraft on the fog, you typically observe colored rings around the shadow as shown in Figure 5.5. These rings are caused by the diffraction of light on the drops of mist, which in such a fog are relatively small and mono-disperse with radii of somewhere around 10–50 μm, depending on atmospheric conditions (in each individual fog, they are very mono-disperse with deviations of only a few percent). This means that these distances are of the order of the wavelength of the light and interference can occur. Such a droplet has a constant density, up to the radius R, that is, the intensity distribution is obtained from the Fourier transform of a constant up to a certain size. We have already looked at this, and the result was a distribution $A(\vec{q}) = \sin(qR)/q$. This distribution is also shown in Figure 5.5 and will continue to concern us when we treat diffraction in the next section. There is a minimum in the intensity distribution at an angle given by $qR = \pi$, which is $\alpha = \lambda/(2R)$. For a radius of $R = 10$ μm and green light ($\lambda = 500$ nm), this corresponds to an angle of about 6 degrees. That is, at different wavelengths (or different frequencies, i.e., colors), the minimum lies at different angles. At certain angles, therefore, certain colors are missing from the intensity distribution, which leads to the colored rings.

In the following section, we will look at a few more examples of interferences in nature before we deal with the structure determination by X-ray scattering.

5.3.3 Interference Patterns in Nature

Perhaps the most striking example of an interference pattern is seen in the introduction by the color of the butterfly morpho menelaus. Only the blue portion of the incident light is reflected, which is caused by interference from the nanostructures on the scale of the wing. Such a butterfly and an electron micrograph of the surface structures is shown in Figure 5.6.

In order to make this fact clear to us quantitatively, we consider the phase shift of a light beam, which is reflected on a chitin layer in the butterfly. To get an interference, we need coherent light beams, that is, as a second beam, we consider the one that was reflected at

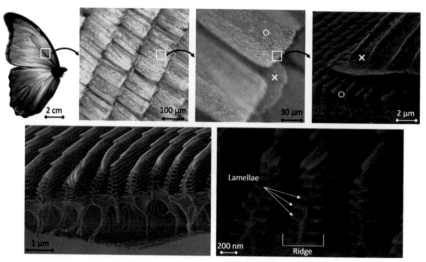

the back of the layer. Thus, the two beams have a path difference of $\Delta = 2d$, where d is the thickness of the layer. The corresponding phase difference is then $\delta = k_n \Delta = 4\pi nd/\lambda$, where n is the refractive index of chitin and λ is the wavelength in vacuum (or air to a very good approximation). In addition, a further phase shift occurs because the reflections have occurred at different boundaries. As we have seen in the treatment of reflections of waves, a phase shift of π occurs when the reflection is from the denser medium. In that case, the reflected wave goes opposite to the incoming wave. This phase shift does not occur at the reflection toward the thinner medium. In the situation of the butterfly wing, we have a reflection at the air–chitin boundary (i.e., a phase shift of π) and the second and the boundary chitin–air (i.e., no phase shift). That is, the total phase shift of the reflected beams on a thin layer in air is as follows:

$$\delta = \frac{4\pi nd}{\lambda} + \pi$$

In order to obtain a constructive interference, this phase shift must be a multiple of 2π so that the two waves are in phase. Therefore, $\delta = m2\pi$, where m is an integer. This number is also called the order of the interference maximum. With the phase shift we have derived, we thus get the condition for constructive interference if the following is true:

$$2m - 1 = \frac{4nd}{\lambda}$$

That is, with a given thickness of the layer, constructive interference occurs at a certain wavelength given by $\lambda = 4nd/(2m - 1)$. Thus, when we look in the first order ($m = 1$), we obtain the largest wavelength at which a constructive interference occurs. This is the case with a wavelength of $\lambda = 4nd$. Destructive interference occurs when the phase shift is an odd multiple of π, i.e., $\delta = (2m + 1)\pi$. With the phase shift of the two beams, we again get a condition, this time for destructive interference:

$$m = \frac{2nd}{\lambda}$$

This again means that at a corresponding wavelength of $\lambda = 2nd$, there is an intensity minimum (of the first order), i.e., no reflection takes place at this wavelength!

The spectral dependence of this reflection intensity is shown in Figure 5.7 (green curve) and shows a relatively broad distribution.

The effect of increasing reflection of certain spectral parts can be amplified by repeating such a reflection several times on multiple layers of the same thickness. Only in case of full constructive interference no intensity is lost, whereas even a small fraction lost at another wavelength is increased exponentially. This leads to a very striking coloration in the butterfly, as seen in Figure 5.7 for the case of 8–10 layers. The color selectivity increases the more interfering rays there are. An example of a single thin layer showing such a coloring you can observe at home when looking at a soap film, which shows exactly these properties.

In the treatment of reflection, we have also seen that we have to use the refractive index of the reflective layer and that the wavelength at which an intensity maximum occurs is proportional to the refractive index. If the refractive index of the environment is increased, e.g., by the addition of alcohol, the wavelength of the maximum is increased accordingly. We have seen this in the example of *Morpho menelaus*, which appeared green after the addition of alcohol.

Other frequently encountered examples of the same effect can be seen in oil or water films (see Figure 5.8), as well as in the antimirror coating of glasses or spectrally separating mirrors used in fluorescence microscopes. In these systems, it is important to note that sometimes the thin layer is located on an optically thicker medium, such that both

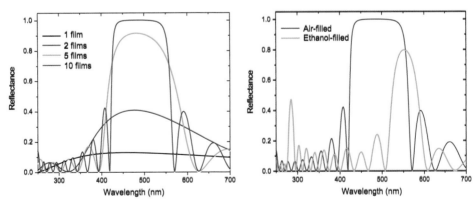

Figure 5.7 The reflectivity of layers of chitin 80 nm thick interspersed by layers of 125 nm of air (right). With increasing number of layers, the reflectivity maximum becomes sharper and is centered around blue light. When the interspersed layers are filled with alcohol for 10 layers (left), the reflectivity maximum shifts to the green.

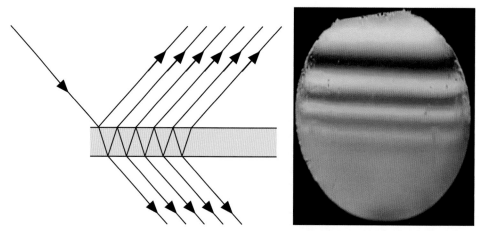

Figure 5.8 | Illustration of the reflection of light from a thin film. Light paths reflected on the top surface of the film and those reflected on the bottom surface of the film have experienced a path-length difference. If these different light paths are in phase, i.e., interfere constructively, the corresponding wavelengths are enhanced in the reflection and give rise to a coloration of the thin film, as seen in soap bubbles or soap films (right).

reflections occur with a phase shift of π. The interference phase shift then becomes $\delta = 4\pi nd/\lambda$. The conditions for constructive and destructive interference change accordingly.

5.3.4 More Complex Interference Patterns

If we superimpose not two but several (N) waves with the same relative phase shift $\delta = k\Delta$ to its neighbor, the corresponding total amplitude is given by the following:

$$u(x, t) = \sum_{m=1}^{N} u_m = \sum_{m=1}^{N} u_0 \sin(kx - \omega t + k(m - 1)\Delta) =$$

$$= u_0 \frac{\sin Nk\Delta/2}{\sin k\Delta/2} \sin(kx - \omega t + (N - 1)k\Delta/2),$$

which gives an intensity of
$$I(\Delta) = I_0 \frac{\sin^2(Nk\Delta/2)}{\sin^2(k\Delta/2)} .$$

For which we obtain main maxima when
$$\sin(k\Delta/2) \to 0 , \frac{\sin(Nk\Delta/2)}{\sin(k\Delta/2)} \to N,$$

i.e., at $\dfrac{k\Delta}{2} = 0, \pm\pi, \pm2\pi, \ldots ,$ $\dfrac{\pi\Delta}{\lambda} = \pm n\pi$ (n integer) $\Rightarrow$ $\Delta = n\lambda$, $I_N = N^2 I_0$.

Again, for constructive interference the path-length difference is an integer multiple of the wavelength. Moreover, the principal maxima are the more intense and sharper the greater the number N of the interfering waves (see Figure 5.9). Between the maxima,

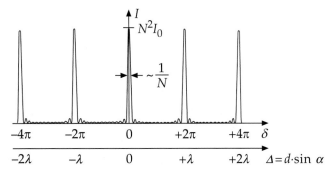

Figure 5.9 Intensity distribution following a diffraction grating with N slits.

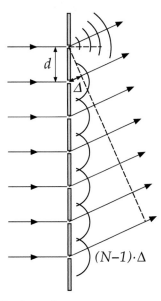

Figure 5.10 Construction of path differences in a diffraction grating.

there are zeros at $\sin(Nk\Delta/2) = 0$ with $\sin(k\Delta/2) \neq 0$ and small side maxima at $\sin(Nk\Delta/2) = 1$.

If a plane wave is incident perpendicular to a set of N equidistant slits, a so-called *line grating*, this acts as N coherent sources (see Figure 5.10). The path difference is then dependent on the observer angle α as in the double slit ($\Delta = d \sin \alpha$). The main maxima are found at

$$\sin \alpha_{max} = \frac{\Delta}{d} = m\frac{\lambda}{d} , \; m \text{ integer.}$$

If the gap distance d, also called *lattice constant*, is known, then the wavelength λ can be determined from the position of the maxima. This is the principle of the *grating spectrometer* and also gives the coloration to a CD (see Figure 5.11).

Since all wavelengths coincide in the zeroth order, $m = 0$, in the grating spectrometer (see Figure 5.12), we have to observe at the least the first order, $m = 1$. Since in that case

The grooves on a CD or DVD containing the information stored on it are usually read out with a laser and the difference in reflection of the laser spot corresponds to a digital bit. This means that the grooves are of a size comparable to the wavelength of light and hence the CD as a whole can act as a grating spectrometer. If you watch different light sources with a CD, you can find different kinds of spectra.

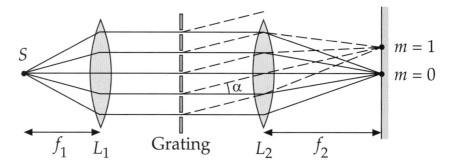

Construction of a grating spectrometer: The lens L_1, at whose focal point the source is, ensures that the wave falling on the grating is a plane wave (i.e., we have parallel beams). The lens L_2 causes parallel rays at the angle α to converge and interfere at a point on the observation screen (rather than infinity). The interference conditions are not changed by the lenses since the differences of the light paths remain the same. If the incident light contains different wavelengths, the corresponding maxima lie in different places. We thus get a split according to colors.

$\lambda = d \sin \alpha < d$, the lattice constant must be greater than the wavelength to be measured. In the visible range ($\lambda = 400 - 750$ nm), scratched gratings are used, while for X-rays ($\lambda \simeq 0.1$ nm) one uses a crystal lattice. Conversely, if the wavelength of the X-rays is known, the intensity distribution in the interference pattern can be used to infer the lattice constant or even the structure of the crystals or macromolecules.

With a crosshatch grating, equidistant points in two dimensions are obtained as interference patterns (see Figure 5.13). A wavelength-dependent splitting of the interference pattern is found not only in lattice spectrometers.

White light illuminating a diffraction grating. Due to the different wavelengths, the different colors are differently diffracted and hence split into different angles. This happens at every point of the diffraction image.

5.3.5 X-Ray Crystallography

As we have already alluded, an atomic lattice can be considered as a lattice spectrometer for electromagnetic waves with short wavelengths, which we know as X-rays. So when we send X-rays through a crystal, they will form an intensity pattern, as discussed earlier in the case of the lattice. This represents nothing but a Fourier transform of the lattice of the atoms. This scattering of the X-rays can thus be used as a method to determine the structure of the crystals. This is one of the most common techniques used in the determination of the structure of macromolecules, and one of the best-known examples in biology is the X-ray scattering data of Rosalind Franklin on oriented DNA molecules that Crick and Watson used to determine the structure of the DNA. Such an intensity pattern is shown in Figure 5.14 and represents, accordingly, a Fourier transform of the DNA double helix. The geometric variables of the DNA can be read directly on this scattering image. The opening angle of the characteristic X-shape indicates the slope of the DNA strands, or the helical pitch. The vertical distance between the respective intensity maxima in the X corresponds to the inverse distance between the maxima of the helix, such that this information together with the helical pitch gives the width of the DNA molecule. Finally, the two large spots vertically above and below correspond to the inverse distance of the base pairs.

5.3.6 Diffraction

So far, we have assumed that the openings in a screen are so small that they can be regarded as point sources. This is true, as long as its diameter s is small compared to the wavelength λ. Expanded openings consist of a continuum of points from which secondary waves emanate, the phases also being continuously distributed.

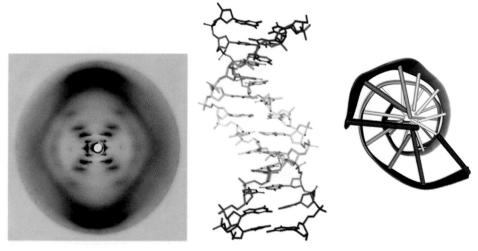

Figure 5.14 Left: diffraction pattern of oriented DNA molecules obtained by Franklin and Gosling (1953), used with permission. The characteristic arrangement of the diffraction in the shape of an X allows the identification of the structure as a helix. Further details of the diffraction pattern (such as a missing spots on the X, as well as the spacing between spots) led Watson and Crick to conclude that the structure of B-DNA consists of a double helix with base pairing for storing genetic information (right). Data from Drew et al. (1981).

We derive the expression for the intensity distribution from a wide single slit (intensity $I_0 \propto u_0^2$), where we represent the slit as a sequence of N points with distance $d = s/N$, from each of which a wave with an amplitude u_0/N ($I = I_0/N^2$) originates. If we observe at the angle α, the path difference incurred for waves from adjacent points is again $\Delta = d \sin \alpha$. The intensity distribution from the overlap of the different parts of the N points is, as we know from the previous section, the following:

$$I = \frac{I_0}{N^2} \frac{\sin^2(Nk\Delta/2)}{\sin^2(k\Delta/2)} \;\; \rightarrow \;\; (k\Delta \text{ small}) \; I = I_0 \frac{\sin^2(Nk\Delta/2)}{(Nk\Delta/2)^2} \; .$$

Here we replace Δ by the slit width s and the observation angle α:

$$\frac{N}{2} k\Delta = \frac{s}{2d} \frac{2\pi}{\lambda} d \sin \alpha = \frac{\pi}{\lambda} s \sin \alpha \equiv g(\alpha) \,, \;\; \Rightarrow \;\; I = I_0 \frac{\sin^2 g(\alpha)}{g(\alpha)^2} \; .$$

The resulting distribution is shown in Figure 5.15. As can be seen, the main maxima (denominator $= 0$) are at this location:

$$\sin g(\alpha) = g(\alpha) = \sin \alpha = \alpha = 0 \,, \; I_{max} = I_0 \; .$$

Minima occur, when the nominator is zero, but the denominator is nonzero, i.e., for

$$\frac{\pi s \sin \alpha_{min}}{\lambda} = \pm n\pi \; (n \neq 0 \,, \text{ integer}) \; \rightarrow \; \sin \alpha_{min} = \pm \frac{n\lambda}{s} \; .$$

Side-maxima, whose intensity decreases quickly with increasing n, then can be found at

$$\sin \alpha_{max} = \pm \left(n + \frac{1}{2} \right) \frac{\lambda}{s} \,, \; I_{max,n=1} \simeq 0.05 \, I_0 \; .$$

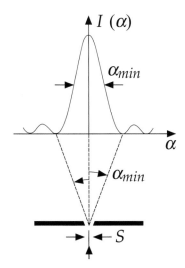

Figure 5.15 The intensity distribution following diffraction through a slit.

Figure 5.16 Diffraction of white light on fungal spores of lycopodium. Similar to the case of the glory in mist, diffraction rings occur, which appear at different angles for different colors, hence giving rise to colored rings.

The width of the main maximum is given by $\sin \alpha_{min,1} = \lambda/s$, i.e., it increases with decreasing width of the slit. When $s \rightarrow 0$, we actually obtain a spherical wave in all directions as embodied in Huygens' principle. The exact functional form of the intensity distribution is again obtained by a Fourier transform of the slit, which we have already discussed in the treatment of the Glory. This distribution can also be seen in Figure 5.16, where white light is scattered off fungal spores from lycopodium, which are spherical in shape and about 20 μm in size. Since the angular position of the diffraction minima depends on the wavelength, i.e., the color, differently colored rings appear similar to the glory.

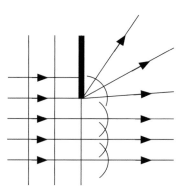

Diffraction around an opening.

Of great practical importance are diffraction effects at circular openings (iris, lens, etc.). The interference pattern will be circular for symmetry reasons. The first minimum appears at the angle of

$$\sin \alpha_{min,1} = 1.22\frac{\lambda}{D} \quad (D = \text{diameter of the opening}).$$

Since the intensity pattern otherwise is similar to that of the single slit, the secondary maxima are hardly visible, resulting in a bright circular disc (the *Airy disc*).

Diffraction effects always occur when a light wave is laterally constrained, e.g., at edges in front of apertures and lenses (see Figure 5.17). They are the more pronounced the smaller the apertures and the longer the wavelength of the light. Since all optical instruments work with limited light bundles, diffraction is inevitable and limits the resolving power.

Resolving power is given by the minimum distance at which the diffraction images of two adjacent points are still separable, i.e., the corresponding principal maxima are separated by the angle α_{min}. We consider a few optical systems as examples:

Pinhole Camera

Probably the simplest imaging system is the pinhole camera. The entrance hole cuts out a small number of beams emanating directly from the object of interest, each of which produces an image on the screen (see Figure 5.18). How big should the hole be chosen?

If the hole is large, the image of a point is not a point, but a bright disc with the diameter D of the hole (geometric shadow) that increases with size of the hole. If, on the other hand, it is chosen to be smaller and smaller, the size of the Airy disc increases in diameter inversely proportional to the size of the hole given by

$$L \sin \alpha_{min} \simeq \frac{L\lambda}{D}.$$

The image diameter becomes the smallest, i.e., the image is the sharpest when both contributions are approximately equal, that is,

$$D \simeq \frac{L\lambda}{D} \text{ or } D \simeq \sqrt{L\lambda}.$$

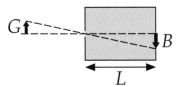

Figure 5.18 Image construction in a pinhole camera.

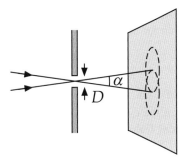

Figure 5.19 Angular distribution of diffracted light of two points to be separated.

Actually, you can use the principle of a pinhole camera in order to see clearly even if you have lost your glasses. As we have just discussed, if we send light through an opening of a size $\sqrt{L\lambda}$, we get a sharp image. If we want to have this on our retina, the L becomes the size of your eyeball, which is roughly 2 cm in diameter. Thus for visible light of a wavelength of roughly $\lambda = 500$ nm, an opening of $D \simeq 100$ µm will do the trick. If you roll up your index finger and hold it with your thumb, there will be a very small opening left, which is only a few hundred microns in size. Thus viewing the world through this hole, every object makes a sharp image on your retina, even without glasses. The only drawback is that you throw away a lot of light this way and therefore the images can be quite dark.

Simple Lens

Also a lens cuts out a bundle from the incident light wave with the lens diameter D.

Thus, if we want to make an image of a certain point with a microscope (for example, a fluorescently labeled molecule), then the image lies in the focal plane; however, it is not a point (see Figure 5.19), but again an Airy disc with the following angular diameter:

$$\alpha_{min} \sim \sin \alpha_{min} = 1.22\frac{\lambda}{D} \text{ , or the size } d_m \simeq 1.22\frac{\lambda f}{D} \text{ .}$$

The images of two such points can be separated or distinguished if their angular distance is equal to the diameter of the Airy disc. Otherwise, they become blurred and are perceived as a single point. The resolution limit is therefore α_{min}. The ratio of the lens diameter D to the focal distance f describes the opening angle viewed by the lens, which is also called the numerical aperture. The larger this numerical aperture, the better the resolution of a lens, with a limit at half the wavelength, which also appears in the preceding equation. This is Abbe's limit for the resolution of a lens, which specifies a limit for optical microscopes that is less than roughly 250 nm. Smaller structures cannot be resolved with optical microscopes.

However, recent developments in the field of microscopy have somewhat blurred this fundamental limit, and today fluorescent structures can actually be imaged with better resolution. In order to understand these, however, we first need to look more closely at fluorescence processes, which we will do in Section 5.6.4.

The human eye has an index of refraction of approximately $n = 1.34$ and a pupil diameter of $D = 4$ mm. For a wavelength in the visible ($\lambda = 600$ nm) we therefore obtain a resolution limit for the eye of

$$\alpha_{min} = 1.22\frac{\lambda_{vac}}{nD} = \frac{1.22 \times 6 \times 10^{-7}}{1.34 \times 4 \times 10^{-3}} = 1.37 \times 10^{-4} \text{ rad} = \frac{1'}{2}.$$

At a distance of 1 km, two luminous points can be resolved at a distance of 14 cm. If the focal distance of the eye is $f = 2.3$ cm, then the minimum distance between two resolved image points is as follows:

$$r_{min,1} = \alpha_{min} f = 3.1 \times 10^{-4} \text{ cm} = 3.1 \text{ μm}.$$

This value corresponds approximately to the spacing between adjacent cones of the retina.

5.4 Polarization

5.4.1 Polarized Light

Everything we have said so far about electromagnetic waves (reflection, refraction, diffraction, interference) can occur for all types of waves. Polarization, on the other hand, is a property that can only be observed in transverse waves. Since light can be polarized, as we can see using polaroid filters, we know that light actually is a transverse wave. This can also be obtained by studying how an electromagnetic wave (and thus light) is generated. Here, we basically only need to know that electric charges can be moved by electric fields, and a time-varying electric field can therefore lead to a vibration of an electric charge. Since such charges are sources of the electric field, a vibration of a charge leads to a time-varying electric field. We will see where these fields come from more closely in Chapter 9.

A dipole transmitter antenna emits an electromagnetic wave where the intensity depends on the angle θ as shown in Figure 5.20: $I(\theta) = I_0 \sin^2 \theta$. It is at its maximal value perpendicular to the antenna and equal to zero parallel to the antenna.

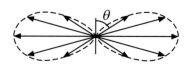

Figure 5.20 A dipole transmitter antenna emitting an electromagnetic wave.

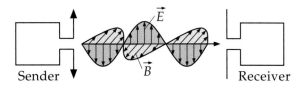

Figure 5.21 The emitted electric field vector is parallel to the vibrational direction of the electrons.

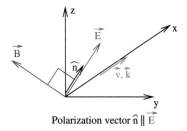

Polarization vector $\hat{n} \parallel \vec{E}$

Figure 5.22 A linearly polarized wave.

The emitted electric field vector is parallel to the vibrational direction of the electrons (see Figure 5.21). In order for the electrons of the receiving antenna to be excited to vibrate, this must again be parallel to $\vec{E}$, that is, to the transmitting antenna. If it is rotated by 90 degrees, it receives no signal.

The described wave is *linearly polarized*, that is, the $\vec{E}$-vector has a fixed direction in space, the so-called *polarization direction* ($\hat{n} \parallel \vec{E}$). *Circularly polarized* waves are those whose $\vec{E}$-vector rotates around the propagation direction at constant angular velocity (see Figure 5.22).

Electromagnetic waves, e.g., light, which are emitted from independent, nonaligned sources, e.g., atoms, are unpolarized. This means that all transversal $\vec{E}$ directions occur equally frequently and are statistically distributed. For such light to be polarized, it must pass a *polarizer*, where it goes through a process that prefers certain $\vec{E}$ directions with respect to others. Examples of this will be discussed in the following section and include reflection, propagation in anisotropic media, and scattering. Since our eyes cannot distinguish polarized light from unpolarized light, we need to use an *analyzer* with the same properties as the polarizer. As shown in Figure 5.23, depending on the relative position of polarizer and analyzer, differences in intensity are obtained.

Longitudinal waves cannot be polarized by definition, since there is only one direction of vibration, which is the direction of propagation.

5.4.2 Polarization by Scattering, Reflection, and Refraction

Electromagnetic waves can be polarized both by scattering at the atoms of a medium as well as by reflection and refraction at the interface between two media.

If a light beam hits a glass plate, the normal plane on the plate containing the beam is called the *incidence plane*. The intensities of the reflected and refracted beams are different for the two components of the electric field parallel ($E_\parallel$) and perpendicular ($E_\perp$) to the incidence plane.

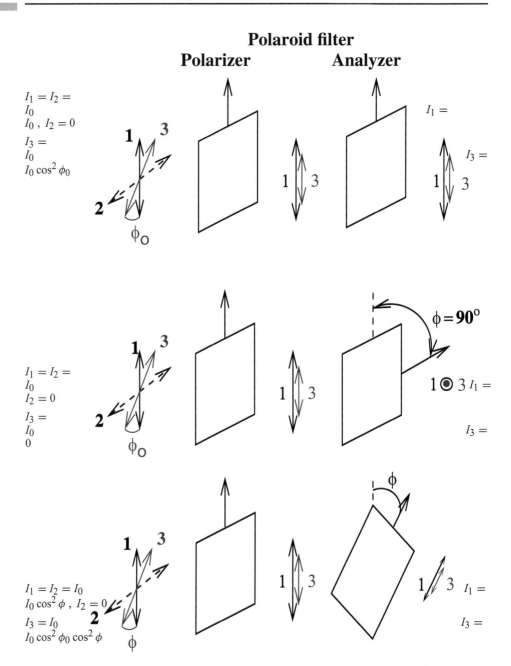

Figure 5.23 Polaroid filters as polarizers and analyzers of light. Three polarized waves fall on the polarizer: I_1 with $\vec{E} \parallel$ polarizer, I_2 with $\vec{E} \perp$ polarizer, and I_3 with $\vec{E}$ at an angle ϕ_0 to the polarizer. In addition, there are three positions of the analyzer relative to the polarizer: parallel, perpendicular, and at the angle ϕ. The intensity behind the analyzer is indicated.

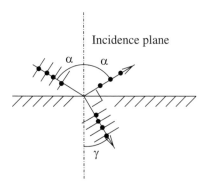

Figure 5.24 Geometry of polarized reflection under the Brewster angle.

If the angle between the reflected and the refracted beam is exactly $\pi/2$ ($\alpha + \gamma = 90°$) (see Figure 5.24), i.e., if we have

$$\frac{\sin \alpha}{\sin \gamma} = \frac{\sin \alpha}{\sin(\frac{\pi}{2} - \alpha)} = \tan \alpha = \frac{n_2}{n_1} \, ,$$

then the reflected beam does not have any component in the direction $E_\parallel$, meaning that it is completely linearly polarized. At adjacent angles, the reflected beam is partially polarized. If it runs through an analyzer afterward, it is weakened. An application of this effect is found in polarization filters for suppressing reflections in photography.

If electrons in atoms or molecules are excited by an incoming wave, they emit electromagnetic waves like the dipole antenna. Light scattered under 90 degrees is therefore linearly polarized as there is no component perpendicular to the dipole in that case. The light from the sky is sunlight scattered by the air of the atmosphere. Its direction of polarization depends on the position of the sun. The polarization pattern in the sky, which one can observe, is used by certain polarization-sensitive animal species for navigation (bees, ants). We discussed this in more detail in Section 1.1.2. The scattered intensity strongly depends on the frequency of the incident light:

$$I(\omega) \sim \omega^4 \, .$$

High frequencies, i.e., short wavelengths, are scattered much more strongly. Therefore the clear sky appears blue. The unscattered, direct sunlight is therefore reddish in the evening, when the path through the atmosphere is long and all of the high-frequency (blue) light has been scattered away.

The optical properties of anisotropic crystals cannot be characterized by a single refractive index. This depends on the $\vec{E}$ direction with respect to the optical axis of the crystal. An incoming light beam is generally split into an *ordinary* and an *extraordinary* ray. The two beams are polarized perpendicular to each other and are geometrically separated from one another at certain crystal positions. Such birefingent properties may also be introduced in isotropic bodies when they are subjected to mechanical stresses (*stress birefringence*) or electric fields (*Kerr effect*).

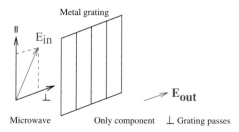

Figure 5.25 A metal grid and a dipole antenna.

Also, the *optical activity* is based on anisotropy in the molecular structure of certain substances. It describes the fact that the plane of polarization of a light rotates when passed through a medium. For example, the angle of rotation in a sugar solution is proportional to its sugar concentration and the thickness of the layer.

The most common polarizers are the so-called *Polaroid filters*. These films were developed by Land in 1938 and consist of a plastic in which the hydrocarbon chains are strongly elongated in one direction. In an iodine solution, one attaches to the chains of macromolecules to iodine, thus obtaining electrons free to move along a linear direction, the direction of absorption.

The physical principle used can be demonstrated, for example, with a metal grid and a dipole antenna (see Figure 5.25). The component of the electric field perpendicular to the grating passes through, while for the component parallel to the grating, the incident wave excites vibrations of electrons in that grating and is thus *absorbed*.

Also, in the case of the polaroid filter, if an unpolarized wave is passed through the polarizer, the component polarized in the absorption direction is absorbed, while the perpendicularly polarized component passes through the filter. Thus, for our eyes, such a filter appears gray and only by rotating the filter can we determine whether the incident light is polarized.

5.4.3 Stress Birefringence

As indicated in the preceding section above, materials are birefringent when a mechanical stress is applied to them. This is because the refractive index is changed differently in the direction of the stress and perpendicular to that direction. The polarization of the light sent through such a birefringent material thus changes relative to the axis given by the direction of the stress. If the incoming polarization is set at 45 degrees to the direction of the stress, then when we add a crossed analyzer behind the sample, there will be no intensity passed through the setup where there is no stress. Where a stress is applied, the linearly polarized light is converted to an elliptic polarization, which results in a component perpendicular to the original polarization. Thus light will then pass through the analyzer and lead to a signal. This is illustrated in Figure 5.26, using the stresses frozen in during production of a plastic tube, mainly along its length due to the production by pulling the plastic. When the tube is rotated relative to the polarizers, the signal becomes smaller or larger depending on the direction of the stress relative to the set of polarizers. This method can be used to measure stresses in materials or to simulate loads incurred in machines.

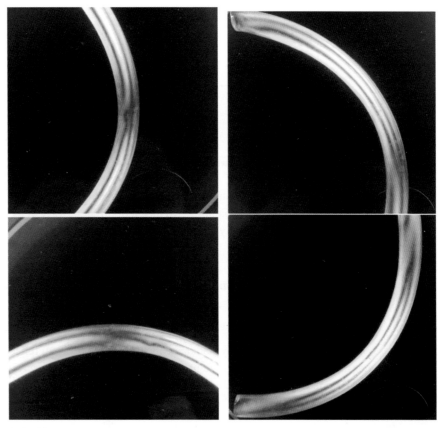

Figure 5.26 Illustration of stress birefringence on a plastic tube with frozen in stresses due to its production. The tube is placed between crossed polarizers, acting as an analyzer. Depending on the direction of the polarizer and analyzer relative to the main direction of the stress, the birefringence (as seen by the light passed through the arrangement) increases or decreases.

5.5 Geometric Optics

5.5.1 Light Rays

While we know that light is a wave from the polarization and interference effects we have described in the preceding sections, we also know that we can produce seemingly sharply defined light beams with the help of, e.g., an iris. These light rays are decisive in designing optical imaging systems such as microscopes, which is why we will be looking in more detail at geometric optics in the coming sections.

Such beams are reflected on well-defined surfaces (polished glass or a calm water surface). Rays are also refracted into the other medium. We have the law of refraction, which we have established earlier from the wave properties of the light, also known as Snell's law. We will now however formulate it generally when we consider two materials, each with a refractive index n_1 and n_2 respectively.

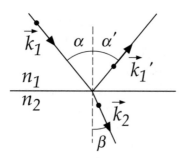

Figure 5.27 Refracted and reflected light rays following Snell's law.

If a plane wave impinges onto a surface in the direction of $\vec{k}_1$ with an angle α relative to the normal of said surface separating two media of different wave velocities v_1 and v_2, it is partially reflected and partially refracted (see Figure 5.27). The *law of reflection* says

$$\alpha = \alpha' .$$

In addition, we have *Snell's law of refraction:*

$$\frac{\sin \alpha}{\sin \beta} = \frac{v_1}{v_2} = \frac{c}{n_1} \frac{n_2}{c} = \frac{n_2}{n_1} .$$

These laws do not make any statement about the proportions of reflected and transmitted intensity. Only in the special case of the so-called *total internal reflection* conservation of energy implies that the reflected intensity is equal to the incident intensity. This case can occur for the transition from the optically denser to the less dense medium ($n_1 > n_2$). From

$$\sin \alpha = \frac{n_2}{n_1} \sin \beta \leq \frac{n_2}{n_1} < 1,$$

we know that there is a maximum angle of incidence α_T. If $\sin \alpha > \sin \alpha_T = n_2/n_1$, there is no refracted part anymore and all of the incoming intensity is therefore reflected.

If the speed of light depends continuously on position, i.e., there is no well-defined boundary between two materials, this also leads to a continuous change in the propagation direction. This can for instance be seen when a light beam is passed through a liquid with a large density gradient (e.g., a sugar solution with stratified concentrations). Such phenomena are well known in the atmosphere, for example, aerial reflection on hot surfaces (fata morgana), and "reflection" of radiowaves in the upper atmosphere, which allows the transmission over great distances. Also, the sun rays can be bent so that the sun is visible, even if it is under the geometrical horizon, something that can give rise to the phenomenon of the "green flash," where the last rays of the setting sun are green because red light is refracted slightly more in the atmosphere.

So far, we have not talked about the edge of the light beam. Of course, the electromagnetic fields that oscillate do not suddenly become zero at the edge, but rather decrease in amplitude over a certain distance, which is roughly given by the wavelength. The treatment of light as geometrical light rays is thus limited by the fact that the width of the beam has to be large compared to the wavelength, while it also has to be small compared to the optical instruments we consider. Because we are thus considering objects that are very large compared to the wavelength, we can therefore safely neglect effects of diffraction and interference in geometric optics, as these are only relevant when the objects are comparable in size to the wavelength.

Another way to obtain the laws of refraction and reflection, but now for rays, is given in the principle of Fermat. This states that a light beam takes the path that uses the shortest time relative to any adjacent paths. This can also be considered in terms of optical path lengths, which are given by the geometric path multiplied by the respective refractive index.

For the accompanying sketch (see Figure 5.28), Fermat's principle implies that the light on the way $A \rightarrow B \rightarrow E$ and $A \rightarrow B \rightarrow C$ takes longer than on $A \rightarrow D \rightarrow E$ and $A \rightarrow D \rightarrow C$, respectively, which exactly corresponds to the laws of refraction and reflection. This means in particular that

$$t_{ABE} = \frac{\overline{AB}}{v_1} + \frac{\overline{BE}}{v_2} = \overline{AB}\frac{n_1}{c} + \overline{BE}\frac{n_2}{c}$$
$$> t_{ADE} = \frac{\overline{AD}}{v_1} + \frac{\overline{DE}}{v_2} = \overline{AD}\frac{n_1}{c} + \overline{DE}\frac{n_2}{c}.$$

The importance of the refractive index can be illustrated by the fact that it is due to differences in refractive index that we actually see things at all. If the refractive index of two substances is equal, light will not change its direction or other properties at the boundary between these two substances and will continue unimpeded. Therefore, the boundary between two substances of the same refractive index is invisible. This can be seen when looking at a turbid window. This has been ground and hence has a rough surface, where light is scattered in all directions, thus completely scrambling the light. However, if adding a layer of oil of the same refractive index to the rough side, the rough boundary between the oil and the glass becomes invisible and only the smooth boundary between the oil and the air leads to reflections and refraction. Hence the turbid window has become a clear window, as seen in Figure 5.29.

Very many optical phenomena, in particular most optical instruments, can be understood solely by means of light rays, boundary surfaces, and refractive indices. If the description of the light is limited to the index of refraction and the direction of propagation, we are doing geometric optics. Let's see how this works in practice.

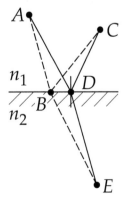

Figure 5.28 Different path lengths for light rays show that the one following Snell's law takes the shortest time.

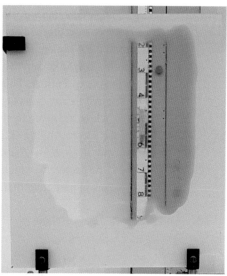

Figure 5.29 The importance of the refractive index can be shown by making a turbid glass transparent. Due to the rough surface of the ground glass, light is scattered in all directions and hence the glass becomes turbid. However, when adding a layer of oil with the same refractive index, the voids in the rough surface are filled and the surface becomes smooth again, such that the glass becomes transparent. Because of the equal refractive index of the oil, the boundary between the glass and the oil will not lead to scattering of light.

5.5.2 Images

In microscopes, we image processes, e.g., in a cell, from the object we study onto a camera (film or charged coupled device [CCD] chip). How a microscope makes such images can be understood with the help of geometric optics. Mathematically speaking, an image can be treated as some sort of function between two abstract spaces (or collection of points), specifying how each point in one set corresponds to a point in the other set, and vice versa. In optics, we can be more specific and say that the set of object points is mapped to the set of image points. In other words, each point of an object is transferred to exactly one point of the image, which therefore is a faithful representation of the object.

If the light rays emanating from one point of the object converge again at a single point in the image after passing through the imaging system, a *real* image point is obtained. However, if the rays leave the imaging system as a divergent bundle, then the point from which they appear to come is a *virtual* image point.

The difference between real and virtual images is illustrated in Figure 5.30. As shown in this figure, the human eye cannot distinguish between real and virtual images. On the other hand, only real images can be captured on a screen or a film.

Ideally, we would require that the assignment $G \leftrightarrow B$ applies to all object points equally given the imaging system and also that the geometric arrangement of the image points is similar to that of the object points. These conditions cannot be strictly fulfilled in real life. We will always have to live with some extent of several possible imaging aberrations (see Section 5.5.7).

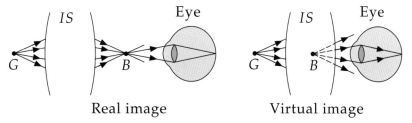

Real image   Virtual image

Figure 5.30 The two different types of imaging systems (IS) produce either a real (left) or a virtual (right) image points B of the object point G.

5.5.3 Mirrors

When an object is viewed in a plane mirror, the mirror acts as an imaging system, which produces a virtual image without distortion or change in size. The virtual image can be seen with the eye. Examples are shown in Figure 5.31. Real or virtual images are also obtained with spherical mirrors. Enlargements or reductions occur as shown in Figure 5.32.

For the construction of an image, we need to know some geometry and the law of reflection. If we look at similar triangles in the construction shown in Figure 5.32, and the definitions of G = object size, B = image size, g = object distance (distance to the mirror, positive on the front of the mirror), b = image distance, and f = focal length, we have two different sets of similar triangles shaded in red and blue in Figure 5.32. From the red triangles, we thus know that

$$-\frac{G}{g-f} = \frac{B}{f}$$

whereas the blue triangles imply

$$\frac{B}{b-f} = -\frac{G}{f}.$$

We can now rearrange both of these equations to describe $-\frac{G}{B}$, which of course has to be the same from both ways of describing it. We therefore obtain an equation relating the different distances as

$$-\frac{G}{B} = \frac{g-f}{f} = \frac{f}{b-f}.$$

Rearranging this to get rid of the fractions, we get $(g-f)(b-f) = f^2$ or $gb - fb - fg = 0$ and $gb = fb + fg$. This basically is what we are looking for, namely a relation between the image distance and the object distance via the focal length, but it becomes still somewhat nicer, i.e., more obvious in what the effects of the different parameters are, if we divide the entire equation by f, g, and b, which gives us the following:

$$\frac{1}{b} + \frac{1}{g} = \frac{1}{f}.$$

We will look at the structure of the equation and how different situations are described by it more closely in the next section, where we deal with lenses, as their images are described

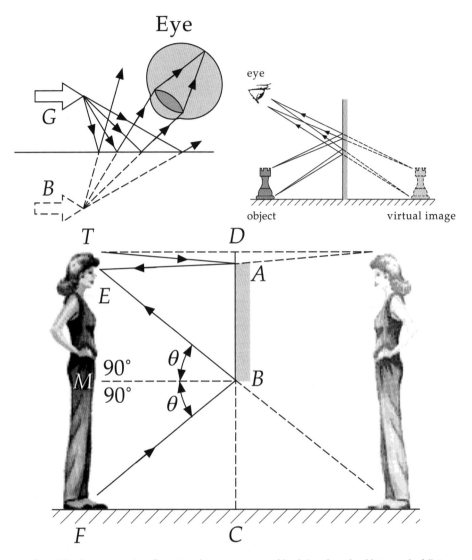

Eye

eye

object   virtual image

Figure 5.31 Images formed by plane mirrors. A uniform, virtual image is generated (top). In order to be able to see the full size, only a half-size mirror is necessary, as the bottom sketch illustrates.

by the same equation. The preceding equation is actually only valid as long as the object is not too large compared to the radius of the mirror ($G < r$), since otherwise the larger of the blue triangles is no longer similar to the smaller blue triangle. Inserting this relation into that for the magnification, we also find the following:

$$\frac{B}{G} = -\frac{b}{g}.$$

Finally, one finds that $f = r/2$ for the focal length of the curved mirror, at least as long as the preceding assumptions are valid.

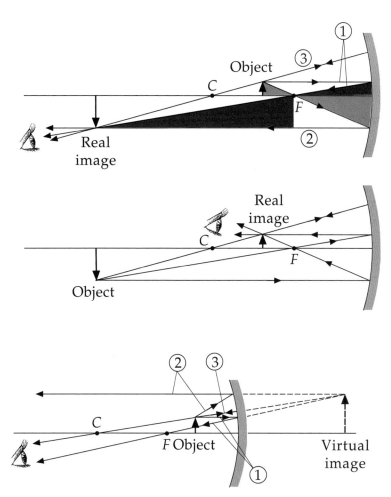

Figure 5.32   Image constructions for spherical mirrors: C marks the center of the sphere (distance to the mirror surface r). F is the so-called focal point at half distance r/2. If the distance of the object from the mirror surface g is varied, the image distance b changes accordingly. Top: object between center of sphere and focal point: real, inverted, and enlarged image. Center: object outside the center of sphere: real, inverted, and reduced image. Bottom: object between focal point and mirror: virtual, upright, and enlarged image behind the mirror.

In any imaging system however, we define the focal point F as the point where parallel rays meet.

The following recipe can be used for constructing an image. First, the parallel ray from the object passes through the focal point after the imaging system. Second, the ray passing from the object through the focal point becomes a parallel ray after the imaging system. Where the first and the second beam intersect, the image is located.

This rule applies to all imaging systems.

As mentioned earlier, spherical mirrors only form an imaging system for the case where $G \ll r$. A hollow mirror that is to produce an exact image for any object needs to have the shape of a paraboloid. This means the surface must be described by the following equation:

$$y = \frac{1}{4f} x^2,$$

where x is the extent of the mirror in the vertical direction and y is the coordinate of the optical axis. The parabolic mirror thus makes a spherical wave out of a plane wave and vice versa(!). Spotlights and mirror telescopes have parabolic mirrors.

5.5.4 Lenses

To understand the image formation with a lens, we begin by looking at one side of this lens, i.e., the formation of an image through a refractive, curved surface. Normally, this surface will be spherical to a reasonable approximation – we will treat the spherical case here as it can be well described mathematically. The only two properties of the lens we need are a curved and refracting surface. Therefore, we can already expect that the properties of the lens solely depend on the refractive index difference between the lens and the outside as well as the radius of curvature of the spherical surface. For the sake of simplicity, we place the object point G on the optical axis (symmetry axis).

The axial ray is perpendicular to the lens, i.e., the incident angle is $\alpha = 0$ and thus $\beta = 0$. The beam is not refracted. Figure 5.33 implies the following for the other rays:

$$\alpha = \phi_1 + \phi_2 \text{ and } \beta = \phi_2 - \phi_3.$$

For near-axial rays, the corresponding angles in the law of refraction are small and hence we can approximate the sines by the angles themselves. Thus we obtain the following from Snell's law:

$$\frac{\sin \alpha}{\sin \beta} = \frac{\sin(\phi_1 + \phi_2)}{\sin(\phi_2 - \phi_3)} \approx \frac{(\phi_1 + \phi_2)}{(\phi_2 - \phi_3)} \cong \frac{n_2}{n_1}, \Rightarrow n_1\phi_1 + n_2\phi_3 = (n_2 - n_1)\phi_2.$$

With the same accuracy of approximation, we can also set

$$\phi_1 \simeq \frac{h}{g}, \quad \phi_2 \simeq \frac{h}{r}, \quad \phi_3 \simeq \frac{h}{b},$$

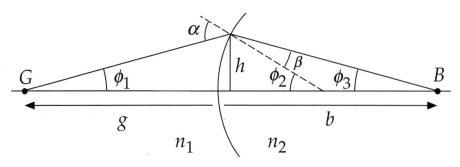

Figure 5.33 The construction necessary to describe the focal length of a curved surface and hence the basic property of a lens.

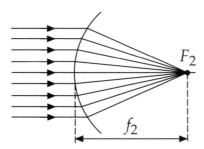

Figure 5.34 Image-side focal point.

which yields the following imaging equation:

$$\frac{n_1}{g} + \frac{n_2}{b} = \frac{n_2 - n_1}{r} .$$

g is again the object distance and b the image distance, which now only depends on g and not on ϕ_1. This implies that all near-axial rays ($\phi_1 \ll 1$) converge at the same point B, which is therefore the image point.

As usual, we want to make sense of this equation by looking at some limiting cases that will allow us to get a more intuitive feel for what the equation describes as well as simplify image constructions for any type of lens. There are three basic cases we want to discuss:

(i) The image of an object infinitely far away, or in other words, the point where incoming parallel rays meet, is called the image-side *focal point* (F_2) (see Figure 5.34). By setting $g = \infty$, in the image equation, we find

$$\frac{n_1}{\infty} + \frac{n_2}{b} \equiv \frac{n_2}{f_2} = \frac{n_2 - n_1}{r}$$

and hence

$$f_2 = \frac{n_2 r}{n_2 - n_1} .$$

The focal point is a real image point if $f_2 > 0$, that is, $n_2 > n_1$ and $r > 0$ (surface convex).

(ii) The object-side focal point F_1 is the point whose image is infinitely far away, or in other words rays emerging from this point leave as parallel rays (see Figure 5.35).

If we set $b = \infty$, we obtain the following:

$$\frac{n_1}{g} = \frac{n_1}{f_1} = \frac{n_2 - n_1}{r} , \quad \Rightarrow f_1 = \frac{n_1 r}{n_2 - n_1} .$$

The two focal distances are related like the two refractive indices, since one is material 1 whereas the other is in material 2, i.e.,

$$\frac{f_1}{f_2} = \frac{n_1}{n_2} .$$

(iii) If the refractive surface is curved the other way, i.e., it is concave rather than convex, or in an equation, if $r < 0$, one finds for the image-side focal point (see Figure 5.36), i.e., for $g = \infty$ and $n_2 > n_1$:

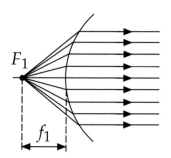

Figure 5.35 The object-side focal point.

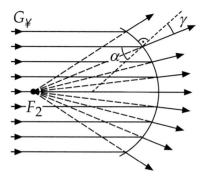

Figure 5.36 Virtual image of a faraway object.

$$f_2 = \frac{n_2 r}{n_2 - n_1} < 0 \,.$$

Hence F_2 is where a virtual image of an infinitely faraway object forms.

A real *lens* actually consists of two curved, refractive surfaces, one in front and the other in the back, which are described by two radii of curvature r_1 and r_2 (see Figure 5.37). These are counted positively when the relevant centers of curvature lie to the right of the refracting surfaces (light incident from the left). We can again derive an image equation for this situation by performing the previous calculation twice.

For the transition n_1 to n_2, at the first surface, we have

$$\frac{n_1}{g} + \frac{n_2}{b'} = \frac{n_2 - n_1}{r_1} \,,$$

where, b' is the image distance for the intermediate image B'. With respect to the second surface with the transition from n_2 to n_1, B' becomes the object point G'. For thin lenses, the two refracting surfaces are basically at the same position, such that $g' = -b'$ is a good approximation, which we can then insert in the image equation for the second surface:

$$\frac{n_2}{g'} + \frac{n_1}{b} = \frac{n_1 - n_2}{r_2} \,, \quad \Rightarrow \quad -\frac{n_2}{b'} + \frac{n_1}{b} = \frac{n_1 - n_2}{r_2} \,.$$

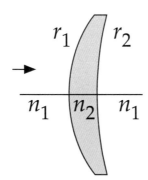

A real lens with two different radii of curvature.

If we add the image equations from the two surfaces, we obtain the following:

$$\frac{n_1}{g} + \frac{n_1}{b} = (n_2 - n_1)\left(\frac{1}{r_1} - \frac{1}{r_2}\right), \quad \Rightarrow \quad \frac{1}{g} + \frac{1}{b} = \frac{n_2 - n_1}{n_1}\left(\frac{1}{r_1} - \frac{1}{r_2}\right).$$

From this, we can again determine the object- and image-side focal lengths by setting $g = \infty$ or $b = \infty$. Because we assumed the refractive index to be the same, n_1, on both sides, actually both of these focal lengths will be the same and are given by the following:

$$\frac{1}{f} = \frac{n_2 - n_1}{n_1}\left(\frac{1}{r_1} - \frac{1}{r_2}\right).$$

This equation is called the *lens equation*, which describes how large the focal length of a lens becomes given the refractive indices and the radii of curvature of its surface. An important special case of this is when the lens is symmetrical, i.e., both sides have the same (but opposite) curvature, $r_1 = -r_2 = r$. In that case, the focal length is given by the following:

$$\frac{1}{f} = \frac{n_2 - n_1}{n_1}\frac{2}{r}.$$

The image equation for such a thin lens then has the form

$$\frac{1}{g} + \frac{1}{b} = \frac{1}{f},$$

which we have already seen in our discussion of curved mirrors! The variable $1/f$ is the refractive power of the lens. It is specified in *dioptries*, where f is measured in meters. A lens in air with 5 dioptries thus has the focal length $f = 0.2$ m.

If f is positive, one speaks of a converging lens; f is negative of a diverging lens.

In the practical application of the previous image equation, the following sign rules must be observed:

1. Light impinges from the left. The direction of incidence determines the signs of g, b, and r.
2. Object distances are positive if G is to the left of the lens.
3. Image distances are positive if B is on the right side of the lens.
4. r_1 and r_2 are positive, if the centers of curvature are to the right of the refracting surface.

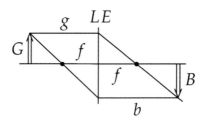

Figure 5.38 Image construction for a symmetric, thin lens.

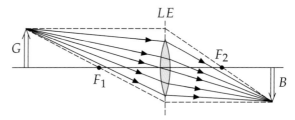

Figure 5.39 Illustration of the image bundle.

Here is an example of a converging lens. The image B of any object G is constructed graphically as follows. The lens is replaced by the lens plane LE.

On the axis of the lens, the two focal points lie symmetrically to its plane. The image point B is the intersection of the special rays that pass through the two foci (see Figure 5.38). It is not necessary that they actually run through the lens for the construction. They are geometric auxiliary lines whose intersections G and B have axis sections g and b that satisfy the image equation.

Only the rays that pass through the lens contribute to the actual image. They form the imaging bundle (see Figure 5.39). The lateral magnification m can also be read from the figure. It is as follows:

$$m = \frac{B}{G} = -\frac{b}{g} \ .$$

The negative sign of m takes into account that the real image is inverted. If m is positive, the image is upright (as it was for a spherical mirror).

Using the image equation, we can, however, also get all of this information without carrying out these constructions. So now let us see how we can get a qualitative understanding of the images formed by a lens (as well as by a spherical mirror) by looking again at the image equation qualitatively.

If we are interested in where an image forms, we can rearrange the image equation to have b alone on one side:

$$\frac{1}{b} = \frac{1}{f} - \frac{1}{g}$$

If we have a converging lens with a positive f, the right-hand side will be positive as long as g is larger than f. This means that an object outside the focal point always forms a real image for such a lens (eye, camera). If, on the other hand, g is smaller than f, the right-hand side will be negative, meaning that a virtual image is formed for an object inside the focal length (magnifying glass).

Similarly, we can look at a diverging lens, i.e., a lens with negative f, where the rays are scattered. Now, irrespective of the value of g (it cannot be negative), the right-hand side will be negative. Thus such a lens will always give a virtual image (glasses for nearsighted people, door spy).

We can also use this formulation to see what has to happen in an autofocus. In our eyes as well as in a camera, the image distance b is fixed, basically by the size of the eyeball in the case of the eye. This means that if we want to image objects at different distances g, the focal length has to change in order to keep $\frac{1}{f} - \frac{1}{g}$ constant. Decreasing the distance g (from very far away, for instance, where $1/g$ is virtually zero) implies that f also needs to be reduced, because we are now subtracting something from $1/f$, which thus has to increase.

If we consider the lens equation again qualitatively, we see the essential elements that we had already expected at the outset. The focal length of the lens depends on the difference in the refractive indices between the lens and the environment (see Figure 5.40). If there is no difference (i.e., $n_1 = n_2$), the focal length is infinite. That means we might as well have no lens. If n_2 is less than n_1, the focal length is negative for the same curvature. That is, if we make a lens from a material with a smaller refractive index than the environment, a converging lens can be changed into a diverging as one can see, for instance, if a water-filled glass (which in air acts as a converging lens) is put into oil, where it acts as a diverging lens. In addition, the focal length depends on the radius of curvature. The smaller the radius of curvature, the smaller is the focal length. Therefore, if we want to have a lens as short as possible, we must have a radius of curvature as small as possible, or a very large curvature, as well as a large difference in refractive index. Since the refractive index difference between biological materials and water is not very great, and there is not much room to increase it, the lenses in a fish eye are quite good spherical lenses since this corresponds to the minimum radius of curvature for a given opening of the lens. This is

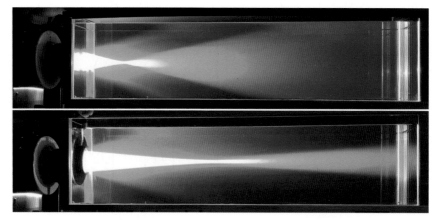

Figure 5.40 The focal length of a lens is not just determined by its curvature, but also by the refractive index contrast. This can be shown by examining a converging lens made of glass once in air (top) and once in water (bottom). As can be clearly seen, the focal length increases in water due to the reduced index contrast.

what the fish wants to have as large as possible, since it only sees light passing through this opening.

If you want to do that for a big lens, you need a lot of material, on the one hand, and on the other hand, the lens is so thick that we cannot make some of the assumptions that we have made. Since, however, the curvature of the surface is what counts, the lens can also be composed of pieces of high curvature placed side by side. Such Fresnel lenses are used in lighthouses or overhead projectors.

5.5.5 Optical Imaging Systems

Imaging systems composed of several thin lenses as well as thick lenses can be completely characterized by two principal planes H_1 and H_2 with the corresponding focal lengths f_1 and f_2.

The points P of H_1 are mapped with the lateral magnification $m = +1$ into points P' of H_2 and vice versa. The focal lengths f_1 and f_2 are as defined before, but they are counted from the respective principal planes. The image is reconstructed with the help of two rays, which pass through the two focal points (see Figure 5.41). If image and object media have the same refractive index, $f_1 = f_2 = f$, then we again have

$$\frac{1}{g} + \frac{1}{b} = \frac{1}{f},$$

where g and b are calculated starting from H_1 and H_2 respectively. The lateral magnification then is

$$m = \frac{B}{G} = -\frac{b}{g}.$$

We obtain the case of the thin lens when H_1 and H_2 coincide with the lens plane.

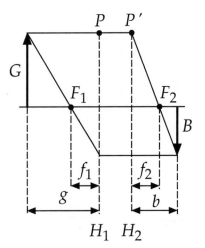

Figure 5.41 Image construction in a thick lens.

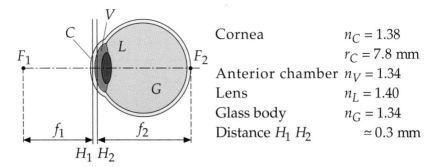

Cornea	$n_C = 1.38$
	$r_C = 7.8$ mm
Anterior chamber	$n_V = 1.34$
Lens	$n_L = 1.40$
Glass body	$n_G = 1.34$
Distance $H_1\,H_2$	$\simeq 0.3$ mm

Figure 5.42   Buildup of the human eye.

The Human Eye

The human eye is a composite optical system. However, the refractive indices of the different components differ so little that the incident rays are refracted mainly at the boundary air $\rightarrow$ cornea (see Figure 5.42). The radius of curvature of the lens can be reduced by a ring muscle so that the total focal length f_2 is shortened. This so-called *accommodation capacity* allows the normal-sighted eye to image objects at distances between the far point $a_R = \infty$ ($f_2 \simeq 23$ mm) and the near point, depending on the age $a_p \simeq 10 - 40$ cm ($f_2 \simeq 18$ mm), sharply on the retina. As previously mentioned, if we want to image a close object onto the retina, the focal length needs to be reduced. This can only be achieved to a certain limit of about two to three times the radius of the lens, hence the limit is about 18 mm. This reduction becomes worse with age, due to the aging of the muscle. Therefore, the near point $1/g = 1/f - 1/b$ moves outward. If we use the diameter of the eye of 23 mm for b and 18 mm for f, we obtain a near point of a little over 8 cm. Reducing the accommodation to $f = 19$ mm already gives a near point of 11 cm and 24 cm for $f = 21$ mm.

In a myopic eye, the initial focal length is too short for the eye, and even in the relaxed state, the image forms in from of the retina. Thus, objects far away cannot be properly imaged and this corresponds to nearsightedness. This can be corrected by means of an additional diverging lens. On the other hand, in a hyperopic eye, the eyeball is too short, such that objects close at hand cannot be properly imaged. This is corrected by an additional converging lens (Figure 5.43).

The size of the retinal image does not depend on the size of the object, but on the visual angle under which it appears to the eye. A determination of size is only possible once the distance to the object is known either from the setting of the accommodation muscle or by the use of binocular vision.

Optical instruments (magnifying glass, microscope, telescope) are generally used in such a way that the (usually virtual) image is produced at a standard distance from the eye, the so-called distinct visual range $L = 25$ cm. In the following applications of geometric optics, we use thin lenses. They are uniquely characterized by the focal length $f_1 = f_2 = f$.

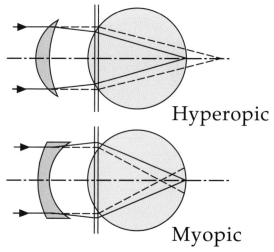

Hyperopic

Myopic

Figure 5.43 The correction of near- and farsightedness by spectacles. In the myopic eye, a blurred image results in long-distance accommodation without glasses, a sharp picture for nearby objects in near-accommodation. Here a diverging lens helps. In the hyperopic eye, one can see distant objects with additional near-accommodation; for nearby objects, the accommodation is not sufficient, so we need an additional converging lens.

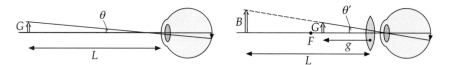

Figure 5.44 Magnifying glass.

Magnifying Glass

If a small object is placed closer to the eye in order to enlarge the viewing angle, we lose the sharp image on the retina. A magnifying glass is used to produce an image at a distance of $L = 25$ cm from this object, which is seen by the eye (see Figure 5.44).

The object G must lie within the focal length of the biconvex lens. The image distance is negative. If we hold the magnifying glass directly in front of the eye, then $-b \cong L$. The lateral magnification $m = -b/g$ in that case is the same as the angular magnification θ'/θ, where θ' is the angle of the image and θ the angle of the object at distance L. It then follows that

$$m = -\frac{b}{g} = -b\left(\frac{1}{f} - \frac{1}{b}\right) = 1 - \frac{b}{f} \simeq 1 + \frac{L}{f} \simeq \frac{\theta'}{\theta}.$$

Lens Doublet

Two thin lenses with focal lengths f_1 and f_2 are placed one after the other at distance d. The first lens produces the intermediate image B_1, which is then transferred to the final image B_2 by means of the second lens (see Figure 5.45).

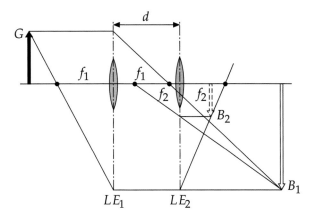

Figure 5.45 Lens doublet.

From the image equation, we obtain the following for the first image:

$$\frac{1}{b_1} = \frac{1}{f_1} - \frac{1}{g_1} \ , \ m_1 = -\frac{b_1}{g_1} \ .$$

The object distance for the second image therefore is $g_2 = d - b_1$. B_1 lies behind the second lens, such that g_2 is negative! This implies

$$\frac{1}{b_2} = \frac{1}{f_2} - \frac{1}{d - b_1} \ , \ m_2 = -\frac{b_2}{g_2} = \frac{-b_2}{d - b_1} \ .$$

The total magnification therefore becomes

$$m = m_1 m_2 = \frac{b_1 b_2}{g_1(d - b_1)} \ .$$

Microscope

The microscope is also, in principle, a lens doublet. The results from before can therefore be used. The two lenses are called *objective lens* and *eyepiece* (see Figure 5.46). The lens has a short focal length f_1. The object is located just outside of the focal length ($g_1 > f_1$) and is imaged by the lens into a greatly enlarged, real intermediate image B_1. The eyepiece is used as a magnifying glass and creates a virtual intermediate image B_2. This is seen by the eye approximately in the distinct visual range L.

In commercially available microscopes, the distance between the objective lens and the eyepiece is fixed and the eyepiece is also fixed in the microscope. This means that the (adjustable) properties of the microscope are primarily determined by the objective lens. In contrast to what we have discussed, they are usually not characterized by their focal length and size, but by the numerical aperture (NA), the magnification and the working distance (WD). We have recognized the numerical aperture in the discussion of the resolving power, where the resolution limit is given by $x = \lambda/2NA$. With the size of the lens aperture, the focal length is obtained from the numerical aperture by $f = nD/2NA$, where n is the

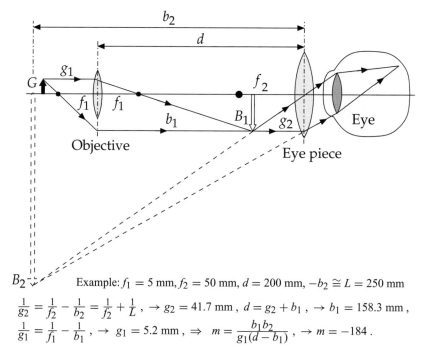

Example: $f_1 = 5$ mm, $f_2 = 50$ mm, $d = 200$ mm, $-b_2 \cong L = 250$ mm

$$\frac{1}{g_2} = \frac{1}{f_2} - \frac{1}{b_2} = \frac{1}{f_2} + \frac{1}{L} , \rightarrow g_2 = 41.7 \text{ mm} , d = g_2 + b_1 , \rightarrow b_1 = 158.3 \text{ mm} ,$$

$$\frac{1}{g_1} = \frac{1}{f_1} - \frac{1}{b_1} , \rightarrow g_1 = 5.2 \text{ mm} , \Rightarrow m = \frac{b_1 b_2}{g_1(d - b_1)} , \rightarrow m = -184 .$$

Figure 5.46 Schematic design of a microscope and typical numbers for a roughly 200-fold magnification.

refractive index of the medium outside the lens. The working distance corresponds to the object distance of the objective, i.e., the distance between object and lens (surface), in which a sharp image is produced in the microscope. Using the image equation for a given magnification, we obtain that $WD = nD(m-1)/(2NAm)$. The working distance is therefore smaller the larger the numerical aperture, or the smaller the opening of the lens. For large magnifications, the WD does not depend on the magnification.

5.5.6 Dispersion

Dispersion describes the fact that the refractive index of most substances depends on the wavelength of the radiation:

$$n = n(\lambda).$$

This is due to the fact that the dielectric constant of a material depends on the wavelength, or on the frequency of the incident radiation. The dielectric constant describes how the molecular (electrical) dipoles of the material react in the oscillating field of the electromagnetic radiation. This is closely related to how the frequency of the radiation is related to the resonance frequency of these molecular dipoles. Depending on whether an increase of the frequency shifts the frequency closer to or farther away from a resonance leads to an increase or decrease of the dielectric constant, that is, of n with increasing frequency.

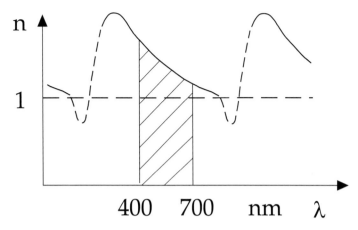

Figure 5.47 Schematic illustration of the variation of the refractive index with wavelength.

In the visible region, n usually decreases with increasing wavelength (or decreasing frequency; see Figure 5.47). In normal glasses, n(blue) is greater than n(red). Blue is therefore refracted more strongly than red. This is called normal dispersion. This basically originates from the fact that the resonance frequency for the electric dipole is in the UV part of the spectrum, such that the frequency has to be increased to reach the resonance.

In certain wavelength ranges, the inverse can also be true: n increases with increasing wavelength. This occurs when the frequency of the incident radiation is close to a natural frequency of the transmitted material, i.e., when resonance occurs.

Since the n occurs in the law of refraction, the light can be decomposed into its different wavelengths with the aid of, for example, a prism.

A rainbow is created by refraction in the water droplets, as we will treat in detail. Because of the dispersion, the rainbow is colored.

5.5.7 Aberrations

Optical systems cannot produce perfect images, since aberrations occur, which can never be completely eliminated. However, they can be partially suppressed with suitable lens combinations in such a way that they do not disturb the image too much. Among the most common aberrations are the following:

Spherical aberration: In the case of lenses with spherical surfaces, axis-parallel beams do not all meet at the focal point. The rays distant from the axis are more strongly refracted than the near-axis rays.

Astigmatism: A point lying laterally from the axis is not exactly represented by a spherical surface in a point B, but in two mutually perpendicular, successive line elements. The astigmatism of oblique bundle is the aberration most difficult to fix. It can, however, easily be observed with a square grid as an object. If oblique beams are used for imaging, either the horizontal or then the vertical bars are sharply imaged in two different image planes.

Figure 5.48 A sample image of a contrast poor image in bright-field microscopy in order to illustrate the advantages of different contrast mechanisms discussed in Section 5.5.8. Figure from Sanderson, 2000, used with permission.

> **Chromatic aberration:** Because of dispersion, the focal length and hence the position of the image will depend on the wavelength. This can lead to colored edges. If lenses of glass with different dispersions are glued together, the chromatic aberration can be partly compensated (*achromatic lenses*).

5.5.8 Modern Microscopy

We have looked at the possible resolutions of a microscope, as well as some of the errors in imaging that need to be taken in to account when designing a microscope objective. Microscopy, however, is not only about resolution, but also very much about having enough contrast between the surroundings and the object of interest (see Figure 5.48). One means of increasing contrast that has proved to be very useful is that of fluorescence, which we will discuss in more detail in the next section. The essential point is that a fluorescent dye (a fluorophore) emits the light at a different (longer) wavelength than it absorbs the light. A fluorescence microscope is then set up in such a way that light of short wavelength is used for illumination, but only long wavelength light is observed. This is usually achieved by passing the illumination over a coated mirror that only works as a mirror for the illuminating wavelength. Such mirrors are produced according to the same principle as the butterfly wings, which we have discussed earlier. The fluorescence emitted is not reflected by the mirror but passes straight through to the observing camera or eye. In this observation, all the illuminating light is absent and therefore only the fluorescent structure is seen. By the discovery of fluorescent proteins, which can be combined genetically with interesting structures, one can thus obtain a very high contrast for the process of interest. There are, however, also other means of increasing the contrast,

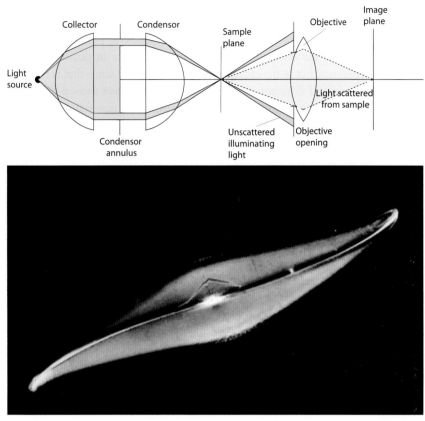

Figure 5.49 Illustration of the light path in dark-field microscopy (top) and a sample image taken in dark-field illumination. Microscopy image from Sanderson, 2000, used with permission.

as well as increasing the resolution in modern microscopes, some of which we will briefly introduce now.

Dark field Microscopy

As we have seen in the preceding discussion of lenses, the main mechanism of achieving contrast in imaging is by making use of differences in refractive index between two different parts of a sample. However, in normal samples, changes in refractive index between different parts are rather small, such that the corresponding changes in the imaging are covered by the unscattered background illumination, and thus there is no discernable imaging contrast. One way of surmounting this is to use dark-field microscopy, where the straight-through illuminating light is blocked; see Figure 5.49. This means that in the absence of any scattering, the image will be completely dark (hence the name), thus greatly enhancing the boundaries of samples in terms of contrast. The main body of a sample, however, is not necessarily imaged with greater contrast, since there the scattering may be small because of a homogeneous structure. Thus it is usually useful to take both bright-field and dark-field images in order to maximize the information gained from the imaging.

Phase Contrast

A similar technique to dark-field microscopy is posed by phase-contrast microscopy; however, the contrast here is not in the change in direction due to a refractive index difference, but in the change in phase of the light wave that is incurred by passing through materials of different refractive index. As we have seen, a difference in refractive index corresponds to a difference in light speed in the material. Thus if light is passing through different materials, the phase of the light, i.e., the times at which the amplitude of the light wave shows maxima, is changed slightly. This change of phase largely enhances changes in the refractive index and therefore gives a large increase, in contrast even for relatively homogeneous samples and also works, for instance, to increase the contrast of X-ray microscopes. Similar to dark-field microscopy, in phase-contrast microscopy the straight-through light has to be blocked and separated from the rest. Therefore, the illumination of the sample is similarly done from an annulus, which occurs, however, before the image plane is sent through an annular phase plate rather than being blocked. Such a phase plate is a piece of glass of a specified thickness, such that the phase change after passage through the plate corresponds to exactly 90 degrees. This means that the thickness of the plate has to fulfill the condition that $d = \lambda/(4n)$. Now the background light has incurred a phase shift similar to the light passing through the sample, such that these parts of the light interfere constructively, thus enhancing the sample. Finally, the unenhanced background light is partially blocked by a gray filter. The schematic buildup of a phase-contrast microscope is illustrated in Figure 5.50.

Differential Interference Contrast

Another contrast enhancement technique based on the interference of differently polar-ized and phase-shifted part of a sample is present in differential interference contrast (DIC) microscopy (see Figure 5.51). As in phase-contrast microscopy, the wave nature of light is used as an enhancement of contrast, specifically the fact that light can be polarized and that waves in different polarization states do not interfere. In order to achieve a contrast enhancement in this way, the illuminating light is passed through a birefringent prism, thus separating the two different polarization states of the ordinary and extraordinary beams. Due to the birefringence of the prism, these will pass the sample at slightly different positions, which means that especially at the edges of objects, the different polarization states will incur different phase shifts. When the two beams are recombined in another birefringent prism after the sample before image formation, these phase shifts lead to constructive and destructive interference at the edges, thus enhancing them.

Confocal Microscopy

Apart from increasing the contrast using different optical methods, the resolution can also be an issue. In particular, three-dimensional samples lead to a loss of resolution due to signal originating from other layers also being imaged, even in fluorescence microscopes. In order to solve this problem and obtain three-dimensional data from

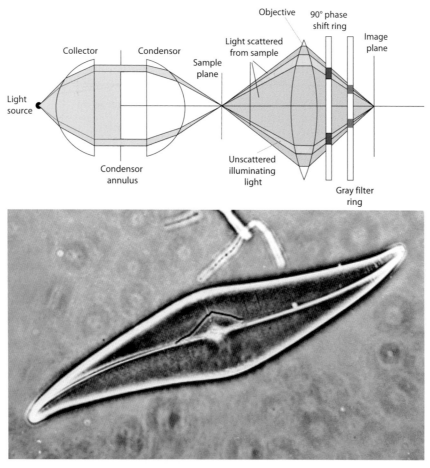

Figure 5.50 Illustration of the light path in phase-contrast microscopy (top) and a sample image taken using phase contrast (bottom). Microscopy image from Sanderson, 2000, used with permission.

images, confocal microscopy has become widespread since the late 1990s. Here, the main point is to reject light in the image that originates from outside the focal plane. Using a point source illumination as shown in Figure 5.52, the imaging optics actually imply that only a corresponding point should be excited and thus reflect light. If this is taken into account in the imaging optics by adding an additional pinhole that images the same point from the objective side, this can be taken into account and thus all out-of-focus light is rejected.

If the imaging is carried out in reflection, such that the illumination and imaging objectives are the same, the rejecting pinhole has to be in the same geometric position as the illuminating one. For a fluorescence microscope, where the illuminating light can be separated from the sample response spectrally by use of a dichroic mirror, the rejection pinhole has to be positioned at the right place with respect to this mirror (see Figure 5.52). Since this only images a single point, the detection can be a photomultiplier that is highly sensitive and can also detect faint fluorescent signals. On the other hand, this means that

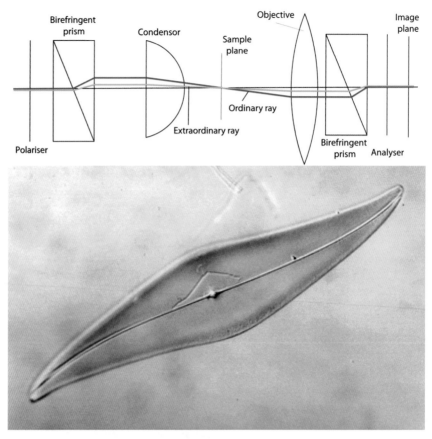

Illustration of the light path in differential interference contrast microscopy (top) and a sample image taken using DIC (bottom). Microscopy image from Sanderson, 2000, used with permission.

the point of illumination has to be scanned across the sample in all three directions, which makes confocal imaging somewhat slow. Nevertheless, good three-dimensional resolution is obtained; see Figures 5.53 and 5.54.

5.5.9 The Rainbow

We have seen in the introduction that one of the very first scientific experiments has served to explain the appearance of the rainbow. At the same time, we have seen that the essential element for creating a rainbow is a ball-shaped water drop, no matter how big it is. In contrast to the "glory" we described previously as an interference phenomenon, the rainbow can be explained with geometrical optics.

So if we look at a single raindrop, which is much larger than the wavelength of light (typical raindrops are about 1 mm in size), and illuminate it from the sun, then the situation shown in Figure 5.55 can be found. For simplicity, the light is parallel to the x axis on the drop. Depending on the height at which the light ray hits the drop, the angle α changes between the drop surface and the incident light. What ultimately interests us is at what

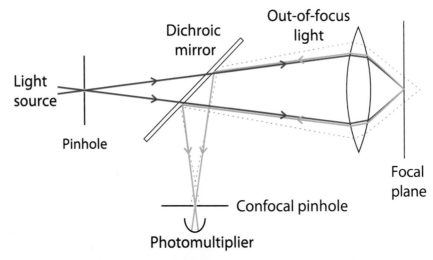

Figure 5.52 Light path in a confocal microscope. The pinhole aperture of illumination is mirrored by a pinhole in front of the detector, which rejects all light reflected away from the focal plane. With a scanning of this pinhole, a three-dimensional image of a sample can be taken with good depth resolution.

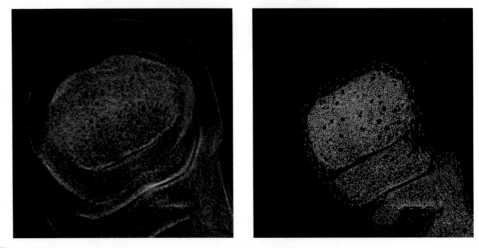

Figure 5.53 Comparison of a wing imaginal disc once imaged using a wide-field microscope and once using a confocal microscope. Due to the rejection of out-of-focus light in a confocal microscope, the image section obtains a much higher resolution in depth.

angle (relative to the incident angle) the light beam leaves the drop again. This angle is indicated in Figure 5.56 with ϕ.

Let us look more closely at the situation. The light beam that hits the drop at the angle α is refracted according to Snell's law and continues at the angle β in the drop. The light beam is reflected at the rear end of the droplet. Since the drop is a sphere, the angle of incidence is also β and thus also the angle of reflection. This reflected beam again hits the surface of the drop where it is again refracted. Since the angle of incidence is β because of the spherical shape, the refracted beam again occurs at the angle α, since the difference in

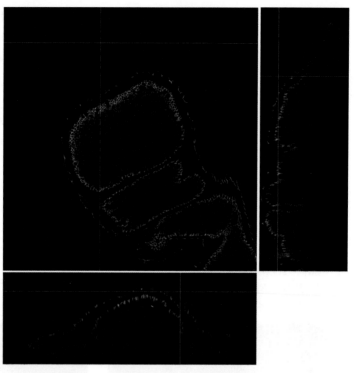

Sections through the wing disc above in all three spatial dimensions. The three-dimensional, folded structure of the wing disc can thus be investigated.

refractive index is the same as in the case of entry and we just have to invert Snell's law. If we consider the two isosceles triangles, each having an angle β to the surface at the base, we see that the angle must be $2\pi - \phi = 2\pi - 4\beta + 2\alpha$. In other words, the angle at which the light ray exits again is given by the following:

$$\phi = 4\beta - 2\alpha,$$

where α and β are connected by Snell's law: $\sin\alpha = n_W \sin\beta$. That is, the exit angle is only dependent on the refractive index of water and the angle of incidence. If we consider different rays, arriving at different heights of the drop in Figure 5.34, then we see that the angle of incidence is small at small angles and increases with increasing angle of incidence. However, the beams close to the one indicated all reemerge at a very similar angle, and the angle does not exceeded a certain value. This angle is exactly 42 degrees for red light ($n_{red} = 1.33$). This maximum angle depends only on the refractive index. This can also be seen in Figure 5.35, where the exit angle is given as a function of the entry angle. The different curves correspond to blue, green, and red light. Since, as stated previously, the refractive index of a material, i.e., also that of the water, is changed with the wavelengths (i.e. the colour), the maximum possible exit angle is changed with the color of the light. Since the index of refraction for blue light is somewhat larger than for red (more precisely, $n_{blue} = 1.34$), a slightly smaller maximum angle of about 39 degrees is obtained (see Figure 4.35).

A rainbow showing its characteristic size, coloration, as well as a secondary rainbow with an inverted color sequence.

This means that the incident light from the sun, which is essentially parallel to a falling raindrop, is predominantly reflected by an angle of 42 degrees for red light, or 39 degrees for blue light. Thus, a rainbow is produced when light from the sun can fall directly onto the raindrops, with an angle of 42 degrees between the direction from the sun to the drop, and the direction from the drop to the eye (see Figure 5.57). Raindrops for which the angle is greater do not reflect light in the observer's eye and thus no sunlight is seen in those areas. For smaller angles, all colors are reflected approximately equally, resulting in a white shimmer in the center of the rainbow. At viewing angles of 42 degrees, only red light is reflected into the eye, so a red arc is seen. At somewhat smaller angles, green or blue light are reflected predominantly, which is why the rainbow appears in those respective colors. We see, therefore, that due to the geometric-optical properties of the water droplet of the falling rain, the rainbow is red at the top and blue at the bottom, has an opening angle of about 42 degrees, and is dark on the outside and bright on the inside.

If we take a closer look at the process, we also see that at the angle of the rainbow the reflection angle inside the droplet is very close to the Brewster angle. This means that the light in the rainbow is reflected in a polarized fashion! The rainbow is tangentially polarized. We can test this by looking at the rainbow with a polarizing foil.

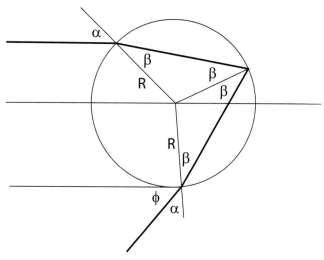

Construction of a light path through a circular drop of rain. The different angles are determined by Snell's law and thus yield a dependence of the outgoing angle in response to the incoming angle.

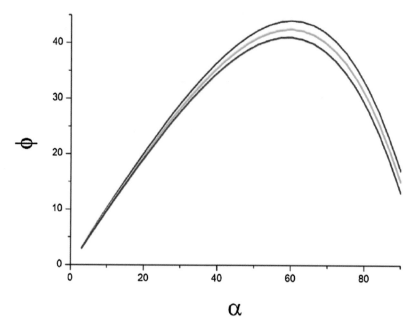

The distribution of outgoing angles as a function of the incoming angles for different refractive indices of water corresponding to red, green, and blue respectively. There is a maximum outgoing angle, which changes slightly for the different colors, thus giving rise to the intensity distribution and coloration of the rainbow.

In addition, we can understand why rainbows are mostly observed in the morning or in the evening. At noon, the sun is usually at an angle of more than 42 degrees above the horizon. Then it is not possible for the sun to illuminate a raindrop, which is at an angle of at most 42 degrees to the observer, since those raindrops have already fallen to the ground.

In the image of the rainbow above there is also a second rainbow visible, which has the inverse color sequence and is slightly larger. This can also be understood with a similar argument. If we look at this section's treatment where the ray emerges from the drop after being reflected once more, we can make the same argument. We again get a maximum angle, for the beam reflected twice, which is now about 51 degrees. Since we now have two reflections, the order of colors is reversed. Also this rainbow is polarized.

5.6 Absorption, Colors, and Fluorescence

We have seen in the treatment of interference that many colors in nature are caused by interference processes. However, there are also very many colors that arise differently. Likewise, the fact that we can see colors at all is based on other physical effects. These depend on how light of different wavelengths (or frequencies) is absorbed by materials. This is why we will now deal briefly with absorption processes. To this end, we introduce atoms as nuclei-bound electrons that can be excited to vibrate. For this, we must know with which energies the electrons are bound to the nuclei, which we have already briefly discussed in the treatment of wave mechanics in Chapter 4. However, the treatment of absorption here will rely mostly on classical physics rather than quantum mechanics, which already gives a good intuitive picture of what happens when light interacts with matter. It can also be applied to systems that are more difficult to treat quantitatively, such as complex molecules. Here we do not imagine the electrons as a vibrating system, but the individual components of a molecule. In this case, the masses of the atoms and the covalent bond energies are decisive for the absorption properties.

5.6.1 Absorption Processes

When we consider the electrons bound to the nucleus, we can apply the restoring force as a harmonic potential so that the binding energy is given by $E_B = ka^2/2$, where a is the distance of the electron from the nucleus and k is the effective spring constant of the bond. The binding of the electrons to the nucleus is essentially electrical in nature, the bond distance being quantum mechanically justified (see Section 4.8). This results in a natural frequency with which the electron can be excited to a vibration of $\omega = \sqrt{k/m}$, where m is the mass of the electron. Thus, when light is irradiated with this frequency, the electron can be excited to a resonant vibration. This is exactly the same as the resonance process we have observed in mechanics, and there will be a similar resonance spectrum of frequencies that can lead to resonant vibration. This means that we can estimate which frequencies of light lead to increased absorption from the properties of the binding energies and the vibrating masses.

We will see in Section 9.3 that the bonding energies of electrons in atoms, or of covalent bonds in molecules, are of the order of $E_B = 10^{-19}$ J. The atomic distances are of the order of $a = 0.1$ nm. This gives an effective spring constant of about $k = 20$ N/m. Depending on the type of oscillation we wish to excite by the electromagnetic wave, the natural frequency of the oscillation changes with the mass of the vibrating object. The mass of an electron is approximately $m_e = 10^{-30}$ kg, which leads to natural frequencies of electrons in atoms of the order of $\omega = 4.5 \times 10^{15} s^{-1}$. An electromagnetic wave of this frequency has the wavelength $\lambda = 2\pi c/\omega = 400$ nm. This means that at these wavelengths the electrons are excited to vibrations so that almost all materials absorb wavelengths of this magnitude (and smaller). At lower wavelengths, the energies are high enough to ionize atoms, a process that is not resonant; we'll discuss this process in the next section. When oscillations of molecules are to be excited, the oscillating masses are correspondingly greater by a factor of 2,000, which results in correspondingly about 40 times larger wavelengths for the resonant absorption. Such wavelengths of 10–20 µm correspond to the infrared heat radiation in the electromagnetic spectrum. This explains why molecules such as CO_2 or water reflect back the heat radiation of the earth and therefore act as greenhouse gases. Higher harmonics of such vibrations provide additional absorption at lower wavelengths, e.g., can completely explain the absorption spectrum of water. This results in water being transparent in the range of about 900–400 nm. This spectral range corresponds fairly closely to the range of visible light. This makes sense, since our eyes have evolved in water, such that only in this range of wavelengths there is a signal for which a detector (the eye) must be developed.

5.6.2 Pigments

Since these absorption processes are caused by resonance, they occur only in a small wavelength range, and only certain wavelengths of the electromagnetic radiation are actually absorbed by the substance. Which frequencies determine the appearance or colour of the fabric. This is the type in which pigments are used for colouring.

If we have hitherto considered the color of light, we have always limited ourselves to the wavelength (or the frequency) of the electromagnetic radiation. Red light has a wavelength of about 600–650 nm, blue one of 400–480 nm, with the rest of the spectral colors have wavelengths in between. If light of all colors (or wavelengths) is present, we have white light. This is called additive color mixing. If we mix pigmentary paints, as in a paint box, for example, then we do not get white, but black. This is called subtractive color mixing. The difference is in the way a pigment produces a color. Since this is caused by absorption, the color we see in a pigment is just that color of the light that is *not* absorbed. That means a red pigment absorbs everything except the red light. The subtractive color mixing thus consists of an aggregation of absorptions so that when all pigments are mixed together, all the colors are absorbed and therefore the pigment appears black.

The coloration of a pigment and its correlation with the absorption behavior is illustrated in Figure 5.58, which shows the absorption spectra of hemoglobin (left) and chlorophyll (right). Hemoglobin is the oxygen transporter in the red blood cells, which also gives the blood its red color. When oxygen is bound to the iron group, the absorption behavior changes, and oxygenated blood has a somewhat harsher red color than nonoxidized blood. This is seen in the difference between the coloration between venous and arterial blood. Chlorophyll is the essential active ingredient of photosynthesis in plants, which gives the leaves a green color.

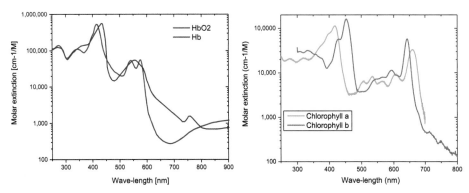

Absorption spectra of oxygenated and deoxygenated hemoglobin (left) (Prahl, 1998) and chlorophyll a and b (right) (Strain et al., 1963).

These absorption spectra show that for hemoglobin the red light is not absorbed, resulting in its red coloration. The comparison of oxy- and deoxyhemoglobin also shows that in the case of deoxyhemoglobin, there is an additional, weak absorption in the red part of the spectrum, which is why the venous blood appears darker and shows a less rich red color. The observation of the absorption spectrum of chlorophyll shows that absorption bands are present in the red and the blue. Thus the blue and red light of the sun is used for photosynthesis, although the greatest light intensity of the sun lies in the green spectral range.

5.6.3 Color Vision

As we have just seen, colors are formed by selective absorption of light by pigments. If we define a color only by the wavelength of the light, then there are problems with our experience of additive color mixing. If we take two lamps, e.g., green (520 nm) and red (620 nm) light, then we see yellow, even if no light with the corresponding wavelength (570 nm) is emitted. However, we see that the color we see is spectrally about halfway between the colors that meet our eye. To understand how this works, we have to look at the eye first. The light in the eye is absorbed by the retina and this absorption leads to an electrical nerve impulse that is guided into the brain. Various types of cells with different pigments serve as absorption centers. On the one hand, there are the rod cells, which have a relatively broad absorption spectrum. These cells are mainly responsible for seeing at night. They are therefore more sensitive, but (almost) do not pass on any color information because of their broad absorption spectrum. The cones are present in three different variants (see Figure 5.39), which have quite different absorption spectra.

The intensity of the respective excitation of the different cones is also reflected by color perception. Thus, red light (620 nm) is absorbed exclusively by the red cones. This means that only these give a signal and the brain concludes from this information that red light has hit the eye. In blue light (450 nm), the situation is similar, only the blue cones give a signal. However, if, for example, yellow light (570 nm) hits the eye, the green and red cones will respond approximately equally. The situation is the same as when we only shine green and

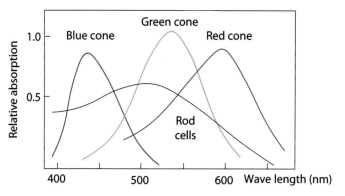

Figure 5.59 The basis of color vision in the human eye. The different absorption spectra of cone and rod cells are shown. The three different types of cone cells lead to a distinction between blue, green, and red light. This is the basis of the RGB colors in computer screens.

red light onto the eye. The absorption signal in the brain is the same, which is why the color perception is the same, and we see yellow light in the mixture. This is also how computer screens create all kinds of colors from only red, green, and blue pixels.

Since color perception depends on a neuronal signal, adjacent colors can also interfere with the color perception by saturation effects in the different cones. Depending on the ambient color and brightness, a different color impression results. These relative intensities are responsible, e.g., for the fact that the veins appear bluish even though the venous blood is red.

Likewise, color sensations can result from time-varying light-dark patterns that lead to different excitations of the different suppositories, as can be shown by Benham's disc, which is completely black and white, but when rotated seems to show colours.

5.6.4 Fluorescence

When we treated the absorption of light, we have described it as a resonant excitation of the oscillation of an electron in an atom or molecule. However, if such an electron is in a vibrating state, it also generates an electromagnetic wave, as we have seen in the dipole transmitter. The oscillation leads to a time-varying electric field, which at the same time makes a change in the B-field, which in turn induces an E-field. We have neglected this emission in the preceding discussion. Normally, this emission takes place at the same frequency as the excitation, so that the emitted light is also absorbed again. However, if the excitation of the electron is attenuated, or if different modes of oscillation are possible, the frequency will change accordingly. Since the energy can only be delivered to other parts, the change in frequency will be such that the frequency is reduced by this relaxation. Then the pigment will emit light of a different (longer) wavelength. This process is called fluorescence and the pigment in question is called a fluorophore. In nature, one can find many fluorescent structures and in recent years it has become possible to manipulate them in a highly controlled fashion using genetics. Some examples of typical biological fluorophores and their absorption and emission spectra are given in Figure 5.60.

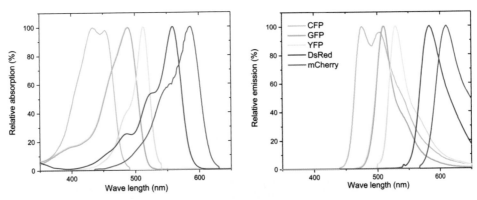

Figure 5.60 Excitation and emission spectra of standard fluorophores. Data from Thermo-Fischer, 2017.

The gene sequences of these fluorescent proteins can be introduced into model organisms and placed under the transcriptional control of morphogens or other interesting structures. This allows the study of development processes with good contrast (fluorescence microscope!). Alternatively, it is also possible to fuse the gene sequence with the sequence of a protein so that the protein of interest is fluorescently labeled and its transport can be followed.

In addition, it is now possible to produce fluorophores that can be switched on or off by the irradiation of light of specific wavelengths. In essence, the light absorption leads to a chemical or physical structure change of the fluorophore and thus to the switching on or off of the fluorescence. If interesting structures are labeled with such fluorophores, the fluorescence can be switched on by irradiating light of a certain wavelength and the fraction of molecules that is switched on can be controlled by the activation light intensity. These molecules then give a fluorescent signal, which is limited to diffraction, but if we know that this signal is due to only one molecule, the position of the molecule can be determined with greater accuracy if the intensity is large enough. This is similar to the difference between the variance of a distribution (diffraction limit) and the error of the mean value (accuracy of the position determination). Thus, if only a subset of the fluorescent molecules in the field of view is switched on such that their images do not substantially overlap with each other, then switched off and other fluorescent molecules switched on, an image of the fluorescent structure can be generated from the positions of these fluorescent molecules, which avoids the diffraction limit and allows resolutions of down to 10 nm or better. This method of fluorescence microscopy is called Photoactivated Localization Microscopy (PALM) or Stochastic Optical Reconstruction Microscopy (STORM) (see Figure 5.61).

Another way to achieve a similar increase in spatial resolution is to use an intrinsic property to all fluorophores to switch off the fluorescence using a process called stimulated emission (which by the way is also the basis for how a laser works). Here, shining in light of exactly the same frequency as energy difference between the highest ground state and the lowest excited state stimulates the emission of exactly the same light, therefore depleting the excited fluorophores in this region. Scanning this depletion beam and having it such that it only depletes a ring around the basic excitation, a scanning confocal microscope with a

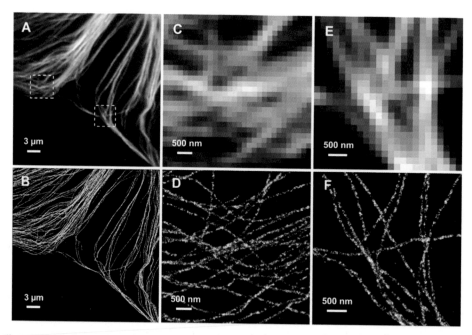

Figure 5.61 Illustration of the increase in resolution beyond the diffraction limit when using STORM as a microscopy technique. The top shows the microtubule network using standard conventional diffraction-limited microscopy, whereas the bottom shows the same situation using STORM. Figure from Bates et al., 2007, reprinted with permission from the American Association for the Advancement of Science (AAAS).

resolution down to 10 nm and below can be obtained. This technique is called STimulated Emission Depletion (STED) and is illustrated in Figure 1.10.

5.7 *Absorption of Radiation

In nuclear transformations, which we have not discussed, several high-energy rays are emitted, of which the so-called γ rays are directly comparable with light, since they are also electromagnetic waves. When these are absorbed in a body, comparable processes occur as in the absorption of light, but they are intensified according to the energy, so that the absorption of the high-energy rays can also result in strong chemical (and biological) changes. We will now look at the consequences of the various types of radiation.

5.7.1 Alpha and Beta Decay

In alpha decay, an atomic nucleus is released by excess protons. This is done by emitting a helium nucleus (two protons and two neutrons). Alpha particles are correspondingly heavy and doubly charged. For this reason, they interact very strongly with matter. Since they have a large mass, they move relatively slowly at a given energy, which means that the charge of the alpha particle can interact with the electrons of the shell over a long time. Likewise, the interaction is stronger since the alpha particle is doubly charged. For this reason, alpha particles are absorbed very well. This can be an advantage and a disadvantage with respect to the health effect of the radiation of alpha particles. One advantage is that external sources of alpha emitters can be shielded very well. Only a sheet of paper stops most alpha particles on their way.

On the other hand, this strong absorption is a disadvantage, since internal sources have a very strong effect on the body. So if you have an alpha-active source absorbed by dust in the air or in food, it can be very harmful. Alpha rays are typically emitted from heavy elements.

In beta decay, a neutron in the nucleus turns into a proton. It sends out an electron, which we can measure as radiation. The electrons of the beta rays are also charged and therefore interact with the electrons in the atomic shells of matter. In addition, electrons are relatively light (compared to nuclei), and therefore the beta particles move very fast. Therefore, they interfere with matter as alpha particles, but also beta emitters can be shielded quite well with relatively dense material; aluminum foil is usually sufficient to shield beta rays. Particles that are internalized were also stopped in the body.

5.7.2 Gamma Decay

In the gamma decays, electromagnetic waves are emitted from the nucleus after the particles of the nucleus have moved from an excited state to the ground state. Corresponding to the energies in the nucleus of $E \sim 10^{-13}$ J, the corresponding photons have a very high energy. This energy also determines its absorption properties. The frequencies of the wave are many times higher than the resonance frequencies, and therefore the absorption is not particularly strong. The principal part of the absorption takes place by direct collisions of the photons with the electrons, the photon emitting some of its energy as kinetic energy of the electron. The processes occur randomly and the absorption rate is therefore given directly by the number of photons. The intensity therefore decreases exponentially with the thickness of the shielding material, and very heavy elements are necessary for good shielding. To protect against gamma rays, several centimeters of lead or thick concrete walls are necessary.

5.7.3 Natural Radiation

There are very many sources of radioactive materials in nature, and we are all constantly exposed to certain rays (see Table 5.1). A large part of the natural radiation comes from radioactive uranium or thorium found in rocks. The heat generated by the absorption of this radioactivity deep inside the earth is responsible for the fact that the interior of the earth is still liquid. Without this source, the earth could not be older than a few million years, which was calculated by Lord Kelvin in the nineteenth century using heat conduction. Until the

Table 5.1 Typical doses of natural and artificial radiation sources experienced by humans

Source	Dose per year (mSv)
Cosmic radiation	0.3
Inhalation, mostly radon	1.2
Medical diagnostics	2.0
Terrestrial radiation	0.4
Ingestion of terrestrial sources	0.3
Emissions from nuclear power plants or fallout, etc.	0.02

discovery of radioactivity, this was the most important argument against Darwin's theory of evolution, and thus even though we know the argument is flawed, we shall look at it again when we look at heat conduction in Chapter 8.

On the other hand, the sun sends high-energy radiation to the earth, which in the upper atmosphere stimulates radioactive decay, which leads to radiation exposure. This load becomes higher the higher you are. A transatlantic flight is approximately the same as an X-ray image and makes up a considerable part of the mean radiation load a person experiences. Flight personnel typically have 10–20 times the radiation exposure of a normal person. The Sievert (Sv), which essentially corresponds to the energy that the radiation has given to the body per kg, is equivalent to an Sv of an energy absorbed by J/kg. The type of radiation is also important, since its absorption properties, as discussed previously, make the radiation more or less dangerous. This is achieved by a weighting factor for the various beam types, but in the case of gamma rays this is one.

On average, the radiation load is approximately 1–3 mSv. This, however, fluctuates strongly, depending on how often one is X-rayed or takes transatlantic flights, because each X-ray image gives an additional 0.2–0.5 mSv. Also, where one lives can be important; for instance, in the Swiss Alps, the dose is much higher due to the combination of increased cosmic radiation (height) and radiation from rocks. Here, the load can easily go to 10 mSv (and sometimes even higher).

The damaging effect of the radiation comes from the fact that electrons can be released from the atoms and the binding properties of the molecules can be altered. This can happen with proteins that, after absorption, can no longer fulfill their function and therefore different processes of the cell are altered. More important is a change in the DNA, since this is also passed on to daughter cells. Such mutations also produce the natural radiation, but there are cellular mechanisms that can correct such changes to a certain extent. In any case, it has not been shown that even small doses of radioactivity harm one's health. From the atomic bombs in Japan in the Second World War, one knows that a dose of 100–200 mSv leads to an increased cancer risk. It is also known that from a dose of 500–1,000 mSv an acute radiation disease comes, from which one normally recovers, but then has a correspondingly increased cancer risk. Finally, about 3–6 Sv are fatal after a very short time.

Exercises

5.1   Interference

You are sending red light of a wavelength of 628 nm through two thin slits with a distance of 18 μm between the slits. At what angle do you observe the third-order maximum?

5.2   Interference 2

Determine the thickness of a water film on a glass plate. You are illuminating the film from the top and observe the direct reflection with a spectrometer. There you find a maximum in spectral intensity at a wavelength of 600 nm and a minimum at a wavelength of 450 nm, with no other minima in between. Water has a refractive index of 1.33 and glass has a refractive index of 1.5.

5.3   Diffraction

You are sending light of a wavelength of 500 nm through a slit with a width of 2 μm. You are observing the resulting diffraction pattern at a distance of 1 m. What is the distance of the first side peak from the main maximum?

5.4   Reading glasses

Between the ages of 40 and 50, the human eye typically loses the ability to accommodate fully. This means that the lens can no longer shorten its focal length fully and we can only see object at a greater distance. This is why your parents are holding papers at arm's length or are using reading glasses. Determine the strength of such glasses in dioptries for a reading distance of 25 cm. Also, draw the ray construction for imaging an object far away, which can be imaged sharply.

5.5   Magnifying glass

You would like to observe an object in a magnifying glass with a magnification of 10 times. The image should be 25 cm behind the lens.

(a) Sketch the image construction.
(b) Will this be a real or a virtual image?
(c) What focal length of the lens do you have to choose?

5.6   Lens

What is the uncertainty of the magnification of a single lens with a focal length of $f = 10.0(1)$ cm for an object at a distance of $g = 15.0(5)$ cm?

5.7   Lens 2

You have the lens of a fish eye, which is a good approximation for a spherical lens. The refractive index of the lens is a little lower than the of glass, $n = 1.45$, and the lens has a diameter of 1 cm. What is the focal length of this lens when used in air ($n = 1$), water ($n = 1.33$), and oil ($n = 1.5$)?

5.8   Microscope

(a) Determine the magnification of a microscope: the eyepiece has a focal length of $f_2 = 18.0(1)$ cm, the distance between objective and eyepiece is $d = 20.0(2)$ cm, and the objective has a focal length of $f_1 = 5.0(5)$ mm with a working distance of $g_1 = 6.0(5)$ mm.

(b) What is the uncertainty of this magnification?

5.9 Spherical mirror

What radius of curvature does a spherical mirror need, such that it produces an upright image at a distance of 50 cm from the mirror that is magnified by a factor of two? Sketch the situation.

5.10 Resolution limit

The resolution limit of a microscope is determined by diffraction effects at the opening of the objective with a diameter D.

(a) How does the minimum angle between two points that can just be separated depend on the wavelength and the diameter of the objective opening? Think of the distribution of intensities behind a slit of width D.

(b) To achieve maximum magnification, the object is placed as closely as possible to the focal point of the objective, such that we can assume $g = f$. What is the distance between two points that can still be resolved in this case?

(c) The focal length of a lens cannot be smaller than the lens radius in practice. Thus in the optimal case, we will have $D = 2f$ relating the focal length to the objective opening. In this case, what is the minimum distance between two points that can still be resolved?

(d) The resolution of a microscope can be increased somewhat by filling the space between object and objective with immersion oil (typically having an index close to that of glass, i.e., 1.5). Why does this help?

5.11 Resolution limit 2

(a) How far away from a pointillistic painting do you have to be such that you cannot resolve the individual points making up the painting? A human pupil is roughly 2 mm in diameter and the dots are about 1.5 mm in size. Suppose that you are using red light with a wavelength of 628 nm for your calculations.

(b) What is the distance of the dots on the retina when you can just keep them apart? The eye has a diameter of about 2.5 cm, i.e., the image is formed at a distance of about 2.5 cm.

5.12 Resolution limit 3

(a) The angular resolution of the eye is about one arc minute (i.e., 1/60 of a degree). How far away can a car be at night such that you can tell its a car (distance between the head lights of about 1.8 m) and not a motorcycle?

5.13 Radio protection

(a) Lead has a shielding length for gamma rays at 1 MeV of about 1 mm. How much lead do you need to make sure that only 1 ppm (part per million) of the radiation is passing your shield?

(b) Neutrinos, the elusive particles that hardly interact with anything, have a very long shielding length. Estimate the shielding length in water, if you know that you are detecting about 100,000 per year in cubic volume of 10^3 m^3, when 10^{11} particles from the sun hit every square centimeter of the earth every second.

Quiz Questions

5.1 Double slit

Light hits a double slit and forms an interference pattern as shown in Figure 5.62. If you add a glass plate in front of one of the slits, such that the phase of the light passing through this slit is changed by 180 degrees (because of the lower speed of light in the glass), what happens to the interference pattern?

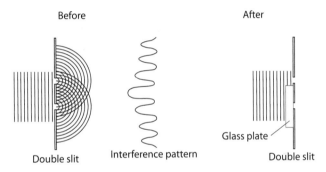

Figure 5.62 Interference pattern formed by light hitting a double slit.

    A The interference pattern disappears.
    B The interference pattern moves to another position as a whole.
    C Maxima move closer together.
    D Maxima move farther apart.
    E Maxima and minima are interchanged.
    F Nothing happens.

5.2 Interference

Two glass plates at a with a small opening angle to each other are illuminated from above with monochromatic light. In reflection, you observe an interference pattern of light and dark stripes (see Figure 5.63). What happens to this interference pattern if you decrease the opening angle between the two glass plates?

    A Nothing changes.
    B Maxima move farther apart.
    C Maxima move closer together.
    D The interference pattern disappears.

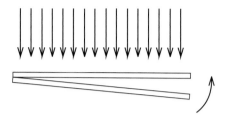

Figure 5.63 Interference pattern formed by two glass plates illuminated from above.

5.3   Interference colors

In a soap bubble, you observe a yellow structure. Which color(s) have been removed due to destructive interference?

A Blue
B Green
C Red
D Magenta
E White

5.4   Polarization

Three polarizers with different polarization directions are held in different sequences. In which of the situations in Figure 5.64 can light pass through all of the polarizers?

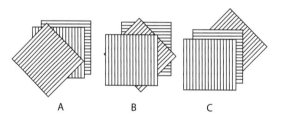

Figure 5.64 Three polarizers with different polarization directions held in different sequences.

5.5   Refraction

If a light beam travels from one medium into another, it is refracted, i.e., the angle of the beam relative to the surface changes. Which other properties change at the boundary of one medium to another?

A The speed of light
B The index of refraction
C The frequency of the light
D The wavelength of the light
E The energy of the light
F The intensity of the light

5.6   Polarization

Which of the following waves can be polarized?

A Light
B Radiowaves

C Seismic waves

D Sound waves in a gas

E Waves on a rope

F Microwaves

5.7 Diffraction grating

A diffraction grating is illuminated with green light. This gives an interference pattern showing three points: one in the middle (straight through), and two each at an angle of ±45 degrees. Now you add a red light source in the same direction as the green one. What pattern do you observe?

A Yellow points in the middle and at ±45 degrees.

B A yellow point in the middle, green points at ±45 degrees, and red points farther out.

C A yellow point in the middle, green points at ±45 degrees, and red points closer in.

D A green point in the middle and yellow points at ±45 degrees.

5.8 Geometric optics

How large does a mirror have to be at least such that you can observe your full length in it?

A This depends on the distance between myself and the mirror.

B The same height as myself.

C Half the height of myself.

D Twice the height of myself.

5.9 Geometric optics 2

A fish is swimming in a pond (water with a refractive index of 1.33) and you want to point at it with a laser pointer. Where do you have to aim to hit the fish?

A Exactly onto the fish

B A little above

C A little below

5.10 Geometric optics 3

A fish is swimming in a pond (water with a refractive index of 1.33) and you want to catch it with a spear. Where do you have to aim to hit the fish?

A Exactly onto the fish

B A little above

C A little below

5.11 Mirror

You are standing in front of your bathroom mirror in a distance of 1 m and want to observe the back of your head. For this purpose, you hold another mirror 1/2 m behind your head. At what distance behind the bathroom mirror do you see the reflection of the back of your head?

A 0.5 m

B 1 m

C 1.5 m

D 2 m

E 3 m

F 4 m

5.12 Rainbow

Which property leads to the color formation in a rainbow?

A The size distribution of the raindrops.

B The falling speed of the raindrops.

C The dispersion of the refractive index of water.

D Brewster's angle.

E Diffraction of light.

F There is no special property leading to the coloration of a rainbow.

5.13 Colors

What is the color of the light that hits your eye when you look at a red rose?

A Yellow

B Blue and green

C Blue and red

D Red and green

E None of the above

5.14 Colors 2

An LCD shows yellow. Which colors in the pixels are lit up?

A Yellow

B Blue

C Red

D Green

E White

F Black

5.15 Radioactivity

What happens to an atom emitting either an alpha or a beta particle?

A It becomes heavier.

B It becomes another element.

C It becomes another isotope of the same element.

D Nothing changes.

6 Forces and Newton's Laws of Motion

We have so far dealt in detail with the mathematical description of natural phenomena and have also looked at some very important biological applications in optics. In particular, in our treatment of waves, we have cursorily spoken of forces, energies, tensions, and mechanical stresses. Many of these concepts are familiar to us from everyday life, and this is what we have alluded to in the discussion on waves. In the coming two chapters, however, we will make these concepts more explicit and. First, let us now consider forces as the cause of movements in detail. The principles necessary for this were first described by Isaac Newton and therefore bear his name. But we will also in this chapter look at a different approach to treat the same phenomena, namely, via energy considerations and in particular the mechanical conservation of energy. The more complete version of conservation of energy in the thermal sense we will treat in Chapter 8.

The description of movements according to the Newtonian principles requires the concept of force, which is regarded as an effect for movements (and deformations). So far, we have used the term more intuitively and based in everyday life. We now want to arrive at a clear, quantitative definition, and therefore want to first look more closely at where forces come from and what effects they have. The Newtonian principles will then emerge.

6.1 Forces and Their Origin

6.1.1 Forces Are Described by Vectors

Experimentally, or empirically, we observe that forces have an effect on movements and/or deformations. We see, for example, that a force is necessary to move something that was previously at rest. We must also exert a force in order to deform, for example, a rubber band. In all of this, the movement or the deformation is in the direction of the force, therefore a force has the properties of a vector – it not only has a magnitude, but also a direction. The direction of the force is the direction into which the movement or the deformation takes place. When we apply several forces, we see that the effect of all of these forces together behaves as their vector sum as long as they act on the same point. Thus, for two forces acting on the same point of attack, we obtain a resultant force:

$$\vec{F} = \vec{F}_1 + \vec{F}_2.$$

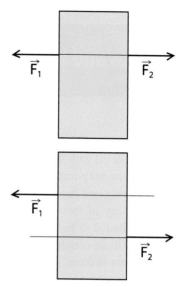

Figure 6.1 Forces must act on the same point or in the same line of action.

We can then treat the problem as if only the resulting force $\vec{F}$ is at work. If, therefore, two forces act that have the same magnitude but are opposed, then this vector sum gives the following:

$$\vec{F} = \vec{F}_1 + \vec{F}_2 = 0.$$

The two forces cancel each other and nothing happens. The situation is the same as if no force were acting at all. This can be seen, for instance, in a tug of war, where both teams are pulling strongly but hardly anything actually moves.

This statement is not at all trivial, and it is important here to note that the forces have to act on the same point, or in the same line of action, as shown in the top sketch in Figure 6.1. Only then is their resulting effect indeed zero and the sum of the forces is a useful quantity to consider. However, if $\vec{F}_1$ and $\vec{F}_2$ have different lines of action, as in the lower sketch in Figure 6.1, there is a resulting effect: the body starts to rotate. The sum of the forces may be zero, $\vec{F} = 0$, but to describe the rotation we would need to consider something called a torque, which takes the line of action into account. We will deal with this briefly in Section 10.4.1, where we look at rotations. In the remainder of this chapter, however, we will not deal with rotations and only look at cases where the forces do indeed act on the same point.

As an example of how different forces are acting on the same point and added vectorially, we consider the direction sensor in our inner ear. In the inner ear are the maculae, which are used to determine the direction of gravitation relative to the head. They tell us, for example, whether we are standing upright or on our head. They therefore supplement the vestibular organ, which we encountered in Section 3.4, determining the direction of acceleration. The position of the maculae is given in Figure 3.12. In Figure 6.2, the detailed structure of the maculae is shown. On the one hand, the head is held upright (parallel to the direction of

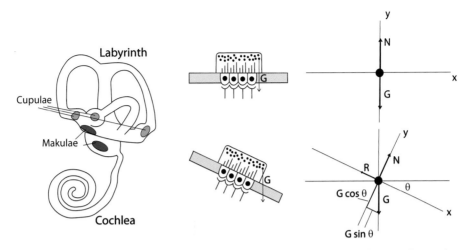

Figure 6.2   Detailed view of the otoliths in the maculae with an upright and tilted head. When the otoliths exert a force on the dendrites of the neuron, it begins to fire. The rate of firing depends on the force, whereby the direction of the head relative to gravity can be determined. On the right, the physical abstraction of the process for calculating the force of the otoliths on the neuron is shown.

gravity), and on the other hand, the head is tilted. The corresponding forces acting, i.e., the physical abstraction, are shown in the right part of the figure. The essential constituents are the otoliths, small calcite crystals, which are retained in a gelatinous substance. Since the otoliths do not move, the force parallelogram is also to be calculated in the inclined case, and the otoliths exert a force on the neurons present in the maculae. These begin to fire according to the force acting, with a force-dependent rate. From the abstraction of the acting forces, the force of an otolith on the dendrites of a nerve can be determined vectorially by means of the equilibrium of forces.

Let's describe the state of the upright head. Since the otoliths remain at rest, the forces acting on them cancel each other out. One force, the weight G, is given by gravity and always points to the center of the earth. The other, from the gelatinous substance to the otoliths, is called the normal force, N (we shall deal with it later in detail), and is perpendicular to the surface. For the upright head, we get $\vec{N} = -\vec{G}$, or if we look at the absolute values in the y direction, $N = G$. For the inclined head (the lower-middle part of Figure 6.2), N and G are no longer parallel. For a balance of forces, we need a third force. This is due to the friction of the otoliths in the gelatinous substance and is parallel to the surface (R in the figure). We will also look at the friction later. Since the otolith remains in rest, the following must hold: $\vec{N} + \vec{G} + \vec{R} = 0$. We now set the axes x and y as shown in Figure 6.2 so that they lie in the plane of the otoliths (x) or are perpendicular to it (y). Then we can consider the forces acting in the x direction. In this case, we have $R = G\sin(\theta)$, where θ is the tilt angle of the head. The force on the gel or the neuron is therefore directly dependent on the angle of inclination with a maximum force per otolith on a neuron, which corresponds to its weight of about 4×10^{-12} N.

In Figure 6.2, we can see that there are two maculae in our head near the vestibular organ. Why is that so? If we look at the possible movements of our head, we see that there are actually two tilting directions that the head can move in, front-rear and left-right. Each of these two different directions of tilt is covered by the force of the otoliths in one of the two maculae only. Thus the two maculae directly correspond to the two different tilt directions of the head.

6.1.2 Origin of Forces in the Muscle

Now that we have seen that our body can measure forces, let us briefly look at how we, and other animals, generate forces in the muscles. In doing so, we will see a few principles of the action of forces, which we will then look at in general. One of these principles is that a force always acts from one object to another. If we look at a muscle more closely (see Figure 6.3), we see that there are different parts that exert forces on one another. On the one hand, the actin filaments, which are arranged between the Z-disks and together with the myosin bundles, form a sarcomere.

The myosin bundles consist of a collection of very many myosin molecules, which are also shown in Figure 6.3 (left). These consist of an elongated chain with a head group whose conformation with respect to the chain changes depending on the binding of an adenosine triphosphate (ATP) molecule. If, as in the muscle, many myosin molecules are arranged in a bundle between two actin strings, a movement of the myosin bundle with

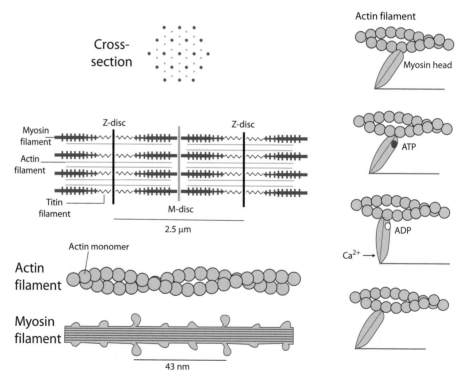

Figure 6.3 Schematic structure of a muscle (left) and microscopic mechanism for the generation of force via a myosin stroke (right).

respect to the actin filament will occur due to the conformational change of the head group. Since the actin filaments are attached to the Z-disks, this movement results in a contraction force, which is given for each individual myosin stroke, and which in sum brings about the force of the muscle. One could thus calculate the force of a single stroke from the force of a muscle and the number and arrangement of the myosin molecules.

The force of a single myosin stroke can also be directly measured. For this purpose, a single actin filament is provided at both ends with a polystyrene sphere (by means of a covalent bond), which is about one micrometer in size, and these two spheres (also known as microbeads) are then held in place by an optical tweezers. An optical tweezer is essentially a strongly focused laser beam, the light exerting a force on the polystyrene due to the different refractive index of the bead and its surroundings. We have discussed this in Chapter 5 in regard to optics. This force is very small, and the applied forces with this method are in the range of a few picoNewton (i.e., 10^{-12} N). If, in addition to this actin filament under tension, we add a further microbead to which individual myosin molecules are bound, we can mimic the action of the muscle in a single-molecule experiment. The myosin molecule on the bead will bind to the actin strand, and if the ATP is present in the solution, it will execute a stroke with its head in order to run along the actin strand. The force of these beats can then be measured in the optical tweezers, because in order to hold the microbead in place when myosin is acting on it, a slightly higher force is needed. The schematic structure of this experiment, as well as the result of a measurement, is shown in Figure 6.4. It can be seen that the action of a myosin stroke is limited in time to the stroke itself, and that the force of a single stroke corresponds to about 5 pN. If the position of the microbead is not left constant, but movement of the actin strand is allowed, one can also measure the step size of a single myosin stroke. This is also shown in the figure, and a step size of somewhat less than 20 nm is found. This step can also be estimated from the geometrical arrangement of the head group and the conformational change. The head group has a length of $L = 21$ nm, and the conformational change makes an angle of about $\alpha = 45$ degrees in the orientation of the head group relative to the chain group. That is, the step length should be given by $\Delta x = L \sin(\alpha) = 15$ nm. It is also possible to estimate the force of the individual impact when we consider that the conformational change is caused by the hydrolyzation of an ATP molecule. Here, chemical energy is converted into mechanical energy, which enables the force. We shall introduce the concept of energy in detail later in this chapter (in Section 6.4) and then also make the corresponding estimate.

The contraction force is thus determined by the number of myosin molecules involved. This will also affect the maximum force that a sarcomere can exert on whether the muscle is prestressed or not. That is, depending on how strongly contracted the sarcomere is, fewer myosin molecules are involved in the contraction. When the sarcomere is contracted too tightly, the maximum force also decreases, since again fewer molecules are involved in the contraction (see Figure 6.5). After the muscle has contracted and thus the actin filament exerts a force on the Z-disks, this force has to be passed on to the outer world. To this end, the muscles are connected to the bones by means of elastic tendons, and the deformation of the tendons gives the force to the skeleton and thereby to the external world. Whether a muscle contracted during contraction depends on the elasticity of the tendons. Since the tendons themselves are elastic, they can be additionally stretched depending on the

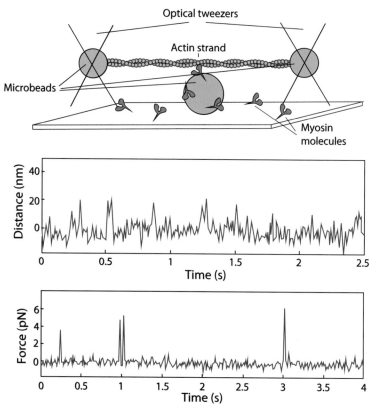

Figure 6.4 Measurement of the force generated by myosin walking on actin in a single-molecule experiment using an optical tweezer. The top panel shows the setup, the middle panel a time trace of the position of myosin, and the bottom panel a time trace of the force exerted by the optical tweezer. Data from Finer et al., 1994.

strength of the acting force. This stretching can cancel the contraction of the sarcomeres, so that the muscle is not contracted but only the muscle tension is increased (see Figure 6.5). The elastic properties of materials will be discussed in detail in Section 7.1. In any case, however, the contraction force that is made by the myosin molecules is passed on to the load. Looking at Figure 6.5 more closely, we see that not only does the muscle exert a force on the load, but also the load exerts a force on the muscle (this is, for example, apparent from the elongation of the tendons). This is a very general principle, which we shall now consider more closely.

6.1.3 The Principle of Reaction

As we have just seen, forces can only act between two distinguishable bodies. When we change our point of view between the two bodies, we see that the other body must return the same force as the one that is exerted onto it. This is summarized in the reaction principle due to Newton.

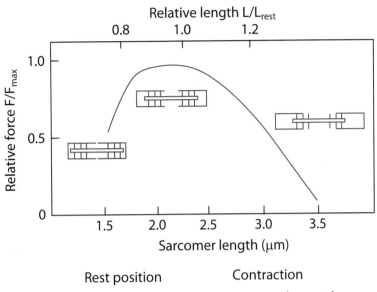

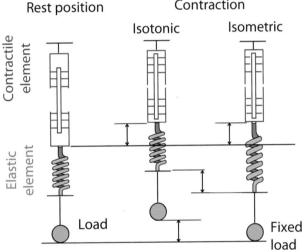

Contraction of muscle fibers leads to different amounts of applied force depending on the extension of the fiber (top). Contracting the muscle fiber can either lead to mechanical work being carried out by lifting a load or by loading the elastic tendons (bottom).

The reaction principle: If a body 1 exerts a force $\vec{F}_{12}$ onto body 2, then body 2 exerts a force $\vec{F}_{21}$ onto body 1, which is equal and opposite $\qquad \vec{F}_{12} = -\vec{F}_{21}$

In this context, it is very important to note that the reaction principle makes a statement about forces acting on *different* bodies. It does not matter whether the two bodies are touching or are any distance apart, whether they have the same or different mass and shape,

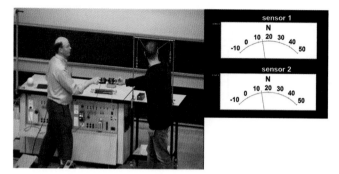

Figure 6.6 Forces in a tug of war. The force of both persons is equal as shown by output of both force sensors. If someone wins in a tug of war, the force of that person on the ground is greater than that of the other one, which leads to an acceleration.

but they have to be different bodies. Then we always have $\vec{F}_{12} = -\vec{F}_{21}$. The essential statement is that forces always act between two bodies, as we have already seen in the muscle. This effect is independent of our point of view, that is, when I say that something exerts a force on something else (for example, the myosin on the actin), then this must also be true the other way around.

Because the two reaction partners $\vec{F}_{12}$ and $\vec{F}_{21}$ act on different bodies, they do *not* cancel out in their effect. It would be pointless to add them up when we want to examine the movements of a single body since they do not act on the same body. $\vec{F}_{12}$ acts on body 2 and $\vec{F}_{21}$ on body 1, or more generally, we abbreviate as follows:

$$\vec{F}_{\text{from } i \text{ on } k} \equiv \vec{F}_{ik}.$$

So if we look at the myosin molecule at the actin thread of the muscle again, we can imagine an abstraction as in a tug of war (see Figure 6.6). One can illustrate this by pulling on two force meters (e.g., two equal springs). No matter what force is exerted, they will always show the same force; the springs will always be extended the same amount. That is to say, even if someone large pulls on someone small, the force of the one is always the same as the force of the other.

How, then, is it possible that someone can win a tug of war? To do this, we must look at the forces that affect the two participants. There is, on the one hand, the force from the rope, which is the same for both. On the other hand, however, there are also forces from the ground, in particular the friction force. As we have seen in the otoliths, these forces depend on the weight (and the coefficient of friction). A smaller (lighter) person will feel a smaller force from the ground. That is, when we look at the total force acting on the smaller person, we get a resulting force in the direction of the larger person. This force results in a movement described by the next Newtonian principle.

6.1.4 Inertia

When resulting forces are applied to a body, its movement changes or it deforms. This fact and its inverse, that the movement of a body remains constant when the forces acting on it cancel out, represents the second basic principle of mechanics: the principle of inertia or action.

> **Principle of inertia:** A body persists in its state of motion when no external forces act on it.
>
> **Principle of action:** If a force $\vec{F}$ acts on a body (a point with mass m), the body is accelerated with an acceleration
> $$\vec{a} = \vec{F}/m$$

This principle therefore gives a relation between the acting force and the mass and change in velocity of the body on which this force acts. In other words, the change of the motion is proportional to the action of the resulting force.

Apart from the vector arrow, we have found the known relationship "force equals mass times acceleration." If we have, in particular, the following:

$$\sum_i \vec{F}_i = 0 \quad \text{so folgt} \quad m\vec{a} = 0 \rightarrow \frac{d\vec{v}}{dt} = 0 \rightarrow \vec{v}(t) = \text{const.}$$

In contrast to the principle of reaction, however, all the forces we are considering here are acting on a single (the same) body! And more precisely, we are considering the (vector) sum of all these forces. So if all the forces on the body cancel out, the motion of the body (the direction and the speed) remains constant. If we want to change the state of motion (either speed or direction), we must act upon the body with a force. This force needs to be larger the larger the mass of the object. This is because the mass leads to an inertia of the object. If we want to exert a force onto a large mass, we must do this slowly, such that the inertial mass can follow. Otherwise, the thing that we use to exert the force may break. You can illustrate this nicely by attaching a heavy sphere using two thin threads above and below on the ceiling. Depending on how quickly you pull, either the upper or lower thread will break. This is because if you pull quickly, the inertia of the ball delays its movement, such that the force mainly acts onto the string, which then breaks. If you pull slowly, you are moving the ball, and therefore the upper string feels the weight of the ball in addition to the force you're pulling with, and thus it will break. You can notice a similar situation if you try to lift a heavy paper bag with groceries quickly. The inertia of the groceries may lead to the rupture of the bag!

Another example of inertia is the air resistance in flight. We have derived the dependencies for the frictional force in Section 2.3.1 from the units of the quantities we want to describe, but we can now do this more precisely. When a fly flies, what it does is move air by the stroke of its wings. This air, however, does have a mass and therefore shows inertia, so to change its movement the fly has to apply a force defined by $F = m \cdot a$, where m is the mass of the air and a is its acceleration. By virtue of the reaction principle, then, the air also exerts a force onto the fly, and the reaction force to this inertial force is the lift that the fly feels and that enables it to fly. Now, how does $F = m \cdot a$ relate to the frictional force $F = \rho A v^2$, which we obtained in Section 2.3.1? The mass of the moving air is given by the density of the air ρ as well as the shifted volume $V = A \cdot \Delta L$, i.e., we shift the cross-sectional area A by a distance ΔL. Thus the inertial force is $F = m \cdot a = \rho A \Delta L a$. The acceleration is given by the velocity change of the air per time interval Δt $a = \frac{\Delta v}{\Delta t}$. Within a distance ΔL, the maximum velocity change within a time interval Δt is given by $\Delta v = \frac{\Delta L}{\Delta t}$. This maximum change is therefore just given by the velocity of the moving air, which gives us the friction force:

$$F = \rho A \Delta L \frac{(\Delta L)/(\Delta t)}{\Delta t} = \rho A \left(\frac{\Delta L}{\Delta t}\right)^2 = \rho A v^2. \tag{6.1}$$

That is, we find the exact relationship that we had already found from dimensional analysis. Since in the preceding derivation we have considered the maximum possible change in velocity, the effective force will be somewhat smaller, which is usually expressed by the drag coefficient c_D. This is dimensionless and depends only on the shape of the object. Typically, c_D is about 0.5, but it can also be smaller or larger, while not exceeding 1.

So now that we know what forces are acting on a hovering fly, we can calculate the frequency at which it has to beat its wings to fly. When a fly hovers, the lift is the same as its weight, $m_{fly}g$, and hence the velocity of the air moved by the flapping wings is given by the following:

$$v = \sqrt{\frac{m_{fly}g}{c_L \rho_{air} A_{wing}}} \tag{6.2}$$

With typical values like $m_{fly} = 3\text{mg}$, $A_{wing} = 6\text{mm}^2$, $g = 10\text{m/s}^2$, $\rho_{air} = 1\text{kg/m}^3$, and $c_L = 0.5$, we obtain a speed of $v \simeq 2.2\text{m/s}$. Given a beating distance of the the wing of about $L = 2$ mm, a fly beats its wings with frequency of about 1,000 beats per second. That this is reasonable you can also check from your everyday experience, since the buzzing of a fly is a tone in the frequency range of 1,000 Hz. This means that a fly must move its wing at least 1,000 times per second. If this flap is associated with a muscle contraction, the muscle should be able to stretch and relax at least within a millisecond. For this, however, the nervous system is much too slow. We will treat this in more detail in Chapter 9, but nerve cells take about a millisecond to fire, so the timing does not work. Flies should not be able to fly! The mistake in this reasoning is that we have assumed that each flap is caused by a single muscle contraction. While this does apply to birds and larger insects, which have a slower beating frequency (see Figure 6.7, top), this is not the case in flies and other small insects. In large insects, such as dragonflies, the wing is attached to a hinge with two muscles each, one for lifting and one for lowering the wing. In the fly, the problem is solved very differently (see Figure 6.7, bottom). The muscles needed to raise and lower the wing do not attach to the wing, but to the thorax of the fly. The muscles therefore only give a momentary pull onto the thorax, which then vibrates at its resonance frequency, which happens to be about 1,000 beats per second. As we have seen in Chapter 4, such a resonance frequency just depends on the masses that are vibrating and their restoring forces. Therefore, the body of the fly has to be built for the proper resonance frequency, but the excitation of this frequency can then simply be achieved by quick application of a disturbance, which we have learned in Section 4.6 contains all frequencies, including the resonance frequency, and will therefore excite the proper wing beat of the fly without needing to control muscles on the time scale of milliseconds.

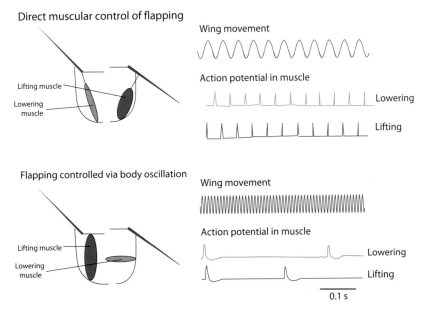

Figure 6.7 Wing beat of a fly with high frequency. In the fly, there is a myogenic control, which means that the frequency of the wing beat is not given by the frequency of the muscle contractions, but by the resonance frequency of the vibration of the thorax. These vibrations are excited by transverse contractions on the body of the fly. Data from Heide, 1968; Pfau, 1986; Suder, 1994.

6.2 Equations of Motion

We have so far seen the basic principles of the physical description of a mechanical system with Newton's laws. We have also seen examples in which we implicitly solved such problems, such as in the treatment of the harmonic oscillator. In the following, we will look at examples of how these principles are applied in several specific cases. We first consider the constant force, which corresponds to the freefall, which we have already seen several times, and second we will consider what happens when we have frictional and normal forces. So, we can for instance also see what happens when we stand on a scale and move. Finally, we consider various types of flow friction, and thus again air resistance, which we have done several times before.

6.2.1 General Considerations

If we start from the action principle and formulated it in the way to describe the acceleration, we obtain an equation for the time dependence of the motion given the forces acting on the body whose motion we want to describe:

$$m\frac{d\vec{v}}{dt} = m\frac{d^2\vec{r}}{dt^2} = \sum_i \vec{F}_i.$$

This version of the action principle is therefore also called the *equation of motion*.

If we are faced with the problem of determining the time-dependent motion $\vec{r}(t)$ for a given set of forces, we must determine the position from the acceleration given by the equation of motion. That is, we must solve the corresponding differential equation. In fact, because the equation of motion describes how vectors relate to each other, we must solve the various equations that represent the components of the vector equation of motion separately. In Cartesian coordinates, they have the following form:

$$m\ddot{x} = \sum_i F_{xi} \quad \left(\ddot{x} \equiv \frac{d^2x}{dt^2}\right) \qquad m\ddot{y} = \sum_i F_{yi} \qquad m\ddot{z} = \sum_i F_{zi}.$$

Sometimes normal and tangential components are more useful, such as if we are treating circular motions with a radius of the circle ρ. Then the equations for the different components are as follows:

$$m\frac{v^2}{\rho} = \sum_i F_{Ni} \qquad m\frac{dv}{dt} = \sum_i F_{Ti}$$

The solving of such equations is mathematical handicraft, with which we do not want to go into too deeply. Most of the equations we will consider here either have an exponential or a harmonic solution, and the essential work of the mathematical solution consists in taking the boundary conditions correctly, i.e., making sure that the initial and final positions and speeds come out correctly. In more complicated cases, the direct analytical solution is not possible and we would have to solve the equations using a computer, as we have seen for the nonlinear population dynamics in Section 3.7. More interesting and important in the description of nature is the setting up of the equation of motion itself for a given situation, which can basically be done by applying Newton's laws.

To do this, we first have to determine all the forces acting on the body to be examined, for example, $\vec{F}_i$. It is useful to do this using an abstracted sketch of the problem in which we draw the acting forces. If several bodies are present, we determine which bodies should be treated separately. Depending on the movement we want to describe, we can also omit certain forces, if they are perpendicular to the direction we want to describe. These do not change the movement in the direction in question.

For this purpose, we first determine which coordinates we want to use for describing the problem. This means we in particular have to determine how many independent directions there are, as this determines the number of coordinates (and hence equations we will have to solve). We decompose the forces into their components with respect to the selected coordinate system, e.g., $F_{iq1}, \ldots, F_{iqf}$, and set up the differential equations of motion in those coordinates, as in the following example:

$$m\vec{a} = \sum_i \vec{F}_i \quad \Rightarrow \quad ma_1 = \sum_i F_{i1}, \ldots, ma_f = \sum_i F_{if}.$$

There is an equation for each component. These must be solved by mathematically integrating twice with respect to time. However, we want to demonstrate the process by means of examples in the next few subsections.

6.2.2 Constant Force

We have already encountered the special case of a constant force several times. If the force is constant, the acceleration is also constant according to the principle of action. This means that we have already solved the equations when dealing with kinematics (Chapter 3), and the results found there are directly the solutions we will be looking for. However, we will study the same case here again, in order to be able to practice the recipe for solutions with an example where we already know the result. To make the situation concrete, we will treat three different cases, all of which correspond to a constant force (see Figure 6.8). These will on the one hand be freefall in the vacuum, which we have already solved when doing kinematics. On the other hand, we will look at the oblique throw (describing the flight of a cannonball) and the migration of ions in an electric field (see Figure 6.9). This latter case is of great importance in electrophoresis, since the separation of the molecules according to their size takes place by a constant electric field. In typical gel electrophoresis, the arrangement of anodes and cathodes is made in such a way that a constant electric field is produced. As we shall see in Chapter 9, an electric field is created between two oppositely charged planar metal plates, which is constant in time and homogeneous in space. Here, we will simply assume that we have such a constant field that then also exerts a constant force on the charged molecules. Thus, although we have very different situations, the mathematical description of the different cases is exactly the same. All we

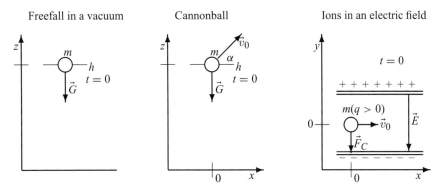

Figure 6.8 The initial situations for the three examples in which an object moves under the influence of a constant force.

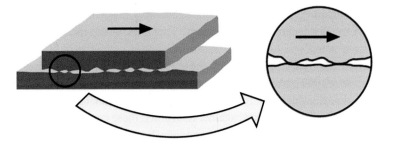

Figure 6.9 The mechanism of sliding friction. The upper surface slides toward the right over the lower surface. In the enlarged section, a region with two potential cold welds is shown.

have to do is interpret the parameters in the equations differently and we can solve all the problems by just solving one of them. This is one of the great strengths of the mathematical description of nature, since the mathematical relationships themselves do not care about what the parameters are, such that they are often transferable from one situation to the other, which makes it possible to solve a large number of problems.

In an oblique throw, we neglect the air resistance against the weight, and in the case of the ions, we neglect the weight against the Coulomb force.

Because the (constant) forces always only pull in a single direction, we obtain the equations of motion as $ma_z = m(d^2z/dt^2) = G_z = -mg$ for freefall and oblique throw, as well as $ma_y = m(d^2y/dt^2) = -qE$ for moving ions in a plate capacitor. In all cases, the solution turns out to be a parabolic time dependence (because of the twofold integration of a constant). The differences in the solution come from the different initial conditions. In freefall, the body is at rest initially at $z = h$, while at the oblique throw it moves into the x and the z direction (with speeds $v_z = v_0 \sin \alpha$ and $v_x = v_0 \cos \alpha$). In the plate capacitor, finally, it moves only in the direction perpendicular to the force (x) with a speed $v_x = v_0$. This means the solutions end up being $z = h - (g/2)t^2$ for freefall $z = h - (g/2)t^2 + v_0t \sin \alpha$; $x = v_0t \cos \alpha$ for the oblique throw; and $y = -(qE/2m)t^2$, and $x = v_0t$ for the plate capacitor.

6.2.3 Surface Forces and Friction

Surface forces microscopically are forces between the individual atoms on the two surfaces (opposite each other). If two very well-polished and cleaned surfaces of the same material are brought together in a very good vacuum, they can no longer be separated; they become welded together into one piece. Even in air, two very well-polished metal surfaces hold together so well that they can only be separated from each other by rotating them relative to each other. Even if the surface atoms are uncharged, the Coulomb forces cause a displacement of the electron shells relative to the nucleus and provide a repulsion at very small distances. At larger distances, a weak and decreasing attraction is obtained. This attraction comes mostly from the van der Waals force between the atoms. This attractive and repulsive interaction is often described by the Lennard–Jones potential, which we will describe more closely in Chapter 9.

On an atomic scale, even polished surfaces look pretty rough, with irregularities of up to a few hundred to a thousand atomic diameters, as can for instance be seen in a scanning tunneling microscope. Surfaces that are exposed to air for longer periods are usually covered by oxide layers. There, the electrons are less mobile, which reduces cold welding.

As Figure 6.9 illustrates, the actual microscopic contact area is typically about a factor of 10,000 smaller than the macroscopic area. The surfaces are connected only at individual points, which are relatively few compared with the total number of atoms. If the two surfaces are displaced against each other, these connections have to break continuously and new ones have to form at other points. For breaking, force is needed, which is usually called friction. In the automobile industry, for example, cylinder pistons were made radioactive by irradiation. After examination of the lubricating oil it was possible to show that the smallest

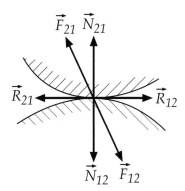

Figure 6.10 Newton's reaction principle enables us to determine that two forces are occurring at the point of contact of two bodies.

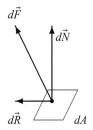

Figure 6.11 If two bodies touch at several points, the forces in one unit of area are combined.

fractions of the metal are transported away from the piston and cylinder during the sliding movement.

Because of Newton's reaction principle, we can in general say that there are two forces occurring at the point of contact of two bodies 1 and 2 – namely, $\vec{F}_{12}$ on the surface of body 2 (exerted by body 1) and $\vec{F}_{21}$ on the surface of body 1 (exerted by body 2), where we know that $\vec{F}_{12} = -\vec{F}_{21}$ (see Figure 6.10). It is customary to decompose these forces into components normal (i.e., perpendicular to) and tangential (i.e., lying within) the contact plane. We call these components *normal force* $\vec{N}$ and *friction* $\vec{R}$. Thus,

$$\vec{F} = \vec{N} + \vec{R},$$

and, because the decomposition is the same for both,

$$\vec{N}_{12} = -\vec{N}_{21} \text{ und } \vec{R}_{12} = -\vec{R}_{21}.$$

If two bodies touch at several points, the forces in one unit of area are combined: in every differential surface element dA, a surface force $d\vec{F}$ is applied. Here again, $d\vec{F} = d\vec{N} + d\vec{R}$. The force per area element $d\vec{F}/dA$ is the *stress* in dA. In particular,

$$\frac{d\vec{N}}{dA} = \vec{\sigma} \quad \text{normal stress}, \qquad\qquad \frac{d\vec{R}}{dA} = \vec{\tau} \quad \text{shear stress}$$

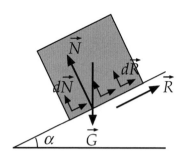

Figure 6.12 Block at rest on an inclined plane.

The surface forces (normal force and friction) are therefore surface forces. They act on every surface element, but not in the interior of the body. (This is in contrast, for example, to the weight representing a volume force.)

Shear and normal stresses play an important role in continuum mechanics (Chapter 7). The behavior of solids, gases, and liquids is described there.

In some cases, distributed surface forces can be combined into a resultant force by integrating them over the connecting surface A.

$$\vec{N} = \int_A d\vec{N} \text{ is the total normal force,} \qquad\qquad \vec{R} = \int_A d\vec{R} \text{ is the total friction.}$$

The effective attack point of this force must be determined by additional conditions. Its position is however irrelevant for translations and only determines rotations.

In the example of a block at rest on an inclined plane (like the otolith in Section 6.1.1; see Figure 6.12), the attack points of $\vec{R}$ and $\vec{N}$ determine whether the block rotates as given by the lever action of the different forces with respect to the center of mass of the block. In order for the block not to fall down, the directions of $\vec{G}, \vec{R}$, and $\vec{N}$ have to meet in a single point. Then the total angular force (or torque) is zero.

It is very complicated and difficult to give a full description of the surface force $\vec{F}$. Since the interatomic force is strongly dependent on the distance, the resulting macroscopic force is very dependent on the surface texture, which in general is rough and complicated. Therefore one can observe qualitative changes in the macroscopic behavior of bodies for changes on the atomic – that is, imperceptible – scale. For example, the interatomic force can change from a repulsion to an attraction. This makes it useful to not try to describe these forces microscopically but macroscopically in the form of materials properties. Here we shall describe the coefficients of friction as such materials properties. In Chapter 7, we will do the same for the elastic properties, and the viscosity in the next section is another such example. Empirically, we find the following relationships for friction on dry surfaces.

Dry Surfaces

The attraction force at the contact surfaces of two dry bodies is generally so weak that it can be neglected. This is due to the fact that microscopically the surfaces are only very close to one another at a few points due to their roughness, as we have mentioned previously.

We will therefore neglect the effects of mutual attraction, i.e., cold welding, and assume the normal force to be repulsive.

On the other hand, we find that there is a relation between the normal force $\vec{N}$ and the friction $\vec{R}$. This is what everyday experience tells us. If we press harder on a body (hence increasing $\vec{N}$), it becomes harder to move (hence $\vec{R}$ has increased). This can be explained microscopically by the increased microscopic contact area when applying a stronger pressure. The stress (i.e., force per area) needed to beak a contact is given by the nature of the material, and hence does not change depending on pressure. However, since the microscopic contact area is increased when applying a pressure, the force necessary to break the connections is increased. This argument also implies that the macroscopic contact area does *not* influence the friction force. In addition, we have to distinguish two different kinds of friction, namely, that of surfaces resting relative to one another (sticktion) and surfaces sliding relative to one another (sliding friction). In this case, fewer contacts can take part in creating the friction force and hence the corresponding coefficient will be smaller. This is illustrated in Figure 6.13.

Analyzing the situation in Figure 6.13 via the action principle gives the following description.

Stiction

As long as the block is at rest on the surface, there is no motion and hence no acceleration, and we have the following:

$$\vec{F}_{tot} = \vec{G} + \vec{F}_a + \vec{N} + \vec{R}_H = 0$$

The weight $\vec{G}$ and the tensile force $\vec{F}_a$ are known. Then what are $\vec{N}$ and $\vec{R}_H$? To introduce perpendicular coordinates (z pointing upward and x to the left), we obtain two equations corresponding to the two components of the preceding equation:

$$F_{z,tot} = N - G = 0 \quad \Rightarrow N = G$$
$$F_{x,tot} = F_a - R_H = 0 \quad \Rightarrow R_H = F_a$$

While here N is constant just as the weight, R_H changes when we change F_a. This must be true until the friction R_H reaches its upper limit, which depends on the normal force N. We find the following:

$$0 \le R_H \le \mu_H N.$$

This upper limit of the friction force is proportional to the normal force. For R_H itself, we only have an inequality. The proportionality factor μ_H is called the coefficient of static friction. The fact that we only have an inequality and no equation between the friction and the normal force is straightforward. If there would be an equation between the forces, pressing on the block would increase N, which would therefore lead to a larger force acting on the block and hence lead to a change in motion according to the principle of action. This is not observed, however, and therefore the normal force gives only a maximum limit in the case of the stiction.

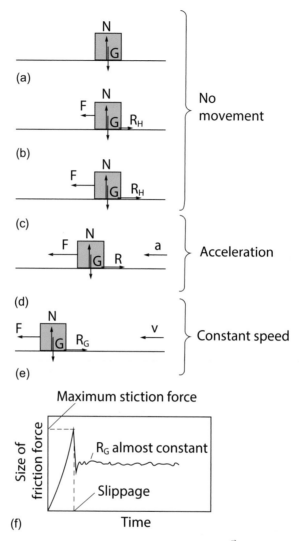

Figure 6.13 Forces on a resting block that is pulled over a table with a variable pulling force. $\vec{G}$ denotes the weight of the block, $\vec{F}$ the pulling (tensile) force, $\vec{R}_H$ the stiction, and $\vec{R}_G$ the sliding friction. While the block rests (a–c), the static friction and the tensile force are compensating each other in the horizontal direction. When the tensile force increases, so does the stiction until it reaches a certain maximum value (d). Then the block starts to move abruptly, i.e., it is accelerated (e). In order for the block to move farther at a constant speed (f), the applied tensile force to overcome the sliding friction is smaller than the maximum value of the stiction. This relation is shown in the bottom graph, where we find the magnitude of the pulling force represented as a function of time.

Sliding Friction

If the externally applied force F_a exceeds the maximum value of the stiction force $R_{H,max}$ in the preceding example, there is a resulting force in the x-direction and the block starts to move, i.e., it is accelerated. After a while, it slides with a constant speed because there is a friction force that is equal to the pulling force. We again expect from experience that this sliding frictional force R_G is directly proportional to the normal force, hence we write the following:

$$R_G = \mu_G N.$$

The sliding friction coefficient μ_G is almost always slightly smaller than the stiction coefficient μ_H for the same surfaces. From the microscopic point of view, this is due to the fact that an arrangement of the surfaces, which allows as many bonds as possible for the given surface, is established during the static friction. In the case of sliding friction, on the other hand, the possible binding sites always move relative to each other, so that an optimal connection cannot be established. In view of what was said at the beginning of the section, it is clear that the material quantities μ_H and μ_G, which only describe the general effects of the interatomic forces, cannot be theoretically determined with high precision since they are depending very critically on the surface shape, roughness, and purity, and this will generally apply to most of the so-called materials properties that we will encounter in Chapter 7 when we look at continuum mechanics.

Nevertheless, such materials properties can be determined experimentally. For instance, the coefficient of static friction (or stiction), one can simply put the object on an inclined plane and determine the angle of inclination, at which it starts to slide. If we consider force balance on the object on the inclined plane, we find for the stiction and the normal force $R_H = G\sin(\alpha)$ and $N = G\cos(\alpha)$ respectively. At the point where the object starts slipping, we can insert the maximum value of stiction into this relationship, i.e., $R_H = \mu_H N$. If we insert this into the preceding relationship and also take into account the relationship between N and G, we obtain $G\sin(\alpha) = \mu_H G\cos(\alpha)$ and therefore have a direct relation between the coefficient of stiction and the angle of inclination of sliding:

$$\mu_H = \tan(\alpha). \tag{6.3}$$

The preceding relations for R_H and R_G imply that the friction does not depend on the size of the macroscopic contact area. Microscopically, this is different and the microscopic contact area does indeed determine the friction force. However, this microscopic contact area is already taken up in the dependence of the friction on the normal force, which depends on the microscopic contact area in the same way. So these descriptions of the friction force are only approximate. For example, they are no longer true in case the surfaces deform under the action of the normal force.

Addendum for Specialists

In the chosen example of the block resting on a support, the normal force and the weight are equal and opposite, but they are *not* reaction partners in the sense of the reaction principle, since the two forces are acting on the same object. The weight is a gravitational force of the earth onto the block, and the corresponding reaction force is the gravitational force with which the block attracts the earth and acts onto the earth at its center. The normal force is a surface force, microscopically of electromagnetic origin, that acts from the support onto the block ($\vec{N} = \vec{N}_{SB}$). Its reaction partner in the sense of the reaction principle is the surface force $\vec{N}_{BS}$

with which the block acts onto the support and also attacks at the support. The mechanism that ensures that the normal component of the gravitational force is compensated by a force of the support on the body consists in an elastic deformation of this support, which, however, is barely measurable. If you act onto the surfaces with an additional force, e.g., by gluing or by putting an iron block onto a magnetic table or vice versa, the normal force can be increased far above the gravitational force felt by the object.

By increasing the microscopic contact area, the adhesion friction can also be increased given the same normal force. This is used in mountain climbing, where people add dry chalk powder to their hands, thus increasing their effective contact area. This is driven to the extreme in the animal kingdom by the gecko, whose fingers contain so-called spatulae, which are nanometer-sized objects, all of which can be so close to even a very smooth surface that the van der Waals attraction comes into play and the friction is large enough to hold the weight of the gecko also on smooth, vertical surfaces.

6.2.4 Viscosity and Friction in Fluids

The situation changes when the space between the two surfaces moving against each other is filled with a liquid. Then the materials property of the viscosity comes to bear at slow rates of displacement.

In this case, not only the friction changes, but also the normal force. Here, the attraction forces between the atoms or molecules are fully effective, since they are now packed as closely as possible, in which case the normal force can also be attractive. We will treat this in more detail when considering surface tension in Section 7.5. Surface tension actually is the property that is mainly responsible for whether a liquid is wet or not, i.e., whether a liquid covers a surface or whether it pearls off. If the liquid is very viscous or solidifies and thus cannot flow away, the bodies adhere to one another; the liquid thus acts as an adhesive. This is why highly viscous liquids such as honey are sticky!

But let's get back to friction. The resistance of a liquid to flow is determined by its viscosity. If a shear stress τ (i.e., a force within a plane; see Section 7.4) is applied to a liquid, it begins to flow. However, the lowest layer of liquid does not flow. It adheres to the confining surface. This results in a change in speed with distance from the bottom, or in other words, a velocity gradient described by $\frac{dv}{dz}$. As it turns out, this velocity gradient is directly proportional to the applied shear stress, and the viscosity η is the constant of proportionality:

$$\vec{\tau} = \eta \frac{d\vec{v}}{dz} \tag{6.4}$$

This defines the unit for viscosity as $[\eta] = \frac{\text{kg}}{\text{ms}} = \text{Pa} \cdot \text{s}$. Microscopically, the viscosity can again be explained by intermolecular interactions. This is illustrated in Figure 6.14. When liquid layers pass along each other, the interatomic bonds must be broken at least briefly between the respective layers. The stronger these bonds, the higher the viscosity.

If a body is moved inside a liquid, the liquid is moved along with the body right next to it, but has to move along the body farther away. Far from the body, the liquid remains at rest

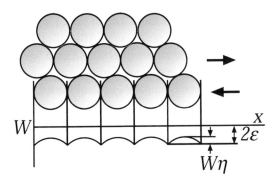

Figure 6.14 Microscopic explanation of viscosity. The atoms or molecules of the liquid must shift in layers with respect to each other when a shear stress is applied. For this purpose, bonds have to be broken, for which a force must be used. The stronger these broken and rejoined atomic bonds, the higher the viscosity of the liquid. In the lower part of the picture, the respective binding energy ϵ, which has to be overcome, is shown. Viscosity is the materials property corresponding to the force, which is necessary for this to happen.

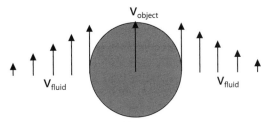

Figure 6.15 Origin of the viscous friction force during movement by a liquid. When a body is moved inside a liquid, a velocity gradient appears in this fluid. The fluid in contact with the body is entrained with the body, thus moving at the same speed, while the fluid far away remains at rest. The force required to uphold this velocity gradient is the viscous friction.

(see Figure 6.15). This results in a velocity gradient in the liquid around the moving body. In order to maintain this gradient, a shear stress is necessary. This can be felt as a flow resistance, as friction, when moving in the liquid, and the corresponding force is called viscous friction. If we are only interested in said force, we can calculate it directly from the shear stress. We obtain $R_V = A\eta \frac{dv}{dz}$, where A is the surface area of the object in line with the applied force. The z-direction is perpendicular to this plane as well as to the velocity of the object. In order to accurately calculate this velocity gradient, one has to expend a lot of effort. We can, however, argue physically that the length at which the flow rate of the liquid changes must be approximately the size of the object perpendicular to the flow direction, i.e., $\frac{dv}{dz} \simeq \frac{v}{r}$. Thus, for a sphere with radius r (i.e., a surface area $A = 4\pi r^2$), which is moved through a fluid with the following velocity v, we obtain the following friction force:

$$R_V = 4\pi r \eta v \tag{6.5}$$

Apart from the prefactor of 4π, this is the same force that we had derived in Section 2.3.1 because of the units involved in the problem. If one does a complete calculation for the gradient, one obtains a slightly different prefactor such that the friction of a sphere becomes

$\vec{R}_V = -6\pi\eta r \vec{v}$. Here we have written the equation in the form of vectors, which is why we had to introduce the minus sign. After all, the friction opposes the motion, therefore it has to point in the opposite direction. As we have already seen, the dependence of the friction force on the velocity will change at high velocities. In that case, the friction is mainly given by overcoming the inertia of the liquid and we have already obtained $R \propto v^2$. The transition between these two regimes is described by the Reynolds number, which is essentially the ratio of these two frictional contributions. We will have a closer look at the Reynolds number in Section 7.17.

Let's return to viscous friction. Using the friction force we have derived previously, we can consider another example for an equation of motion, namely that of a constant external force with friction. This is not only a more realistic form of the description of a falling object, but also a better description of electrophoresis, since the molecules to be separated are typically located in an agarose gel, where friction certainly plays an important role. Let us consider a molecule like DNA. We treat it as a uniformly charged rod and set up its equation of motion in a constant electric field. In that case, the driving, constant force is $\vec{F}_e = q\vec{E}$. The friction force for a rod of length L we can obtain in the same way that we did earlier for the sphere. The surface area of the rod is given by $2\pi L r$, where r is the radius of the rod. The length scale on which the velocity decreases will also be given by the radius if we assume that the rod is pulled in the direction of its length, i.e., with the smallest area facing the fluid. With these assumptions, the friction becomes $\vec{R}_V = -2\pi L\eta \vec{v}$. So we have all of the forces acting on the rod, such that we can set up the equation of motion by setting the sum of the acting forces equal to $m \cdot \vec{a}$. This yields the following:

$$m\frac{d\vec{v}}{dt} = q\vec{E} - 2\pi L\eta\vec{v}. \tag{6.6}$$

Now we place the coordinate system such that we only look at the velocity in the direction of the E-field. If we also say that the rod (i.e., the molecule) is at rest initially (i.e., at time $t = 0$), we can omit the vectors and only look at the motion in the direction of the field. Thus we have an equation in the velocity, which states that the temporal change of the velocity is directly given the velocity itself. The only function that has this property that it is its own derivative is the exponential function, as we have seen before. So, as a solution, we guess that the speed has the form of an exponential given by $v(t) = A + B\exp(-t/\tau)$. Here, A and B are constants that we have to determine from the initial conditions, as we will do later, and τ is the time scale on which a constant speed will set, i.e. where the driving force will be canceled by the friction force. If we insert this ansatz into the equation of motion, where we have to use $\frac{dv}{dt} = -\frac{B}{\tau}\exp(-t/\tau)$ for the acceleration, we obtain the following:

$$-\frac{mB}{\tau}\exp(-t/\tau) = qE - 2\pi L\eta(A + B\exp(-t/\tau)). \tag{6.7}$$

At very long times, the exponential functions approximate zero, so if we wait long enough we get them as close to zero as we want. This allows us to determine A to be $A = \frac{qE}{2\pi L\eta}$. This actually is the speed the rod (or molecule) will have after a long time. The condition that the molecule is at rest initially (i.e., $v = 0$ at $t = 0$) allows a determination of B because at $t = 0$ the exponential is equal to one no matter what the value of τ is. Therefore,

$A + B = v(0) = 0$ or $B = -A = -\frac{qE}{2\pi L\eta}$. Therefore, our guessed solution for the velocity turns out to be the following:

$$v(t) = \frac{qE}{2\pi L\eta}(1 - exp(-t/\tau)).$$ (6.8)

When we insert this into the equation of motion, we obtain the following:

$$-\frac{mB}{\tau}exp(-t/\tau) = -(2\pi L\eta B)exp(-t/\tau),$$

which directly gives us τ. In the preceding equation, the term $B \cdot exp(t/\tau)$ appears on both sides and therefore cancels out, such that we are left with the following:

$$\tau = \frac{m}{2\pi L\eta}.$$ (6.9)

We obtain, therefore, a solution to the problem (Eq. 6.8), which defines a final velocity that is determined by the driving force and that becomes smaller for larger molecules and larger viscosities. In addition, this final velocity is reached in a time τ (Eq. 6.9), which becomes smaller the larger the molecule and the greater the viscosity. That is, in a highly viscous liquid, objects move only as long as an external force acts on them. In such a case, a movement defined by the outside is reversible since a particle reacts only to the external force. Thus, if one moves in one direction and then reverses and makes the same movement in the other direction, one necessarily arrives at the same point. This has important consequences for the movement of bacteria, as we will see in Section 7.17.

If we look at the preceding results in the context of electrophoresis, we conclude that the shorter the molecules are, the higher the final speed of the molecules becomes. Therefore, for a given amount of time, the shortest molecules travel farthest. This is exactly what is observed in electrophoresis. However, we have assumed a constant charge of the molecules. In reality, however, longer molecules are correspondingly more charged. Then the final velocity $v \propto q/L$ is independent of the length of the molecules. However, electrophoresis does work very well in the lab for separating molecules of different sizes. How we can solve this conundrum is discussed in Section 8.5.5, because we first have to understand how long molecules deform due to thermal motion.

6.3 Conservation of Momentum

We will now discuss two alternative approaches to the principle of action. They do not contradict the principle of action, but as we will see, they can lead to a much simpler way to solve some problems. There are problems, thus, that are much easier to solve if one does not set up the whole equation of motion but only considers, for example, the initial and final states. For this purpose, we need two new physical quantities, namely energy and momentum. We deal with momentum in this section and with energy in the next.

To enable us to describe motion's objects, momentum is a property of an object. It is formally defined as the product of the mass and the velocity of the object:

$$\text{momentum } \vec{p} = m\vec{v}$$

Because $\vec{v}$ is a vector, $\vec{p}$ also has to be a vector.

When the mass m is constant, the principle of action says the following:

$$\vec{F} = m\vec{a} = m\frac{d\vec{v}}{dt} = \frac{d(m\vec{v})}{dt} = \frac{d\vec{p}}{dt}$$

Actually, the final part, i.e., that the force corresponds to a change in momentum, also applies if the masses change, and should justifiably be really known as the principle of action. From this more general formulation of the principle of action, it is now possible to directly deduce the conservation of momentum. In general, this is fixed in the set of momenta for a system of particles, if there are no external forces acting on the particles, but only forces acting between the particles. This means that we obtain for the total momentum of all particles, $\vec{P} = \sum_{i=1}^{n} \vec{p_i}$:

$$\frac{d\vec{P}}{dt} = 0 \quad \Rightarrow \quad \vec{P} = \text{const}$$

as long as there are no external forces acting on the system as a whole. This is due to the fact that in this case we always add action and reaction forces from two respective particles that in the sum cancel each other. Therefore, the fact that the total momentum of a system without external forces is constant (or conserved) is a direct consequence of the principle of reaction.

6.3.1 Squids and Rockets

An example where conservation of momentum is important is in the case of rocket propulsion. In this case, the rocket pushes away mass in the exhaust, which carries away momentum from the rocket. Therefore, because momentum is conserved, the momentum of the rocket needs to change accordingly, such that the sum of both remains zero as it had been initially. This propulsion is not just used in rockets, but also squid use the same principle by squirting out water to drive them away from predators quickly (see Figure 6.16). With the help of conservation of momentum, we can find the accelerations reached by a typical squid due to its rocket propulsion. For this purpose, we need a few physiological data, such as the size of the "exhaust" through which the squid squirts out water, the rate at which the water is discharged, and the mass of the squid. When the squid emits water with a speed v, its momentum changes as $dp/dt = vdm/dt$, where dm/dt is the mass change of the squid per time. This change in mass is due to the fact that the squid squirts out water. Therefore, we can calculate this mass change given the speed of the outflow and the size of the opening, as $dm/dt = \rho dV/dt = \rho A dx/dt = \rho A v$, where ρ is the density of water, V is the volume emitted, and A is the area of the opening. We therefore get a recoil force on the squid given by $dp/dt = \rho A v^2$. For a squid, squirting out water with a speed of $v = 15\text{m/s}$ from an opening with a cross-sectional area of $A = 7.5 \text{ cm}^2$, we obtain

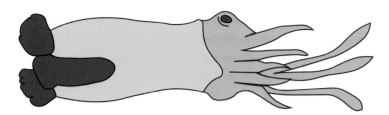

Figure 6.16 A squid moves like a rocket.

Before the collision
Momenta of the two balls:
$$p_{1x} = mv_0 \quad p_{2x} = -mv_0$$
Total momentum:
$$p_x = 0$$

During the collision
Change of momentum:
$$\Delta p_{1x} = 2mv_0 \quad \Delta p_{2x} = -2mv_0$$
Total momentum: $\Delta p_x = 0$
Force on ball 1: $F_{1x}(t)$
Force on ball 2: $F_{2x}(t)$

After the collision
Momenta of the two balls:
$$p_{1x} = -mv_0 \quad p_{2x} = mv_0$$
Total momentum:
$$p_x = 0$$

Figure 6.17 Direct collision of two balls.

a force of $F = 160$N. Assuming a mass of our typical squid of roughly 50 kg, we obtain an acceleration of 3.2 m/s^2, which is about a third of the acceleration due to earth's gravity!

6.3.2 Collisions

Another instance where conservation is useful in the description is the collision of two objects. Typically, it is very difficult to describe what happens exactly during the collision, but if we just want to know how things are before and after the collision, we can limit ourselves to describing the momenta and have a full description. To illustrate this, we consider an idealized example consisting of the elastic collision of two balls (spheres of equal size) that approach each other at the same speed, and whose impact is in the center of the balls. We decompose the time sequence of the entire process of collision into three distinct regions: before the impact and after the impact, where they are moving at equal velocities away from each other; and during the impact, where they are touching and mutually deforming (see Figure 6.17).

If one of the two balls is initially at rest, conservation of momentum can be written mathematically in the following form:

$$m_1 v_1^{before} = m_1 v_1^{after} + m_2 v_2^{after} \tag{6.10}$$

To describe the movements of both bodies after the collision, we need one additional condition. To this end, we consider two special cases of such a collision. First, we consider the case where the two colliding bodies move as one after the impact, i.e., they move with the same speed in the same direction. Then $v_1^{after} = v_2^{after} = v^{after}$, which yields the following for the conservation of momentum:

$$m_1 v_1^{before} = (m_1 + m_2) v^{after} \tag{6.11}$$

In other words, the speed of the two bodies after the collision is given by $v^{after} = \frac{m_1}{m_1 + m_2} v_1^{before}$. This implies that if it is the lighter body that is initially moving, i.e., m_1 is a lot smaller than m_2, the speed afterward is much smaller than before by the same amount as the ratio of the masses. This is because we can neglect m_1 relative to m_2 and we approximatively have $v^{after} \simeq \frac{m_1}{m_2} v_1^{before}$. If the heavier object is moving initially, the situation is reversed and we can neglect m_2 relative to m_1, such that we approximatively get $\frac{m_1}{m_1 + m_2} \simeq 1$, which means that the speed hardly changes. Both of these examples make sense intuitively. A speeding car does not change its speed after collision with a fly. Similarly, if you run into a wall, the wall will not move afterward.

As a second example, let us consider the case in which the first body is at rest after the impact, thus transferring its entire momentum to the other body. This means mathematically that $v_1^{after} = 0$, and in the equation for conservation of momentum, we find the following:

$$m_1 v_1^{before} = m_2 v^{after} \tag{6.12}$$

Thus the speed after collision will be given by $v^{after} = \frac{m_1}{m_2} v_1^{before}$. Thus, when the heavy body was moving initially, the speed of body hit by the incoming ball will be faster than that incoming body has been, again by the ratio of the masses. We will look at such collisions again further later in this chapter, when we have introduced conservation of energy, to see in general how objects behave in elastic collisions.

Another way in which conservation of momentum can be useful in the description of collisions is in the average forces incurred during such a collision. As we have seen, the change of momentum of one particle (in the x-direction) dp_{1x} during the time interval dt is directly given by the time-dependent and deforming force $F_{1x}(t)$ acting during the impact due to the principle of action, as follows:

$$\frac{dp_{1x}}{dt} = F_{1x}(t).$$

Although the detailed time dependence of the momentum or the deformation force $F_{1x}(t)$ is very complicated and maybe not even known, we do know its result, namely, the total momentum change Δp_{1x} during the collision. If we also know something about the impact duration time $t_i \to t_f$, we can use this to obtain the following:

$$\Delta p_{1x} = p_{1x}(t_i) - p_{2x}(t_f) = 2mv_0 = \int_{t_i}^{t_f} \frac{dp_{1x}}{dt} dt = \int_{t_i}^{t_f} F_{1x}(t) dt.$$

The observed change in momentum thus at least allows us to conclude something about the time integral and therefore the average of the acting force. This integral is called the impulse. The unit of the impulse is [Ns], that is, force × time.

In general, the result is an impact, where a (vector) force $\vec{F}$ acts for a short time $\tau = t_f - t_i$:

$$\Delta \vec{p} = \vec{p}_f - \vec{p}_i = \int_0^\tau \vec{F}(t)dt.$$

If the change in momentum is known, we may not know the force $\vec{F}$ itself or how it varies in time, but we do know the average force $\langle \vec{F} \rangle$ acting when the duration of interaction τ is either known or when it can be estimated:

$$\Delta \vec{p} = \langle \vec{F} \rangle \tau.$$

For a given value of $\Delta \vec{p}$, the ensuing force and its corresponding destructive effect can obviously be reduced by intentionally lengthening the duration of the impact. This is well known and used in many different places, ranging from styrofoam packages for fragile parcels, the installation of seat belts and easily deformable crumple zones in automobiles, to the cushioning of a jump by bending the knees.

If, however, $\Delta \vec{p}$ is not given initially but depends on the duration of the interactions, the effects of the impact are small as long as the interaction time is short. This can be seen, for instance, when shooting a gun onto an empty plastic cup. In this situation, the bullet passes right through, simply leaving a hole in the cup, which otherwise stays intact. If we increase the impact duration by filling the cup with water, the change in momentum of the bullet and thus the momentum deposited in the cup increase dramatically, i.e., by much more than the impact time increases. Therefore, there is a much higher force on the cup and it bursts (see Figure 6.18). Another instance of this is the trick of pulling a tablecloth from underneath a set of crockery. In this process, the average force, e.g., the one with which you pull away the table cloth, is given, and therefore the change in momentum is given by the duration of

Figure 6.18 High-speed recording of a gunshot being fired onto an empty (left) and a water-filled cup (right). Due to the different amounts of momentum imparted onto the cup in the two situations due to the difference in interaction time, the force onto the filled cup is vastly greater than that onto the empty cup. Therefore, the filled cup splatters while the empty cup is simply shot through.

the interaction. So for a fast pull of the tablecloth, the change in momentum is small and the crockery remains in place.

6.4 Energy and Its Conservation

In this section, we introduce two new concepts: *work* and *energy*. We will encounter energy in two forms, namely *kinetic* and *potential* energy. We will first define both terms and then show they can help us to gain insights into many aspects of life from them by reformulating the Newtonian principles. In the latter, the central concept was the force. From the knowledge of the force as a function of the time $\vec{F}(t)$, we could basically obtain the entire movement by twofold integration of the motion equation: $\vec{F}(t) \rightarrow \vec{r}(t)$. This can, however, get quite cumbersome for realistic descriptions where several things interact and the force laws are not simple functions. We have found another way of describing the same physics by using the concept of momentum and momentum changes in the previous section. There we did not have to get a full description of the process, but could restrict ourselves to describing specific moments, e.g., before and after a collision, and not have to worry about the exact forces acting during the collision. But still we had to take into account that momenta are vectors and we had to be careful about keeping directions in mind. Work and energy have their role in the same context, but their not being vectors implies that in this case we do not have to care about three-dimensional directions of the process.

The consideration of work and energy also has a deeper background that has consequences outside of pure mechanics. Considerations of work and energy have led to the general principle of energy conservation, stating that energy cannot ever be created or destroyed. This principle is, for example, the basis of thermodynamics, where heat, electrical energy, and chemical energy are recognized as alternative forms of energy that must be taken into account in the conservation of energy, and this is what we will do in Chapter 8. This view of the concept of conservation of energy is of particular importance for biological systems, where energy is usually transferred between the parts of a system, for example, in the exchange with the environment or in organ function, where energy is constantly turned over from one form into another, but always remains constant in the total sum of all its forms (see Figure 6.19).

Conservation of energy is one of the cornerstones of physics. *There is no evidence of any kind that conservation of energy is violated in any system whatsoever.*

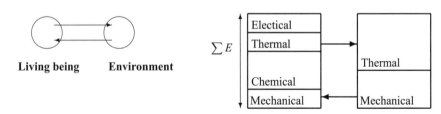

Living being   **Environment**

$\sum E$

| Electical |
| Thermal |
| Chemical |
| Mechanical |

| Thermal |
| Mechanical |

Figure 6.19 The concept of energy conservation.

6.4.1 Work and Kinetic Energy

If there is a force $\vec{F}$ acting on some mass (concentrated in a point) that leads to a movement of this mass by a change in position $d\vec{r}$, we define the work carried out by this force as follows:

$$dW = \vec{F} \cdot d\vec{r} = F\,dr \cos\phi.$$

Note that while both the force $\vec{F}$ and the displacement $d\vec{r}$ are vectors, the work dW does not have a direction. Because of the use of the scalar product between the two different vectors, dW is a scalar. This means that the vectors are projected onto each other, and apart from the absolute values of F and dr, the work also contains $\cos\phi$, where ϕ is the angle between the force and the displacement directions (see Figure 6.20). This has the following consequences on the sign of the work performed:

$$\phi = \frac{\pi}{2} \Rightarrow dW = 0 \qquad \frac{\pi}{2} < \phi \leq \pi \Rightarrow dW < 0 \qquad 0 < \phi < \frac{\pi}{2} \Rightarrow dW > 0$$

This is illustrated in Figure 6.21. When lifting the cat, we do positive work. When we carry it around at the same height, we do not perform work at all, while we "perform" negative work when setting the cat back down.

There are several possibilities that imply that we do not perform work:

(i) the body remains at rest ($d\vec{r} = 0$)
(ii) no force is acting ($\vec{F} = 0$), i.e., the body moves with constant velocity; and
(iii) $d\vec{r}$ is perpendicular to $\vec{F}$, as mentioned previously.

Some examples of this last case are the Lorentz force in electrodynamics:

$$\vec{F} = q\left[\vec{v} \times \vec{B}\right] \quad \vec{F} \perp \vec{v} = \frac{d\vec{r}}{dt} \quad \vec{F} \perp d\vec{r}.$$

The normal force $\vec{N}$ when the support is at rest, and the string tension in the pendulum (see Figure 6.22).

We have defined the work performed via a microscopically small displacement. If we are interested in the total macroscopic work performed, we have to sum up (integrate) the values of dW along the curve of displacement of the body:

$$W_{1 \to 2} = \int_1^2 \vec{F} \cdot d\vec{r}.$$

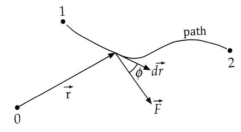

Figure 6.20 Definition of work according to the force exterted along a path.

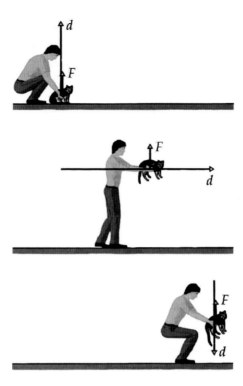

Figure 6.21 Positive and negative work.

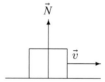

Figure 6.22 The normal force does not perform work.

This is also called the *line integral* of the force $\vec{F}$. The unit of work is the Joule: $[W] =$ Joule $= J =$ Nm. If the force remains constant along the path and is also parallel to that path, we can simply say that work $=$ force $\times$ pathlength:

$$W = \int_1^2 \vec{F} \cdot d\vec{r} = F_x \int_1^2 dx = F \cdot d.$$

Even if we do not perform work macroscopically when we hold an object at a constant height, we still get tired when we do so. Thus, performing work in the macroscopic physical sense does not necessarily mean the same thing as the colloquial expression implies. However, we can make sense of this if we consider that the myosin molecules in our muscles are constantly performing work to create the force that carries the object. In the myosin molecule, the path of contraction is always in the direction of the force, such that it

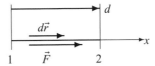

Figure 6.23 If the path and the exerted force are parallel, the work is obtained simply by multiplying the absolute value of the force and the total distance traveled.

is not important whether the force is perpendicular or not, and the thermal motion that we will discuss in Chapter 8 leads to the fact that the actin-myosin bonds must be constantly renewed, such that at least our muscles always carry out work when we hold an object at a constant level or move it perpendicularly to the applied force. Therefore, it makes sense that we are tired after lifting a weight for a long time, even if we strictly speaking have not performed work in this case.

The work performed per unit of time is called the *power* (see Figure 6.23):

$$P = \frac{dW}{dt} = \vec{F} \cdot \frac{d\vec{r}}{dt} = \vec{F} \cdot \vec{v}.$$

Its unit is the Watt $[P] = 1$ Watt $= 1$ Joule s^{-1}. Otherwise put, we can also use the unit of Watt seconds (Ws) or Watt hours (Wh) as a unit of work (or energy).

Next we define *kinetic energy* T of a particle of mass m that moves with a speed v:

$$T = \frac{m}{2}v^2 = \frac{p^2}{2m}.$$

A change in kinetic energy from one state to another then is given by $dT = m\vec{v} \cdot d\vec{v}$. Here we again consider a tiny change in speed dv. But we are interested in conservation of energy and reformulating Newton's laws. How do these definitions help here?

If we take the principle of action for a mass m moving under the influence of a force $\vec{F}$, we have written this as

$$m\frac{d\vec{v}}{dt} = \vec{F}.$$

The work performed by this force on a small path $d\vec{r}$ is as follows:

$$dW = \vec{F} \cdot d\vec{r}.$$

If we use the principle of action in this description of the work performed, we obtain the following:

$$dW = m\frac{d\vec{v}}{dt} \cdot d\vec{r}.$$

Using the definition of velocity, we can rewrite $d\vec{r} = \vec{v} \cdot dt$ and put this into our expression for the work performed:

$$dW = m\frac{d\vec{v}}{dt} \cdot \vec{v}dt = m\vec{v} \cdot d\vec{v} = dT.$$

So the work performed corresponds exactly to the change in kinetic energy that the body has incurred due to the action of that work. We can also sum up all the small changes (i.e., integrate) and obtain a macroscopic change in kinetic energy from a macroscopic work performed:

$$W_{1\to 2} \equiv \int_1^2 \vec{F} \cdot d\vec{r} = \frac{m}{2}v_2^2 - \frac{m}{2}v_1^2 \equiv T_2 - T_1.$$

By thinking in terms of energies, we can often solve mechanical problems more easily than with the full equation of motion using vectors and integration of temporal changes. Since work and kinetic energy are scalar magnitudes, i.e., they do not contain any direction, the considerations are also independent of the specifically chosen path and we do not have to follow all the details that the solution of the equation of motion would give us. We can in energy considerations simply look at starting and end conditions.

6.4.2 Potential Energy: Conservation of Energy in Mechanics

So far, we have encountered the kinetic energy as a quantity associated with the state of motion of the body. If we perform work on a body using a force, this leads to a change in the kinetic energy of the same magnitude. In this section, we introduce the *potential* energy associated with the position of one or more bodies. In connection with conservation of energy, we will also quickly introduce the concept of thermal energy, which is connected with the accidental movement of atoms and molecules in a body. We will deal with this properly in Chapter 8.

If an archer pulls on the string of her bow, then she does the same thing as if we stretch or compress a spring. In the spring, the relative positions of the windings of the spring change, whereas in the bow the relative positions of the ends of the bow to the center change. The windings of the spring resist this change and the result of the work performed by applying the compressive force is an increase in the elastic potential energy of the spring. *Elastic potential energy* is the energy associated with the compression state of a spring (or any other elastically deformable object).

When a weightlifter lifts a dumbbell over his head, he does not increase the kinetic energy of the dumbbell, nor its elastic potential energy (it does not deform due to lifting), but he does increase the distance between the earth and dumbbell, which interact with each other via the gravitational force. His work increases the potential energy of height.

We now formulate these two examples somewhat more quantitatively before we go over to the general definition. In both cases, we are dealing with a system characterized by a single degree of freedom. Things only move in one direction. Our work–energy relation then is as follows (1: beginning, 2: end):

$$W_{1\to 2} = T_2 - T_1 = \int_1^2 F(x)dx.$$

Then the change of *potential energy* U is defined as the negative work performed:

$$U_2 - U_1 \equiv -W_{1\to 2} = -\int_1^2 F(x)dx.$$

For the spring with a force $F(x) = -kx$, we obtain the following:

$$U_2 - U_1 = -\int_1^2 (-kx)dx = \frac{k}{2}(x_2^2 - x_1^2)$$

and for the weight with the force $F(x) = -mg$, we on the other hand obtain the following:

$$U_2 - U_1 = -\int_1^2 (-mg)dx = mg(x_2 - x_1).$$

Potential energy is actually only defined as a difference, i.e., a change. There is no absolute value for any given type of potential energy. In principle, we can add or subtract any constant energy to U, as long as we do this for all points equally. This implies that we can choose at which point we want to have the condition $U = 0$ at will. If we choose $U_1 = 0$ for the spring such that this is the case at position $x_1 = 0$, we can rewrite the potential energy of a spring taking into account that $x_2 \equiv x$ as

$$U(x) = \frac{1}{2}kx^2.$$

If we choose our point of reference for raising the load such that $x_1 = 0, U_1 = 0$, we rename $x_2 \equiv x$ and obtain a potential energy in a constant gravitational field (i.e., close to the surface of the earth):

$$U(x) = mgx.$$

In general, we define the change in potential energy as the negative work performed, as follows:

$$dU = -\vec{F} \cdot d\vec{r} = -dW$$

or as an integral:

$$U_2 - U_1 \equiv -\int_1^2 \vec{F} \cdot d\vec{r}.$$

As such, we can rewrite the work–energy balance equation to show that the change in kinetic energy is given by the negative change in potential energy:

$$T_2 - T_1 = \int_1^2 \vec{F} \cdot d\vec{r} \equiv -(U_2 - U_1).$$

This can again be rewritten to show the following:

$$T_2 + U_2 = T_1 + U_1.$$

Therefore, the sum, $T + U$ always has the same value irrespective of time and space, i.e., it is a constant:

$$T + U = \text{const.} \equiv E_{tot} = \text{total energy.}$$

One can also write this in differential form, which takes into account the changes in energy, where we find that the change in total energy is always zero:

$$dT = \vec{F} \cdot d\vec{r} = -dU.$$

This is the law of *conservation of energy in mechanics.*

$$dT + dU = 0$$

When does this law apply? We have made the energy arguments plausible based on the examples of the spring force and the gravitational force. Obviously, the law applies for these forces, where a reversal of the motion also reverses the whole situation. These are called conservative forces.

When does the law not apply? Let us once again consider a block of mass m sliding on the ground that is brought to a halt by a frictional force. This situation differs from the one we have encountered in weight and spring tension. For there is no way that the block can recover the original kinetic energy by reversing its motion. The reason for this is that energy from the block is turned into thermal energy of the ground and the block. Both heat up. The energy transfer cannot be reversed because the kinetic energy present results in a disordered movement of the individual atoms of the block and ground rather than any kind of potential energy. Sliding friction, like the other friction forces – air resistance, fluid friction, dynamic lift – is not a conservative force.

If we want to include the transformation into heat, we must write the following:

$$dT + dU + dE_{\text{heat}} = 0.$$

Mechanical conservation of energy is only valid for systems with conservative forces.

6.4.3 Collisions

When introducing conservation of momentum, we have shown how we can use this to more easily describe collisions. We have, however, seen there that sometimes conservation of momentum is not sufficient for a complete solution, and we had to make assumptions on the behavior of the final states. Using conservation of energy in addition to conservation of momentum, we can give a more complete description of collisions, which we want to do in this subsection. Such collisions that also conserve energy are called elastic collisions. As we have done previously using momenta, we will consider two balls of masses m_1 and m_2, where ball 2 is initially at rest and ball 1 has an initial velocity $v_{1,before}$. As before, conservation of momentum then implies the following:

$$m_1 \cdot v_{1,before} = m_1 \cdot v_{1,after} + m_2 \cdot v_{2,after}.$$

Assuming a general behavior of the two balls after the collision, we have then been able to describe the motion afterward. But using conservation of energy, i.e., the fact that the collision is elastic, we do not have to assume anything else. Since the balls are not interacting apart from the collision, we only have to look at their kinetic energies for conservation of energy. This means that we have the following:

$$m_1 \cdot v_{1,before}^2/2 = m_1 \cdot v_{1,after}^2/2 + m_2 \cdot v_{2,after}^2/2,$$

which constitutes a second equation describing our problem, such that we can now solve for the speeds of both particles after the collision. Let's look at the speed of ball 2. In order to get at this, we have to express the speed of ball 1 after the collision in the preceding equation by the speed of ball 2 and the incoming speed. For this purpose, we rearrange the equation describing conservation of energy to solve for $v_{1,after}$:

$$m_1 \cdot v_{1,before} - m_2 \cdot v_{2,after} = m_1 \cdot v_{1,after}$$

or

$$v_{1,before} - \frac{m_2}{m_1} v_{2,after} = v_{1,after}.$$

We can then insert this into the equation for conservation of energy, which then gives a relation between the speed of ball 2 after the collision with the two masses and the incoming speed, given by the following:

$$m_1 \cdot v_{1,before}^2/2 = m_1 \cdot \left(v_{1,before} - \frac{m_2}{m_1} v_{2,after} \right)^2 /2 + m_2 \cdot v_{2,after}^2/2.$$

Evaluating the square, we obtain the following:

$$m_1 \cdot v_{1,before}^2/2$$
$$= m_1 \cdot \left(v_{1,before}^2 - 2\frac{m_2}{m_1} v_{1,before} v_{2,after} + \left(\frac{m_2}{m_1} \right)^2 v_{2,after}^2 \right) /2 + m_2 \cdot v_{2,after}^2/2$$

The term with $v_{1,before}^2$ is the same on both sides, such that we can get rid of it and obtain

$$v_{1,before} = \frac{m_2}{m_1} \cdot v_{2,after}/2 + v_{2,after}/2$$

by moving the remaining term containing $v_{1,before}$ to the left-hand side and dividing everything by $m_2 \cdot v_{2,after}$. This can now be solved for $v_{2,after}$ to give the following:

$$v_{2,after} = \frac{2v_{1,before}}{1 + m_2/m_1}.$$

Inserting this into the equation for $v_{1,after}$, one obtains the following:

$$v_{1,after} = v_{1,before} \frac{1 - m_2/m_1}{1 + m_2/m_1}.$$

This shows that the speeds of the two balls basically depend on the ratio of the masses of the two balls, and we can look at three special cases for a more qualitative understanding. If the two masses are the same, we see from the preceding that $v_{1,after} = 0$ and $v_{2,after} = v_{1,before}$, which is a special case that we have already described in Section 6.3.2, where ball 1 transfers all of its momentum to ball 2. If, on the other hand, m_2 is much larger than m_1, we see from the preceding that $v_{1,after} = -v_{1,before}$, meaning that the incoming ball is reflected with the same speed. This is what we would expect for an elastic collision with a wall. Finally, we can look at what happens if m_1 is much larger than m_2. Then we have $v_{1,after} = v_{1,before}$ and $v_{2,after} = 2v_{1,before}$, i.e., the incoming ball continues more or less unimpeded while the small ball is shot away.

6.4.4 Applications

Here we look at three applications: a grasshopper, a spring pendulum, and a myosin stroke.

Grasshopper

Let us look at a model for a grasshopper shown in Figure 6.24. How high can this grasshopper jump? One can solve this problem with the help of the equation of motion, considering all the forces acting while the loaded spring is relaxed and the constant downward acceleration due to gravity. But this is very difficult and complicated and we would have to look at all the details of how quickly the spring is relaxed, etc. On the other hand, we can use an energy consideration, which leads to the same goal much faster and easier. In the legs of the grasshopper, the deformation energy of a spring is

Figure 6.24 The model of a grasshopper discussed in the text. Left: the spring is loaded and the grasshopper is ready to jump. Right: the spring was released and the jump height can be recorded from the meter stick to be about 1.7(1) m.

released, which it releases at the jump and which is converted into potential energy of height. In the initial resting state, we set the potential energy of gravitation to zero, take the jump height as a measure, and begin with zero on the ground. The kinetic energy is also zero because the grasshopper does not move. The energy that is not zero is the potential energy of the loaded spring. This is given by $E_{spring} = kx^2/2$, where x is the distance the spring is compressed. At the maximum level of the jump, the kinetic energy is again zero, because then the grasshopper again does not move (at the maximum jump height, it reverses direction and therefore the velocity is zero). The spring energy is also zero because the spring is now relaxed (actually, this is already the case right after the jump), but now that the grasshopper is at a height h, we have potential energy of height given by $E_{grav} = mgh$. Due to conservation of energy, these two energies must be equal, i.e., $mgh = kx^2/2$. Thus we get the following for the height:

$$h = \frac{kx^2}{2mg},$$

The model shown in Figure 6.24 has the following values for mass and spring: $m = 450(1)$ g, spring constant $k = 550(20)$ N/m, and loading distance $x = 16(1)$ cm. This gives a height of $h = 1.6(2)$ m, which within errors corresponds to what one finds experimentally as shown in Figure 6.24.

Spring Pendulum

In a spring pendulum, potential energy is constantly converted into kinetic energy and back again. The total energy remains constant at all moments of the movement, however, as conservation of energy requires. The potential energy is in the deformation of the spring: $U = kx^2/2$. The kinetic energy is $T = mv^2/2$. That is, the total energy is given by $E = U + T = \frac{k}{2}x(t)^2 + \frac{m}{2}\dot{x}(t)^2$. Since we know the time dependence of the movement, we can show that this total energy remains constant throughout:

$$x = x_0 \cos \omega_0 t \ \text{ mit } \ \omega_0 = \sqrt{\frac{k}{m}} \quad \Rightarrow \quad v_x(t) = -\omega_0 x_0 \sin \omega_0 t$$

$$T = \frac{m}{2}v_x^2 = \frac{m}{2}\omega_0^2 x_0^2 \sin^2 \omega_0 t = \frac{k}{2}x_0^2 \sin^2 \omega_0 t$$

$$U = \frac{k}{2}x^2 = \frac{k}{2}x_0^2 \cos^2 \omega_0 t = \frac{m}{2}\omega_0^2 x_0^2 \cos^2 \omega_0 t$$

$$T + U = \frac{m}{2}\omega_0^2 x_0^2 = \frac{k}{2}x_0^2 \equiv E_0.$$

We could, however, also have proceeded the other way around and have used the fact that the total energy remains constant to show that the motion of the pendulum is given by a harmonic oscillation.

Myosin Stroke

When we looked at the origin of forces in the muscle, we have seen that the movement of a myosin head on an actin filament results in a force of 4–5 pN at each stroke. We had postponed an estimate of this force from the properties of the myosin molecule for later, because we were going to use the concept of energy. Given that we are at the end of the section concerned with energy, we can now deal with this. During the hydrolysis of ATP to adenosine diphosphate (ADP) ADP, which drives the conformational change of the myosin head, chemical energy is released. This energy can be determined experimentally and is about 30 kJ/mole ATP. That is, for a each molecule, an energy of $E_{ATP} = 5 \times 10^{-20}$ J is released. If we now consider the step length of the myosin stroke L_{stroke} as the distance in which a force that consumes this energy is applied, we obtain a force of $F_{myosin} = E_{ATP}/L_{stroke}$. We had already seen that the distance must be between 10 and 15 nm from purely geometric reasons of the conformation of the myosin head. This means we can now give a quantitative estimate of the force we get from converting the chemical energy from ATP to a myosin stroke of somewhere between 3 and 5 pN. That is, the measured force corresponds quite well to a direct conversion of chemical energy in the ATP molecule to mechanical energy in the myosin stroke.

Exercises

6.1   Muscles

(a) Every beat of the head of a myosin molecule exerts a force of about 4.0(4) pN on an actin strand. In human muscles, actin and myosin are arranged in a hexagonal lattice within a sarkomer. Due to the different sizes of the molecules (21 nm for the myosin head and 8 nm for the actin strand), one sarkomer hexagon has a side length of about 30(2) nm. Empirically, one finds that a single sarkomer contains about 200–250 myosin molecules (i.e., $N = 225(22)$). What force can a typical muscle with a cross-sectional area of 5 cm^2 exert?

(b) What is the uncertainty of this estimate?

6.2   Pendulum on a scale

(a) Sketch a spring pendulum on a scale, i.e., a bob (mass $m = 1.00(5)$ kg) suspended vertically on a spring (spring constant $k = 100(5)$ N/m) such that it can oscillate up and down (see Figure 6.25). This contraption is then placed onto a scale. Draw the forces acting on the bob, as well as on the scale. What conditions do you have for the different forces (normal force, force of the spring, gravitational force on bob, etc.)? Consider in particular the force that the scale will indicate as the weight.

(b) Set up the equations of motion of the bob and the scale. What can you conclude for the weight that the scale indicates?

(c) Now assume that the pendulum oscillates harmonically, i.e., that the position of the bob is given by $z(t) = z_0 \cdot \cos(\omega t)$. For a numerical value, assume $z_0 = 10(1)$ cm for the oscillation amplitude. What will be the angular frequency with which the weight oscillates? Calculate the second derivative of the bob's position and

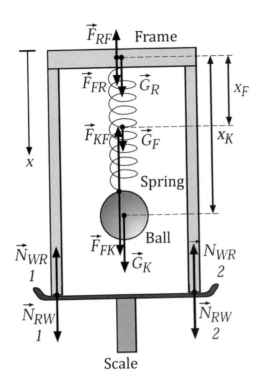

$\vec{F}_{RF}$ Frame

$\vec{F}_{FR}$ $\vec{G}_R$ x_F

$\vec{F}_{KF}$ $\vec{G}_F$ x_K

x

Spring

Ball

$\vec{N}_{WR}$ $\vec{F}_{FK}$ $\vec{N}_{WR}$
1 $\vec{G}_K$ 2

$\vec{N}_{RW}$ $\vec{N}_{RW}$
1 2

Scale

Figure 6.25 Pendulum on a scale.

insert this into the equation of motion. The equation thus obtained can only be solved for a single frequency at all times.

(d) What is the uncertainty of this angular frequency?
(e) What is the amplitude with which the indicated weight oscillates?

6.3   Viscous friction

What is the speed of a falling raindrop with radius r when falling in air? Water has a density of $\rho_{water} = 1,000$ kg/m^3 and the viscosity of air is $\eta_{air} = 1.8(1) \times 10^{-5}$ Pas. The friction force for a sphere in a viscous medium is $F = 6\,\pi\,\eta r v$.

(a) Sketch the forces acting on the raindrop (including the reaction forces) and set up an equation of motion for the raindrop. Solve this equation for the case that the drop falls with constant speed.
(b) Consider two droplets of different radii: $r_1 = 0.45(5)$ mm and $r_2 = 0.045(5)$ mm. Determine the falling speeds and their uncertainties.
(c) Calculate the Reynolds number ($Re = \rho_{air} \cdot r \cdot v/\eta$, $\rho_{air} = 1$ kg/m^3) for the drop with radius r_1. Compare this with the critical Reynolds number for the onset of turbulent flows ($Re_c \simeq 100$).
(d) According to (c), the drop with radius r_1 needs to be treated with friction in turbulent flows. In this case, the friction force is given by $F = 2 \cdot \pi \cdot \rho_{Luft} \cdot r^2 \cdot v^2$. Determine the falling speed in this case for the drop with r_1. Proceed as in (a).

6.4   **Conservation of energy**

Determine the falling time of an object released from a height h. At the moment of release, it is still at rest and thus only has potential energy. When arriving at the bottom of the fall, all of the potential energy has been converted into kinetic energy. From this, you can determine the speed at impact. From this speed and the acceleration acting on the object, you then directly obtain the falling time.

6.5   **Conservation of energy 2**

(a) In a spring pendulum, the frequency depends solely on the spring constant k and the mass m, and the object carries out a harmonic oscillation. Determine the total energy of such a pendulum for all times given the sum of potential energy of the spring and kinetic energy of the bob.

(b) Which parts of the energy in (a) depend on time? If energy is to be conserved, what can you learn from the time-dependent parts of the energy? In particular, look at the prefactor of these terms.

6.6   **Hydro power plant**

In a hydro power plant, water ($\rho = 1,000$ kg/m^3) falls onto a turbine from a storage lake located 200(2) m above and carrying 10.0(1) million m^3 of water, in order to generate electricity. The transported volume through the tube is 400(10) m^3/s. For simplification, we assume that the drop height is constant, i.e., that the level of the lake does not change due to the release of water.

(a) What is the maximum energy in terajoules (TJ) that can be stored in this storage lake? What is the speed of the water just before hitting the turbines, if we neglect friction in the tube?

(b) What are the uncertainties of tis energy and this speed?

(c) Due to friction, the true speed will be about 1% lower than the one determined in (a). How much energy is lost in the tube per second?

(d) What is the electric power generated in, if turbines and generators have an efficiency of 90%? How many kWh can be produced until the upper storage lake is completely empty?

6.7   **Energy and friction**

A block of mass m can slide in a container as shown in Figure 6.26. The side walls are essentially free of friction, but the bottom has a coefficient of sliding friction of $\mu_G = 0.2$ on the length L. The block is released at a height $h = L/2$.

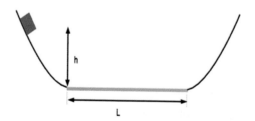

Figure 6.26   A block sliding in a container.

Where does the block come to rest and how many times does it traverse the flat part of the container?

6.8   Elevator

An elevator transports people with a maximum speed of 2 m/s. At full carrying capacity (including people), it has a mass of 1,500 kg. What force does the cable carrying the elevator have to be able to withstand if the elevator needs to be able to stop within a distance of 2 m? Solve the problem using (a) the equation of motion and (b) conservation of energy.

6.9   Bungee jump

A bungee-jumper (mass $m = 100(5)$ kg) is jumping off a platform $100(2)$ m above a river. The elastic tether (considered to be massless) has a rest length of $L = 25(1)$ m and a spring constant of $k = 50(1)$ N/m. Use $g = 10 \text{m/s}^2$ for the earth's acceleration due to gravity.

(a) Which forms of energy appear during the jump? Make a table of all appearing energies for the following situations: (1) before the jump; (2) after a fall of $L = 25$ m, when the tether starts to be extended; and (3) at the lowest point of the jump

(b) What is the speed of the jumper after $L = 25$ m, i.e., when the tether is not yet extended? In other words, what is the highest speed reached by the jumper?

(c) What is the uncertainty of this speed?

(d) What is the lowest height above the river during the jump?

(e) What is the uncertainty of this height?

Quiz Questions

6.1   Forces and friction

A large weight (m = 10 kg) is lying on a table and is connected to a smaller weight (m = 1 kg) via a roll and a string, such that the small weight is suspended from the string next to the table. What is the acceleration of the small weight?

A $a = 0 \text{ m/s}^2$
B $0 \text{ m/s}^2 \le a < 1 \text{ m/s}^2$
C $a = 10 \text{ m/s}^2$
D $a > 1 \text{ m/s}^2$

6.2   Forces

A block is being pulled on a string on a rough surface, such that it moves with constant velocity to the right. The forces shown in Figure 6.27 indicate the direction of the different forces acting on the block, but not their magnitude. What do you know about the magnitudes of these forces?

A $F = k$ and $N = W$
B $F = k$ and $N > W$

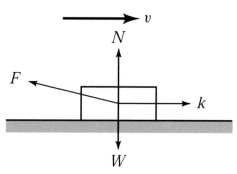

Figure 6.27 A block being pulled on a string on a rough surface.

    C $F > k$ and $N < W$
    D $F > k$ and $N = W$
    E $F < k$ and $N < W$

6.3 Tug of war

A strong man and a small child are engaged in a tug of war (i.e., are pulling on a massless rope at each end). Who exerts more force onto the rope?

    A The man.
    B The child.
    C Both equal.
    D This depends on the surface.

6.4 Reaction forces

A block is at rest on an inclined plane. Which of the following statements concerning the forces and their reaction partners are correct?

    A The normal force is the reaction force of the gravitational force.
    B The normal force is perpendicular to the surface.
    C The sticktion force is pulling the block downward.
    D The gravitational force is pointing toward the center of the earth.
    E The reaction force to the normal force pushes the block upward.
    F The reaction force to the gravitational force is attacking at the center of the earth.

6.5 Friction

Two balls of equal size, one of iron and one of wood, are falling through air. Which experiences the greater friction force?

    A The iron ball.
    B The wooden ball.
    C Both are equal.
    D Cannot be determined.

6.6 Friction 2

You are pushing a wooden crate with constant velocity along the floor. If you turn the crate on its side, such that the contact area with the floor is halved, what happens to the friction force if you push the same crate on the same floor with the same speed?

A It increases by a factor 4.

B It increases by a factor 2.

C It remains the same.

D It decreases by a factor 2.

E It decreases by a factor 4.

6.7 Upward throw

You are throwing a stone having a mass of 1 kg upward with a speed of +5 m/s. What is the force acting on the stone at the top of its curve, i.e., when it reverses its direction and starts falling downward? Do not neglect the action of air resistance and use a value of $g = 10$ m/s^2 for the acceleration due to gravity.

A 0 N

B 10 N

C −5 N

D −10 N

6.8 Energy and momentum

An object ia accelerated, such that its momentum has doubled. What happened to the kinetic energy of said onject?

A It doubles.

B It more than doubles.

C It increases, but less than a factor of two.

D It is halved.

6.9 Conservation of momentum

What can you say about the collision of a bird and an airliner because of conservation of momentum?

A The momentum of the airliner is conserved.

B The total kinetic energy is conserved.

C The change of momentum of the bird divied by the collision time gives the average force on the airliner.

D The momentum of the airliner would be conserved in the absence of friction.

E The total momentum in this situation is zero.

F None of the above.

6.10 Collisions

A moving ball collides with another ball at rest. After the collision, the one initially at rest has a higher momentum than the one arriving had initially. Is this possible?

A No, this violates conservation of momentum.

B Yes, but only if the impacting ball is more massive.

C Yes, but only if the impacting ball is less massive.

Continuum Mechanics

So far, we have looked at what happens when individual particles interact with each other. In this, we have typically neglected what these particles are made of and basically treated them as single points. While many rigid objects can be treated as such very well, we often do have to deal with deformable bodies that interact with each other. Basic examples of this are flowing fluids, which for instance determine how the blood circulatory system distributes nutrients in our bodies. Similarly, the deformation of bones (until they break) or of the cytoskeleton inside cells are others. How we treat this is the task of continuum mechanics, which basically comprises elasticity and fluid mechanics. These two main subcategories reflect the two basic ways a macroscopic body can respond to an external force. The body may deform, but the extent of deformation is limited, and the body returns to the initial shape after the force is released. This is the realm of elasticity. Alternatively, the body keeps deforming as long as there is a force acting, which is the realm of fluid dynamics. Given their qualitatively different response to a force, we will treat these subcategories separately, even though in the course of this treatment we will see that the distinction is not as strict as one may initially think, and we will end the chapter with a description of materials that have characteristics of both liquids and solids.

Basically what we will do is start from the mechanics of individual particles and distribute them in space such that they interact in a prescribed manner. Then we can use Newton's laws to describe these interactions, but only look at a uniform volume filled with particles to describe the extended and deformable bodies. For this, we will need to develop novel concepts, which describe these macroscopic properties of the entire body described by a continuous distribution of particles. These macroscopic descriptions will be based on microscopic properties of the individual interactions of the atoms and molecules that we imagine make up everything. If the motions of these microscopic particles are more or less random, the macroscopic average quantities can have very different properties from the microscopic ones. This is something that we will deal with in Chapter 8, when we look at heat and the laws of thermodynamics, which describe this difference. Here, we are looking at macroscopic bodies that still behave in ways that we can understand from Newton's laws, and we start with looking at solids.

7.1 Elasticity and Materials

If we consider the behavior of many particles, which appear more or less ordered in a solid state, we summarize the microscopic situation (bonds, etc.) in macroscopic quantities, which are the elastic properties of the materials. In doing so, we are primarily concerned with small deformations (elongation, bending, etc.) of the material. Larger deformations of the same type tend to be irreversible: i.e., bonds rearrange or can lead to fracture; i.e., bonds are broken, if the stress is too great. In these cases, we would not be able to describe the properties as those of a solid as mentioned previously, which is why we restrict ourselves to small deformations.

It is the task of elasticity theory to find the connection between the external (macroscopic) forces acting on a body and the internal forces (or stresses) arising from these and their associated deformations. Knowing these connections is of great importance in many fields; building structures must be designed and proportioned in such a way that the maximum permissible stress, as defined by the material, is not exceeded at any point in the interior for a prescribed external stress. At the same time, this typically has to be achieved with as little material as possible. Nature often solves this task optimally, as we have already seen, for example, when looking at the structure of bones and joints in conjunction with muscles (Figure 7.1). In the case of small objects (molecules), the thermal movement constantly exerts forces that have a great influence on the shape and properties of biomolecules and biomaterials, but we will have a closer look at this in the next chapter after we have properly dealt with the concept of heat.

The relationship between the internal forces (or stresses) and the deformations depends on the elastic properties of a body, which we will discuss in the next sections. The connection between external and internal forces is generally complicated. It depends on the geometrical shape of the body and the direction and strength of the external forces. We, however, hope to find a description that leads to materials properties that are independent of the shape of a body and how exactly the body is deformed.

In the case of small loads, the object assumes its original shape once the load has been released. This is the so-called *elastic* range. To give a numerical example, if we have a

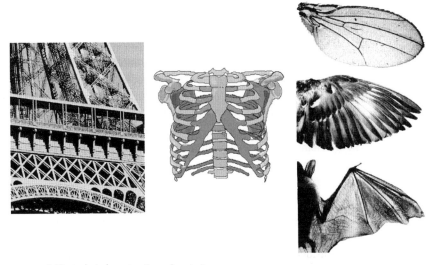

Figure 7.1   Nature as a model for technical construction using struts.

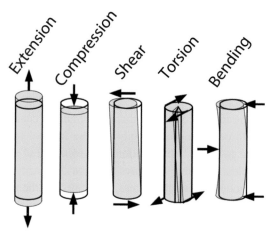

Figure 7.2 The basic types of the deformations of a rigid body: extension (expansion) of a cylinder under tension (left) with concomitant reduction in circumference, shortening under pressure (compression) with enlargement of the circumference, displacement of the cylinder axis (shear) under shear (center), rotation of the shell surface (torsion) under torsional load (center right), and bending (right).

1 m long cylindrical steel rod of 1 cm diameter and we pull on it with the weight of 1 t, the cylinder is extended by 0.5 mm and goes back to its original length when the load is released. If we apply a load of 2 t, it is permanently extended by 1.2 mm. In that case, we speak of *plastic* deformation. At a load of 3 t, the extension is 2 mm, and the rod usually breaks. Extension under *tension* is not the only form of deformation. Figure 7.2 shows that the rod can also deform by *shear*, *torsion*, or *bending* or reduce its volume in all directions by *hydrostatic compression*. In the theory of elasticity, we want to quantitatively describe these properties and define a materials property so that the description does not depend on the specific shape of the rod we are looking at, but depends only on the material itself.

When microscopically examining the structure of a solid body, we will have to more closely consider the attractive, interatomic forces of electrical origin (see Chapter 9). What we will find is that for a pair of atoms there is a certain distance, the equilibrium distance, at which the potential energy of the interatomic interaction has a minimum. This finding can be generalized for many atoms. The mutual interactions force the atoms into fixed positions, leading to regular positions of chains of atoms with high symmetry. With the methods of crystallography, for example with the scattering of X-rays (Section 5.3.5), the positions of the individual atoms can be precisely determined.

In the vicinity of the equilibrium position, i.e., at the distance $r = r_0$, where the position energy is minimal, the curve representing the position energy as a function of the distance has the form of a parabola as in the case of a spring: $U(r) = U(r_0) + c(r - r_0)^2$. Then, of course, there will be a force in response to a deflection, which directs back to the point r_0 and is proportional to the deflection itself, as in the case of a spring $F_r = -dU/dr = -2c(r - r_0)$.

In this approximation, we can actually replace the interatomic forces by such spring forces. This will allow us to get an intuitive understanding of how a body reacts to external

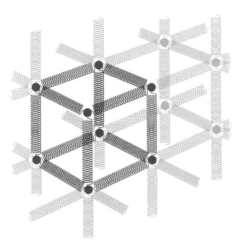

Figure 7.3 Most materials contract in the direction perpendicular to the extension, which can be represented with additional diagonal springs.

stresses. In the case of tensile loading, for instance, an extension occurs in the pulling direction, which disappears again when released.

For most materials, we find that there is also a contraction in the direction perpendicular to the extension. In order to describe this, we have to add additional diagonal springs to the picture (see Figure 7.3). These produce contraction in the direction transverse to the tension. We thus have to know the microscopic structure of the solid to make precise predictions. The deformation depends on the angle between the line of action of the force and the crystal axes.

If there are preferred directions in an object because of the internal structure, these objects are called anisotropic. Think for example of a piece of wood, which has largely different breaking strengths along or across the grain. In the case of an isotropic body, on the other hand, all directions are equal and it does not matter in which direction a force is applied. This can be the case with crystalline substances, if the object consists of many small, arbitrarily oriented crystals. In this case, we can think of the atoms as connected by springs as discussed previously and get an intuitive picture of the different response to external forces. If the springs are overextended, we enter the plastic regime, where the old shape no longer returns. The point at which this happens depends on the material and varies greatly between different types of material. Rubber can be extended by its own length and still return to its original shape, whereas steel or glass deforms plastically already at deformations of much less than a percent of its length. However, as long as we are looking at small enough deformations, all materials are elastic.

7.2 Stress and Strain: Hooke's Law

While thinking microscopically in terms of atoms is useful conceptually to see where the properties of bodies come from, it is not very practical to always go down to the molecular level when looking at everyday forces acting on macroscopic bodies. Thus we will now want to have a description where we do not have to look at forces between individual atoms, and are rather trying to give a macroscopic description, which however is independent of the shape and size of the body. However, the force that gives leads to the deformation of an object does depend on its shape and size. A thick rubber eraser needs much more force to stretch than a thin rubber band, even if they are made from the same material. Thus we need another quantity to describe the force acting on a macroscopic object. For this purpose, let us consider a material that is composed of atoms (or molecules) that are regularly arranged and are connected to the neighbors by springs described by a spring constant. This is a reasonable model for a covalently or ionically bound solid. When we now exert a force F on the body, this force actually acts on single chains of springs. But because these springs are next to each other, all of these chains only incur a fraction of the macroscopic force. In fact, this fraction will depend on the number of such spring chains within the cross-section A, where the force attacks. This number can be obtained from the ration of the cross-section and the area occupied by a single atom, that is, by the square of the interatomic distance a. Each single microscopic spring chain thus senses a force $f = Fa^2/A$. Thus, as a relevant measure of the force, we define the stress given by the forces per unit of area, since this will be constant for all scales, as can be seen by rearranging the preceding relation between microscopic and macroscopic forces $f/a^2 = F/A$. Therefore, for a continuous material, we think of a surface section dA in the interior of the body at a point P, and the stress will be the local force at this point divided by the area dA.

Usually, this local force is separated into the component perpendicular to the cross-section (normal component F_N) and the component within the cross-section (tangential component F_T) $d\vec{F} = d\vec{F}_N + d\vec{F}_T$ (see Figure 7.4). With this separation, we define the following:

$$\text{normalstress}: \quad \sigma = \frac{dF_N}{dA} \qquad \text{shearstress}: \quad \tau = \frac{dF_T}{dA}$$

If $\sigma > 0$, the normal stress is usually called tension, if $\sigma < 0$, it is called pressure.

Note that the pressure in a gas or in a liquid is defined in the same way. However, in such a fluid, the pressure is not dependent on the direction of the cross-section, whereas this may well be the case in a solid.

The tensions within a body are generally not only different from point to point, but also depend on the position of the intersection at each point. They can thus not be defined in one point by only two numbers. If we want to know the complete stress state, we

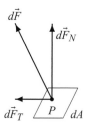

Figure 7.4   A local force separated into the component perpendicular to the cross-section (normal component F_N) and the component within the cross-section (tangential component F_T) $d\vec{F} = d\vec{F}_N + d\vec{F}_T$.

have to specify the normal and shear stresses at every point of the body on the surface elements of all orientations. This is generally a very complex project. We restrict ourselves to describing the stresses by two numbers, so that the stress state does not depend on the cross-section. This is the case in homogeneous or isotropic media. In addition, this description already shows the essential characteristics of a more general theory, which, however, is mathematically complicated, and we do not want to treat this in more detail.

As long as we remain within the linear range, stress and strain are proportional to each other, whereby different constants of proportionality (=elastic constants) can be defined. This is the content of Hooke's law that we will now look at in detail.

7.2.1 Hooke's Law

We look at a slab made from an isotropic, elastic solid with length L_1, L_2, and L_3 in the three different spatial directions (see Figure 7.5). Suppose that we apply a normal stress only in the direction of L_1, i.e., a stress σ_1 (and $\sigma_2 = \sigma_3 = 0$). This stress extends the slab by a distance ΔL_1. In this situation, we define the strain ϵ as the relative change in length $\Delta L/L$ and find that it is proportional to the applied stress σ for small deformations:

$$\epsilon_1 = \frac{\Delta L_1}{L_1} \propto \sigma_1.$$

If we consider the situation again with the microscopic image of atoms connected with springs, we can find a justification for this from the elastic properties of these microscopic springs. We have seen that the chains of springs each feel a force of $f = a^2 \cdot \sigma$. Within the chains, all of the springs feel the same force, because they are connected in series. Each individual spring (with a spring constant k) is thus expanded by a distance of $\Delta x = f/k$. For all springs in a chain taken together, we therefore get for the expansion of the slab $\Delta L = \Delta x L/a$, since there are L/a springs contained in the chain. With $\Delta x = f/k$, this yields $\Delta L/L = f/(ka)$, or when we consider the relationship between the stress and the force on the springs, $\Delta L/L = \epsilon = \sigma \cdot a/k$. The preceding proportionality constant between strain and tension is thus given by a/k, which is only depending on the material as we wanted it to be. This relationship is known under the name of Hooke's law. While we have justified it via the microscopic arrangement and binding of atoms, Hooke's law is, however, generally valid for continuous materials where we simply look at macroscopic stresses and

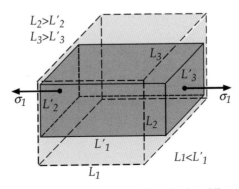

Figure 7.5 A slab made from an isotropic, elastic solid with length L_1; L_2 and L_3 in the three different spatial directions.

strains. So if we have an object of length ℓ, with cross-section A, where a tension force F is applied, there is a normal stress $\sigma = F/A$ and a corresponding strain of

$$\epsilon = \frac{\Delta \ell}{\ell} = \frac{\sigma}{E}.$$

The constant of proportionality $E = k/a$ [Pa $=$ N/m^2] in general is called the elastic modulus or specifically for normal stresses and strain Young's modulus. It depends only on the material (and also its pretreatment), but not on shape and size of the object in question. As we can see from the microscopic justification, its magnitude will basically depend on the binding strength of the atoms in the solid as well as their spacing. This will be important when we look at biological materials as opposed to usual solids. Also, Hooke's law applies only to small strains up to the so-called proportionality limit of the stress σ_p. This corresponds to the region in which the interaction potential between the molecules can be well approximated as a parabola. If it is exceeded, the forces can no longer be described by simple springs, and the linear relationship between stress and strain is lost. This is, e.g., the case for thermal expansion of materials resulting from such effects. At even higher stresses, shifts of the atoms relative to each other can happen. Macroscopically, this corresponds to a plastic deformation, where the deformation remains after the stress is released. That is, after the application of such a high stresses, the material does not return to its starting position, since the atoms have moved microscopically against each other and are therefore bound with other partners. Often, the plasticity limit is associated with the tensile strength σ_B. However, this actually only occurs when the stresses are even higher and the material starts to flow and finally breaks. The extent of these different regions depends greatly on the material (hard – soft, brittle – tough) and its pretreatment (see Table 7.1). Typical metals have only a very small range of elastic strain, but also a large modulus of elasticity, and thus a fairly large area of elastic stress. For polymers and biomaterials, the modulus of elasticity is usually much smaller (the materials are softer), but they also have a larger range of elongation. Many of these materials can be extended to double the length without major problems.

Table 7.1 Elastic constants for different materials. Wood and bones are strongly anisotropic. Therefore, two different values corresponding to the different anisotropic directions are given.

Material	Young's modulus E [10^9 N/m^2]	Yield-stress σ_Y [10^7 N/m^2]	Yield-strain ϵ_Y [%]
Aluminium pure, soft	72	1.3	50
α–Iron	218	10	50
CrV-Steel	212	155	5
Concrete	40	5	
Wood $\parallel$ ($\perp$) to fiber	15 (1.5)	5–20 (0.3–1)	
Compact bone	18 (0.08)	12 (0.22)	
Bone $\parallel$ ($\perp$)	16	8.5 (1)	0.6 (0.2)
Tendons	0.7	6.5 (1.1)	
Human hair	3.6		

For such large elongations, the definition of the strain $\epsilon = \Delta L/L$ has a few problems. The reference length L in that case is comparable to the elongation ΔL, so it is actually unclear whether the proper reference length is the original length or the final length. In all of our considerations, this didn't matter since for small changes ΔL the initial and final lengths are similar and the strain we obtain in both cases will be the same. In order to correctly describe a large strain, what we really have to do is sum up all strains for subsequent small deformations, leading to the large one. Mathematically, this corresponds to integrating over $d\epsilon = dL/L$ between the initial and final lengths. This gives a better definition of a large strain, namely $\epsilon = \ln(L_1/L_2)$. For small changes, this corresponds exactly to the original definition (Taylor expansion), but it also gives a meaningful result for large strains. We should always keep this in the back of our minds, even if we continue with $\epsilon = \Delta L/L$.

7.2.2 Interpreting Stress–Strain Curves

When we consider a stress–strain curve of a real material, we see that it does not always adhere to the ideal of the linear dependence we have seen in Hooke's law. These deviations do, however, give us more information about the properties of the material in question. Consider the (still idealized) curve in Figure 7.6. In this case, we see that the slope of the stress–strain curve initially corresponds to the Young's modulus of the material at small expansions. You can also define an effective modulus if the curve becomes nonlinear or in the plastic regime. These moduli also provide an approximation for the elastic properties, but can only be used in combination with the strain at which it is measured, and the moduli will also depend on the strain state. The end of the curve describes the point at which the material breaks. This can be specified in the required stress (σ_Y) or in the required strain (ϵ_Y). Since a stress–strain curve is never linear until the break point, the two limits do not depend directly on each other, and it may be interesting to know both. Depending on how

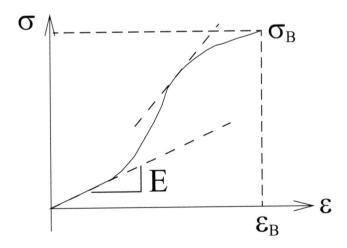

Figure 7.6 Idealized stress–strain curve of a material. Different aspects of such a curve regarding tensile strength, yield-stress, work of expansion, yield-strain, etc., give a characterization of the material in question.

a body is loaded, the force (i.e., the tension) or the strain is given. Think about how the myosin shift makes a force in the muscle. When we look at the area under the stress–strain curve, we get just the work (per volume) we have to perform to pull the material apart. For this, we consider $\int \sigma \, d\epsilon = \int F dL/(AL) = \int F dL/V = W/V$. So when we examine different materials that have different stress–strain curves, we can also be interested in how much work we have to exert for the deformation. A curve that rises steeply in the case of large extensions can be extended with little work but nevertheless be tough when the breaking strength is high. In such a case, the initial slope, i.e., Young's modulus, is relatively small. Rubber is a typical such material, as well as tendons and the material of blood vessels. In Section 8.5, we will look at where these characteristics come from, after we have introduced how heat influences the movement of individual molecules. Metals typically behave just the other way round with a flat stress–strain curve at high strains. This means that in the case of metals, relatively more work of deformation is required for a certain elongation.

7.2.3 *Poisson's Ratio and Compressibility

We have seen that usually, when we extend an elastic body in one direction, there tends to be a simultaneous contraction in the perpendicular direction. This can be clearly seen when pulling on a rubber band. The relative lateral contraction, i.e., the contraction that appears in a direction perpendicular to the acting force, we call q, and find that it is proportional to the strain:

$$q = \frac{|\Delta L_2|}{L_2} = \frac{|\Delta L_3|}{L_3} = -m\epsilon = -m\frac{\sigma}{E}.$$

The constant of proportionality describing the extent of this effect is called the *Poisson's ratio m*. It does not have a unit, i.e., is just a number that depends on the material within the limits of $0.5 \geq m > 0$. This means that the lateral contraction is always at most half of the extension. Why this is, we will see shortly.

For now, let us just see what happens when we compress a cube on all sides instead of just one. Then we have three different normal stresses in the three different directions, σ_1, σ_2, and σ_3. If we now look at the length of the cube in the first direction L_1, it will be extended because of the stress σ_1 by a factor of $(1 + \epsilon_1)$. On the other hand, the stresses σ_2 and σ_3 are perpendicular to L_1 and thus lead to lateral contractions by a factor of $(1 - q_2)$ and $(1 - q_3)$ respectively. Putting all of these changes together, we obtain a new length in the first direction L_1' as follows:

$$L_1' = L_1(1 + \epsilon_1)(1 - q_2)(1 - q_3) \approx L_1(1 + \epsilon_1 - q_2 - q_3).$$

Because the deformations (extensions and contractions) are small ($\epsilon_1, q_2, q_3 \ll 1$), we have left out higher powers (squares and cubes) in ϵ and q in the last step. If we insert Hooke's law and the definition of Poisson's ratio into this relationship, we obtain the following:

$$L_1' = L_1\left(1 + \frac{\sigma_1}{E} - \frac{m}{E}(\sigma_2 + \sigma_3)\right).$$

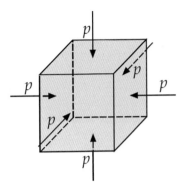

A block compressed from all sides with a uniform hydrostatic pressure p.

In the other directions L'_2 and L'_3, the same will hold when we exchange the indices for the different directions.

If all the (normal) stresses in the different directions are equal, $\sigma = \sigma_1 = \sigma_2 = \sigma_3$, as would for instance be the case in hydrostatic pressure in a fluid (see Figure 7.7), and the stress is compressing, i.e., $-p = \sigma < 0$ (p = pressure), then we can obtain the length change for the three lengths in the different directions $L_i (i = 1, 2, 3)$.

$$L'_i = L_i \left(1 - \frac{p}{E} + \frac{2m}{E} p \right) = L_i \left(1 - \frac{1 - 2m}{E} p \right).$$

The cube of this length gives the new volume, which we therefore can express as a function of the previous volume:

$$V' = L'^3 = L^3 \left(1 - \frac{1 - 2m}{E} p \right)^3 \approx V \left(1 - 3 \frac{1 - 2m}{E} p \right)$$

In the last step, we have assumed that the deformation (p/E) is small, such that we can use a Taylor expansion of the cube, i.e., neglect terms with a square and a cube of this small number. In this case, the relative change in volume becomes

$$\frac{\Delta V}{V} = \frac{V' - V}{V} = -\frac{3}{E}(1 - 2m)p = -\chi p,$$

where χ is called the *compressibility*, i.e., a measure of how well the material can be compressed. If we press on a body from all sides, the volume must decrease, which corresponds to $\chi > 0$ or, according to the preceding equation, $2m < 1, m < 0.5$, as we have already alluded. In case $m = 0.5$, there is no change in volume irrespective of the applied pressure. The compressibility is zero, and any finite pressure will thus always result in a volume change of zero. This is also called in incompressible material (something that only really exists in theory).

7.2.4 *Measuring Elastic Constants

In many solids, the deformations are very small. However, they can be accurately measured with strain gauges. These are fine wire resistors (metals or semiconductors) that can

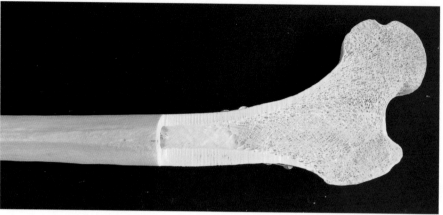

Figure 7.8 The complex, ordered structure of the bone is clearly visible in this frontal section through the proximal part of the human femur (thigh bone). The bone trabeculae in the spongy part of the bone are oriented toward the principal tensile and compressive stresses. In this way, optimum strength is achieved with minimal material expenditure.

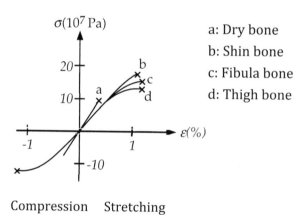

a: Dry bone

b: Shin bone

c: Fibula bone

d: Thigh bone

Figure 7.9 Stress–strain curve for different types of bones. The bone breaks at the strain marked by ×.

be glued or stuck onto a sample and whose resistance is proportional to the strain. Other methods are based on the piezoelectric effect (mechanical deformation → electrical voltage), stress birefringence (the optical refractive index depends on the stress state of the material and the polarization direction of the light and the illumination with polarized light can show stresses as discussed in Section 5.4.3), or the interference of light.

Typical values of different materials and properties are listed in Table 7.1. The ultimate tensile strength, i.e., the yield stress, is the stress at which the material breaks. The permissible stress in engineering is then lowered by a safety margin, which depends on the application. Since wood and bones are strongly anisotropic, two values are given, parallel and perpendicular to the fiber respectively.

From Table 7.1, it can be seen that bones are more elastic than concrete and that the latter breaks first. This allows karatekas to make their spectacular demonstrations of punching through concrete. A fracture can actually occur in response to both tension and compression. For isotropic bodies, the elastic modulus and the tensile strength are the same for all force directions. For most biological materials, this is not the case. These usually have an anisotropic structure. The anisotropy of wood and bone is particularly pronounced. The structure of the thigh bone of a human is shown in Figure 7.8. The stress–strain curve for different bones is shown in Figure 7.9 up to the yield stress (marked by ×).

7.3 Bending a Beam

We have already seen that when we want to study the stability of bones, we must consider their resistance to a bending force. In this way, we had justified the scaling laws of the shape of thigh bones of land animals of various sizes. Likewise, we can look at single, long-chained molecules as beams and ask ourselves what happens to them when they are continually pushed by the thermal motion of the surrounding molecules. This is very important for individual molecules, since they are very small, and therefore the thermal motion can have a great influence, as we shall see later. But for now, let us make a detailed description of the bending of a beam, where we will find the relationships that have led us to explain the scaling of the thigh bones. We will return to this in the next chapter (in Section 8.5) when we consider the flexibility of different biomolecules and how they are directly related to their geometry through bending stiffness.

A bent beam is simultaneously stretched and compressed. As can be seen from Figure 7.10, the top of the beam is stretched while the bottom is compressed. This also means that there is a layer in the beam that is not deformed. This layer is called the neutral fiber. If we want to describe the bending of the beam, we can take this layer as a reference point, which we will do in the following. First, let's consider what energy must be spent to bend a beam by a certain amount. Let us therefore take a bent beam, which may also be bent at different points, as in Figure 7.11, and look at one point of interest, where we want to describe the bending. At that point, we can approximate the shape of the beam by a circle whose radius (the radius of curvature R) is a good measure of the amount of bending. A small radius of curvature corresponds to a large bend, and a large one to a small bend. With

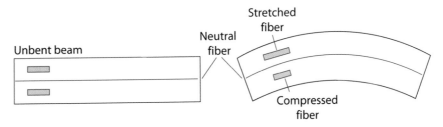

Figure 7.10　When bending a beam, the top is stretched, while the bottom is compressed. Because of the continuity of the deformation, there is therefore a layer, called the neutral fiber, which is not mechanically stressed at all. This neutral fiber is typically in the center, and this is the reason that bones are hollow in the center.

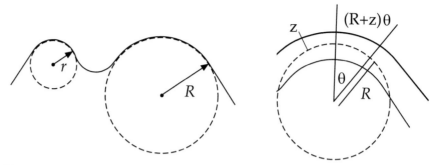

Figure 7.11 Local description of a bent beam by the radius of curvature, i.e., the circle inscribed into the bent structure at the point of interest.

this approximation of the inscribed circle, we can now calculate the strain of the different layers, i.e., the tension of the upper ones, or the compression of the lower ones. As we take the unstrained neutral fiber as our reference, we define the direction perpendicular to the bend, the height z as zero in the neutral fiber and increasing above. Then we know that the change in the length of the rod must be proportional to z, i.e., $\Delta L = az$. The constant of proportionality will have to depend on the bending, that is, on the radius of curvature. The stronger the bending, the stronger the deformation, thus a has to increase with decreasing R. Let's make this more quantitative. From Figure 7.11, we see that any length of the inscribed circle depends on z as $L(z) = (R+z)\theta$, where θ is the angle of the section of the circle that the beam describes. This also applies to the neutral fiber, where the length is just the original length $L(0) = L_0 = R\theta$. This relationship can be used in the preceding relation to eliminate the angle:

$$L(z) = (R+z)\frac{L_0}{R}.$$

In other words, the change in length of the beam as a function of height is given by $\Delta L = L(z) - L_0 = \frac{zL_0}{R}$, which directly gives us the strain of the beam as a function of z as follows:

$$\epsilon = \frac{z}{R}.$$

Now that we know the strain for a bent beam, we can look at the stresses and thus the elastic energy necessary for such a bending. This is what we will do now. From the stress, given by Hooke's law, $\sigma = E\epsilon$, we obtain the force necessary for the tension and compression of the different layers. This is given by $F(z) = \sigma(z) \cdot dA = EdA\epsilon(z)$, where dA is the cross-sectional area of the thin layer considered at height z. We then obtain the energy necessary for such a deformation by multiplying the force with the change in length and integrating over the length of the beam. For a single layer, this change in length is given by $dx = dz\frac{L}{R}$, from the same argument we have used previously. Thus the bending energy for one layer is given by $E_B(z) = \int F(z)dx = \int EdA\epsilon(z)dz\frac{L}{R} = EdA\frac{L}{R^2}\int zdz = EdA\frac{L}{2R^2}z^2$. For the whole beam, we have to sum up (or integrate over) all of the layers making up the beam, such that we obtain for the bending energy of the full bent beam:

$$E_B = \frac{EL}{2R^2}\int z^2 dA = \frac{ELI_z}{2R^2}, \tag{7.1}$$

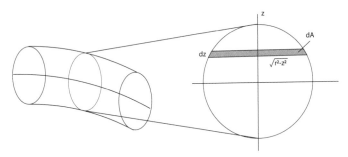

Figure 7.12 The area moment of inertia of a beam is determined by the cross-sectional area at different heights in the direction of bending.

where we have introduced the area moment of inertia, $I_z = \int z^2 dA$, which quantifies the influence of the geometry of the beam's cross-section on the bending energy. This leads to the scaling law for bone sizes, as we will see shortly, and also describes the influence of geometry on the stiffness of biomolecules, as we will see in the next chapter. How the integration in the area moment of inertia is carried out is illustrated in Figure 7.12. We can see how this works in practice for a rectangular beam. Then we obtain the following:

$$I_z = \int_{-h}^{+h} z^2 dA = \int dy \int_{-h}^{h} z^2 dz = \frac{2}{3} Bh^3 = \frac{BH^3}{12} = \frac{AH^2}{12}.$$

This means that the height of the beam enters I_z with the third power! For a round beam, the calculations are a bit more complicated, but we can see just from looking at the units or from the fact that it will be proportional to the cross-sectional area times the height squared that the fourth power of the radius r will be important for a round beam, i.e., $I_z \propto r^4$. A proper calculation gives $I_z = \pi r^4 / 4$. This is exactly the relationship that we needed in the scaling law for bones, i.e., $E_B \propto r^4 L^{-1}$!

The fact that the size of the beam in the direction of the bending is strongly involved in the necessary energy can be understood from the fact that we have to pull the top apart (and compress the bottom) in order to bend the beam. The distance we have to pull (compress) depends on the height of the beam, since we have seen that the strain is directly proportional to z. For the same radius of curvature, we therefore have to pull the material of a high beam much farther apart than with a lower one. This is the deeper significance of the area moment of inertia.

7.4 Flowing and Shear Stress

We had already seen that it can happen that the force acting on some surface does not need to be perpendicular to this surface, even though all of the deformations we had considered so far were for this situation.

In that case, the component within the area was called the shear stress. To see what such shear stresses do, we will look at a model of this, where only shear stresses appear. For this purpose, we look at a cube where there is a shear force F acting within the top surface A of

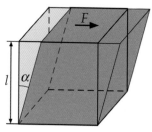

Figure 7.13 Cube with shear force *F* acting within the top surface *A*.

the cube (see Figure 7.13). Thus there is only a shear stress $\tau = F/A$ acting on this cube. This will tilt all of the axes perpendicular to the surface A by an angle α, which is also called the shear strain and which again is proportional to the shear stress τ:

$$\alpha = \frac{1}{G}\tau \quad \text{with} \quad G = \frac{E}{2(1+m)}.$$

Here, G is again a materials property, describing the resistance to shear of the solid, and is called the shear modulus. We will not look into detail at the fact that this shear modulus is connected to Poisson's ratio and Young's modulus and where the previous relation comes from. However, we can justify it somewhat by imagining that instead of applying a shear stress to the edges of a square, we pull on the corners of this square. This will actually lead to the same deformation, but in that case we would describe it as a tensional stress with a corresponding lateral contraction quantified by Young's modulus and Poisson's ratio. If we thus look at the relation between the normal stress on the corner and the shear stress in the edge as well as the normal strain with lateral contraction and the shear strain, we can obtain the previous relation between the shear modulus and Poisson's ratio and Young's modulus.

As we have seen when we introduced viscosity, liquids react very differently to shear stresses compared to solids. We had defined the viscosity by the response of a liquid to a shear stress. In this case, there was no constant shear strain, but a flow, which had a gradient in the speed of the flow. This flow in response to a shear stress is *the* difference between a liquid and a solid. In a liquid, the flow occurs under shear loads even at arbitrarily small stresses. The shear modulus of a liquid is therefore always equal to zero. We will look at this more closely in the final section of this chapter.

7.5 Surface Tension

7.5.1 Adhesion and Cohesion

Cohesion forces are the intermolecular forces between molecules of a single type of material that we have alluded to several times before. There are attractive and repulsive parts of this interaction. The repulsive part is on very short range and basically only acts on direct neighbors. The attractive part in contrast acts on larger distances ranging up to

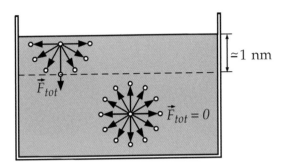

Figure 7.14 Cohesive forces.

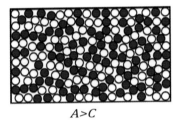

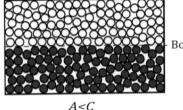

Figure 7.15 Adhesion forces.

roughly 10^{-8} m or 10 nm. In the bulk of a liquid, every molecule is attracted from all sides equally and the resulting force $\vec{F}_{tot}$ is zero (see Figure 7.14). At the surface, however, the situation is different.

A surface molecule has no neighbors outside the liquid and therefore is attracted only by molecules from the inside, which means that there is a resulting force $\vec{F}_{tot}$ toward the inside of the liquid for all molecules located on the surface. These inter-molecular forces toward the inside have the effect of pulling the liquid together and thus lead to an internal pressure p_i, that is ultimately due to the cohesive molecular forces.

Adhesion forces are intermolecular forces between molecules of a different type, i.e., at the boundary of two different bodies (see Figure 7.15). Depending on the kind of molecules involved, the adhesion (A) can be larger than the corresponding cohesion in the bulk of a single material (C), which decides whether two fluids can mix or not:

$A > C$: Two fluids mix. A single fluid will wet a solid body. (Example: vinegar–water.)

$A < C$: Two fluids separate. A single fluid will pearl off a solid. (Example: oil–water.)

If a liquid is in vacuum, there are no adhesion forces. In the absence of any other external forces, the intermolecular distance at the surface will become as small as possible due to cohesion (see Figure 3.16), i.e., the fluid will take the form of a sphere (just like rain drops that are actually spherical, as we have seen in the treatment of the rainbow in Section 5.5.9).

The same result can be obtained from a consideration of the energies involved. Each bond between neighboring molecules provides a negative contribution (the binding energy $-E_B$) to the potential energy. Therefore, the total potential energy is smaller the more such bonds are present. However, there are neighbors missing from the surface molecules, since

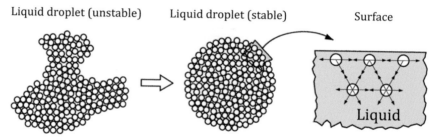

Figure 7.16 Illustration of droplet formation in liquids.

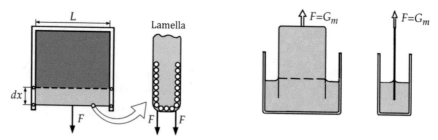

Figure 7.17 How to measure surface tension.

there are no binding partners available, i.e., bonds are missing at the surface, increasing the total potential energy with increasing surface. In order to have a minimum in energy, as few molecules as possible have to sit on the surface, that is, the ratio of surface to volume has to be minimal, which again leads to a spherical shape:

$$\text{Sphere}: \quad V = \frac{4}{3}\pi r^3, A = 4\pi r^2 \quad \Rightarrow \quad \frac{A}{V} = \frac{3}{r} = 3\sqrt[3]{\frac{4\pi}{3V}} = \frac{4.84}{\sqrt[3]{V}}$$

$$\text{Cube}: \quad V = a^3, A = 6a^2 \quad \Rightarrow \quad \frac{A}{V} = \frac{6}{a} = \frac{6}{\sqrt[3]{V}} > \frac{4.84}{\sqrt[3]{V}}$$

If we wish to increase the surface area, we have to add energy to the liquid. This can be demonstrated quantitatively by dunking a wire bracket into soapy water, such that a lamella is formed (Figure 7.17). If the bracket is attached to a scale, this directly shows the force present due to the lamella and hence due to the surface tension. As the surface is increased by the amount $dA = 2Ldx$, the force F due to the increased surface leads to a work performed as $dW = 2Fdx$. The ratio of the applied energy (work done) to surface enlargement is called the *surface tension* γ:

$$\gamma = \frac{dW}{dA} = \frac{2Fdx}{2Ldx} = \frac{F}{L}$$

$$\gamma = \frac{dE}{dA} \qquad \left[\frac{\text{energy}}{\text{area}}\right] = \left[\frac{\text{force}}{\text{length}}\right]$$

Of course, fluids are typically not in contact with a vacuum, but with another medium. If this other medium, however, is a gas, it is sufficiently dilute that the same considerations

Table 7.2 Surface tensions of different fluids.		
Fluid	Temperature [°C]	γ [N/m]
Water	18	0.073
Soapy water	20	0.030
C_2H_5OH (alcohol)	20	0.0223
Hg (mercury)	15	0.407
Oil	20	0.032
Ether	20	0.017

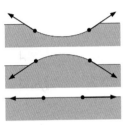

Figure 7.18 Different effects of surface tension.

apply as for a vacuum, at least approximately. Typical values for γ of liquid surfaces in contact with air are listed in Table 7.2. Surface tension strongly depends on temperature, and it can be greatly reduced for water by the addition of surfactants.

On a flat surface, the resulting surface tension is zero. On the other hand, a convex surface feels an inward force, a concave surface one to the outside. In both cases, the disturbed surface is forced in the direction of returning to a flat shape. This restoring force leads to oscillations on a water surface and thus results in the propagation of surface waves on a lake (see Figure 7.18).

7.5.2 Laplace Pressure

For a curved surface, we have seen that surface tension leads to a restoring force toward a flat surface. If a surface is constantly curved, there must therefore be an additional pressure counteracting this restoring force. This is, for instance, the case in a soap bubble (see Figure 7.19). Inside the bubble, there is an overpressure that in equilibrium is equal and opposite to the inward pressure due to surface tension. Via the work necessary to increase the surface and the work due to pressure, we can obtain what this overpressure inside a bubble corresponds to given its radius. In the calculation, there will be an extra factor of two, because in a soap bubble, we actually have two surfaces to consider, one on the inside of the bubble and one on the outside:

$$r \to r + dr \quad \Rightarrow \quad dA = 16\pi r\, dr, \; dV = 4\pi r^2 dr$$

$$dW = \Delta p\, dV = \gamma\, dA \quad \Rightarrow \Delta p = p_i - p_a = \frac{4\gamma}{r}$$

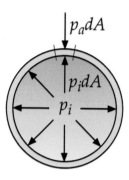

Figure 7.19 Figure 7.19 The pressure distribution for a soap bubble.

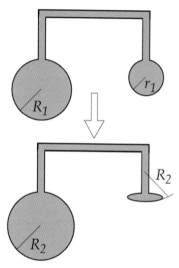

Figure 7.20 A small bubble has a larger Laplace pressure than a large bubble.

Thus the overpressure, called Laplace pressure, increases with *decreasing* radius of the soap bubble. A small bubble will have a larger pressure than a large one! This makes sense based on the previous argument, since the more curved the surface is, the larger the pressure due to surface tension (see Figure 7.20). Think of what happens when you push against a rubber membrane. The more you push, the more it is curved. This can have counterintuitive consequences. If you connect two soap bubbles of different size, they will equalize the pressures inside them. This, however, means that air will actually be flowing from the smaller bubble (larger pressure) to the larger one (smaller pressure), thus increasing the difference in size.

Hence, if two bubbles are connected, the smaller one will shrink indefinitely, or at least as long as it is still a bubble and not a covering meniscus with a radius of curvature equal to the radius of the large bubble. This has important consequences for the stability of our lungs, as we will discuss shortly.

To be general, the surfaces considered for Laplace pressure do not have be of spherical shape, but can be any type of curved shape (see Figure 7.21). However, as we discussed

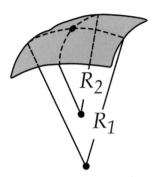

Surfaces conssidered for Laplace pressure can be any type of curved shape.

previously in the bending of a beam, we can describe the curvature at any point P by the radius of the inscribed circle that captures the bent curve. For a surface, we have to do this in two different (orthogonal) directions, such that a curved surface is described by two radii of curvature rather than one. In this case, we can still calculate the corresponding Laplace pressure and obtain the following:

$$\Delta p = 2\gamma \left(\frac{1}{R_1} + \frac{1}{R_2} \right)$$

If $R_1 = R_2 = r$, we have a spherical surface and therefore obtain the same results as for the soap bubble. If we are looking at a tube as our curved surface, which we would have to consider, for instance, for describing the expansion of a blood vessel, one radius of curvature (in the direction along the tube) would effectively be infinite, whereas the other is the radius of the tube itself.

If we do not have a closed surface, but rather a soap film bounded by a frame, the pressure on both sides of the film must be equal, i.e., $\Delta p = 0$. This directly leads to a description of the curvature of this soap film, which either has to be completely flat, described by two infinite radii of curvature ($R_1 = R_2 = \infty$), or one that describes a saddle, where the two different directions are oppositely curved, i.e., $R_1 = -R_2$.

7.5.3 Aneurisms and Collapsing Lungs

The Laplace pressure, which we have just illustrated with a soap bubble, is also of great physiological importance. A natural example of the demonstration of the two soap bubbles we have discussed can be observed in the lungs. In the lungs, there are many small vesicles, the alveoli, at the end of the air tubes, which give the lung as large a surface area as possible to ensure an efficient transport of oxygen into the blood (see Figure 7.22). These alveoli are essentially small soap bubbles. The size of the alveoli can be directly obtained from their surface tension and the pressure in the lungs, which is a problem set in this chapter.

If there are small fluctuations in the size of the alveoli and one of these increases relative to its neighbors, we would expect the same instability to occur that we have observed for the soap bubbles. This means that this one alveolus grows incessantly and the rest of the lung collapses. However, we do not usually observe this. Our lungs are relatively stable even in

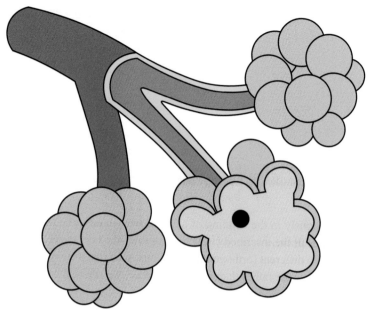

The large surface area of alveoli in the lungs is responsible for efficient gas exchange.

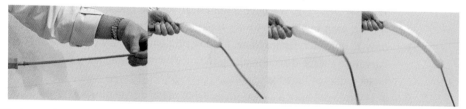

Blowing up a long balloon. Even though the pressure is the same everywhere, the balloon is inflated in the front and uninflated in the back.

the presence of fluctuations. This is, however, only the case because the size of the alveoli is actively regulated by means of a surfactant whose production is actively controlled by a corresponding gland. Thus the fluctuations in size are always actively compensated for by adapting the surface tension of the different individual alveoli. If this mechanism is broken or not yet fully developed, there is a great danger of collapsing lungs due to the aforementioned instability. In the case of premature births, this can actually happen and is known as neonatal respiratory distress syndrome.

If, as a second example, we consider a blood vessel to be an elastic cylinder, the tensile stress in the cylinder is connected with the size of the blood vessel by virtue of Laplace pressure. Basically, we assume the pressure of the fluid to be constant within the tube of the blood vessel, but if there is a pulse of blood, the pressure increases. To model this, let's inflate a long balloon, like those used to create animal shapes at children's birthday parties (see Figure 7.23). If we inflate such a long balloon, it will be blown up very inhomogeneously. The end near the mouth is blown up completely and almost bursts, while the other end is not inflated at all and seems rather limp. If we keep inflating it, the blown-up region simply increases in length, but how much the blown-up region is inflated stays constant. One thing to consider here is that at a given moment, the pressure within the balloon (or within the blood vessel) is the same. There are no walls within and the gas molecules can move around freely. So how can it be that there are two distinct regions of inflation of the balloon? If we rewrite the Laplace pressure for the cylinder ($R_2 = \infty$) to obtain a description of the tensile stress in the walls, we obtain $\sigma = \Delta p \cdot R$, were σ is the tension in the balloon. Therefore, when the pressure is constant, the tension in the walls is actually proportional to the radius of the tube, and we see that where the balloon is under tension, it must also be larger. Thus we see that there is a connection between how big the balloon is and whether it is limp or stiff (the tension). But there could still be many different sizes along the length of the balloon.

To understand where the distinct regions come from, we must first understand what would happen if the balloon were made up of a linearly elastic material. Because the tensile stress is proportional to the radius, a linearly elastic behavior would lead to a disastrous instability. If we imagine that the balloon is inflated somewhat (i.e., has a larger radius) at one point, then there must be greater tension at this point. Thus for an ideal, linearly elastic material, this results in an expansion of the circumference proportional to this tension. Therefore, the balloon will become slightly bigger, which again leads to an increase in tension, which leads to an increase in size, etc. The situation is completely unstable and the balloon would burst almost immediately. However, this does not happen in real life, so there must be more to this. The rubber of the balloon is in fact not linearly elastic, but becomes increasingly difficult to deform with increasing elongation. In other words, the stress–strain curve of rubber is curved upward. In this case, the instability of the bursting balloon only happens up to a certain limit at which point the linear increase in tension due to Laplace pressure is no longer sufficient to lead to an extension of the rubber of the balloon. This is the size up to which the inflated part of the balloon is actually inflated. The rest of the balloon is not under tension and therefore is small and limp. The boundary between the two regions is simply set by the amount of air that has been blown into the balloon, and because of the instability, there are no regions of intermediate inflation and the balloon is either fully inflated or not at all.

The same thing happens in blood vessels. If there is a pulse of fluid or increased pressure, there would be an instability in the blood vessel radius up to the bursting of the vessel. If this happens, this is called an aneurism, and if this happens in the aorta, this is usually deadly. But again, this is rare and only happens in elderly people (actually, Einstein died from this), because the material of the blood vessels have a nonlinear stress–strain curve. This is basically due to the fact that there are two main components with different elasticity making up the blood vessel, the tougher of which only comes into play at larger elongations. With age, this material becomes weaker and therefore blood vessels become more linearly elastic over time, thus leading to an increased risk of aneurisms with age (see Figure 7.24).

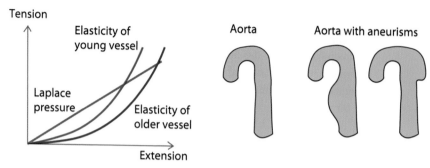

Figure 7.24 Aneurism and its connection to the stress–strain curve of blood vessels.

7.6 Fluids

In the theory of elasticity, we saw that applying a force to something can deform it, but in that case the deformation remained the same, as long as the force did not change. However, we can also observe flows when we apply forces to substances, such that the body changes continuously as long as there is an applied force. In that case, we speak of a fluid and with such deformations, it is actually possible to transport material. An important example of such transport that we will address here is the transport of nutrients through the bloodstream. We will consider how to describe the transport and what requirements the blood vessel system must meet in order to ensure the most efficient transport possible. This is a very directed form of transport that is common on macroscopic scales. In small environments, undirected transport, e.g., by diffusion, is also very common, something that we will deal with later when we cover thermal motion.

To describe how fluids flow, we first have to know what fluids actually are. We have already alluded to this in the theory of elasticity, where we have seen that gases and liquids (both are subsumed in the term fluids) have no shear modulus. This means that no shear stresses are acting ($\tau = 0$) if there are no flows and the fluid is in equilibrium macroscopically.

While in the case of solids the molecules are bound to their equilibrium positions by intermolecular forces (and perform vibrations due to thermal motion around those positions), the molecules of liquids and gases are in a random, unbounded motion. Their average kinetic energy is greater than their binding energy.

The difference between a liquid and a gas is only due to the extent of the intermolecular forces. In liquids, the molecules are tightly packed so that cohesive forces occur. The liquid thus forms drops with a defined free surface and the compressibility is small. Gases, on the other hand, do not form drops, but occupy the entire volume available to them. The compressibility is generally large.

7.6.1 The Equation of Continuity

If we have a macroscopic motion of a group of particles, or an ordered drift, we speak of a flow or a current. This is connected with the macroscopic transport of a quantity, as we shall see again from another point of view in Section 8.3 on general transport processes. Here, we shall only look at the properties of such macroscopic currents in some detail. In general, the flow is illustrated by streamlines (Figure 7.25). The streamlines are the average paths of the many individual particles making up the fluid or the path of a test body that is carried along by the flow. The drift speed is thus tangential to (i.e., along) the current lines.

Mathematically, the velocity vectors of a flow form a vector field $\vec{v}(\vec{r}, t)$, which is to say that they are a collection of vectors that vary in space (and in time). The streamlines are the field lines of this field. We will deal more closely with fields in Chapter 9, when we are studying electric phenomena.

If the streamlines are constant in time, the flow is stationary, which is to say that at a certain point $\vec{r}$, the flow velocity $\vec{v}(\vec{r})$ does not depend on time. This will make many forms of description mathematically much simpler. However, just because the flow is stationary, this does not mean that the velocity of a particle, which moves along with the flow, is constant in time or stationary. The co-moving particle will change its place as time goes on and therefore see different parts of the stationary flow field. Thus the co-moving velocity does not need to be constant in time.

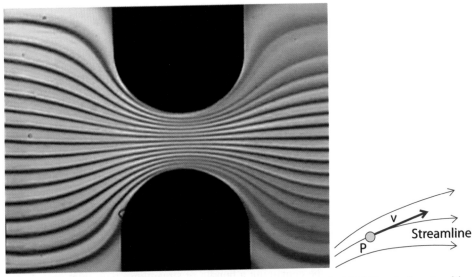

Figure 7.25 Image of a fluid flow through a constriction, which was visualized by coloring the fluid (left). The velocity vector of the flowing fluid is always parallel to the streamline (right).

The speed is the important parameter we look a in a flow field, because the flow we want to describe is ultimately described mostly by the speed. To see that this is the case, let us first quantitatively define the current or mass flux as the temporal change of the mass in one place. Then we obtain the following:

$$I = \dot{Q} = \frac{dm}{dt} = \rho \frac{dV}{dt} = \rho A \frac{dx}{dt} = \rho A v,$$

where A is the cross-sectional area of the current we are looking at and ρ is the (constant) density of the fluid that is flowing. This means that the current, or rather the *current density*, i.e., the current per cross-sectional area

$$I/A = \vec{j} = \rho \vec{v}$$

is directly proportional to the velocity of the flow.

If a flow shows *smooth* streamlines, i.e., when the streamlines do not fold back on themselves, the flow is called *laminar*. If the stream lines are curly, the flow is called turbulent. For any given real gas or liquid, there will always be a transition from laminar to turbulent flow (through or around some obstacle), when a critical velocity v_c is exceeded. Generally, this transition is quantified by the Reynolds number we have seen before, which describes the relative strengths of viscous friction and the friction caused by overcoming of the inertia of the fluid. The continued flows due to inertia are then leading to turbulent eddies. Apart from the speed, the Reynolds number will depend on the size of the obstacle as well as the viscosity and density of the fluid. Therefore, most fluid flows in nature are actually laminar, because we are dealing with fluids of a density and viscosity close to that of water and tube sizes of a centimeter or smaller. A notable exception of this is flying, where we are dealing with air as the fluid, which has a much smaller viscosities and therefore inherently larger Reynolds numbers. In the following, we will thus mainly focus on laminar flows.

Equation of Continuity for Stationary Flows

As we have seen, the current density can also be seen as the number of streamlines per unit area. If this increases, this also corresponds to a higher velocity, because the current density is directly proportional to the velocity. Figures 7.25 and 7.26 illustrate this in the constriction of a river. The physical basis of this observation is the conservation of the total flux: the total amount of water flowing in per unit time must equal to the amount flowing out during that same unit of time. There is neither a source nor a drain anywhere in the problem. Therefore, the flow rate must increase for a smaller cross-section. The mathematical formulation of this fact is the equation of continuity, which we will now derive.

For the concrete mathematical description, we consider a flow tube within a stationary flow, i.e., a tube containing a bundle of streamlines with inlet area A_1 and outlet area A_2 (see Figure 7.27). During the time dt, there are fluid flows through A_1 and A_2 corresponding to the amounts

$$I_1 dt = \rho_1 v_1 A_1 dt \quad \text{and} \quad I_2 dt = \rho_2 v_2 A_2 dt.$$

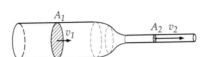

Figure 7.26 In a narrow section under the bridge, the stream-lines converge (right). This also implies a higher flow speed, as long as the river is the same depth everywhere, and the distribution of the streamlines on the river's surface directly describes the process. The same picture and the same effect can be seen when water is flowing through a constriction in a tube (left).

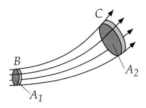

Figure 7.27 A flow tube within a stationary flow.

Since the amount of water remains the same (no sources or drains), the influx at A_1 has to be the same as the outflux at A_2. Thus we have to have $I_1 = I_2$. So for any bundle of streamlines, we have the following:

$$\text{Equation of continuity: } \rho_1 v_1 A_1 = \rho_2 v_2 A_2.$$

In liquids, the density does not change because of a flow, and therefore ρ can be taken as constant, i.e., $\rho = \rho_1 = \rho_2$, which yields a simplified form of the equation of continuity that corresponds to the experience that we have illustrated in the preceding:

$$v_1 A_1 = v_2 A_2.$$

The flux of the $\vec{v}$-field along a flow tube is constant. This is a somewhat special case where the flow is perpendicular to the in- and outflowing areas. For this case, the equation of continuity states the following:

The velocity along a stationary laminar flow is inversely proportional to the tube cross-section.

In general, this will be different and flux can be diverted to different directions. If we want to take this into account, we can consider another (but still special) case, where the area we consider is closed, i.e., the area completely covers a volume. In that case, the total flux (influx minus outflux over the entire area) has to be zero. Again, this is because there are no sources or sinks inside, so fluid cannot appear out of thin air or disappear into nirvana.

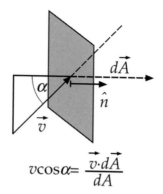

$$v\cos\alpha = \frac{\vec{v}\cdot d\vec{A}}{dA}$$

Figure 7.28 Defining flux.

To be more quantitative, we define the flux through an area A of arbitrary orientation relative to the flow (with $d\vec{A} \equiv \hat{n}dA$):

$$\Phi = \int_A \vec{v} \cdot d\vec{A} \equiv \int_A (\vec{v} \cdot \hat{n})dA = \int_A v_n dA = \int_A v\cos\alpha\, dA$$

Here, the unit vector $\hat{n}$ is perpendicular to the surface at the point considered, dA, thus giving the orientation of the surface, and v_n is the corresponding component of the velocity $\vec{v}$ perpendicular (normal) to the surface (see Figure 7.28).

This definition of the flux (Φ) includes the possibility that the velocity is not equal on all points of the surface A, as well as that this surface does not need to be perpendicular to the streamlines. For $A \parallel \vec{v}$ ($\alpha = \pi/2$), the flux actually is zero, whereas it is maximal for $A \perp \vec{v}$ ($\alpha = 0$). With this definition of the flux, we can now look at a more general version of the equation of continuity, where we do not have to restrict ourselves to stationary flows and incompressible fluids.

For this purpose, let us consider a certain volume V, where we want to look at in- and outflows. So we will have a closed surface A_V that represents the boundary of this volume, and this is the area over which we will have to calculate the total flux Φ. As long as there are no sinks and sources in the volume we consider, we know that mass is conserved and therefore whatever flows in must flow out again. The total flux therefore must be zero:

$$\text{Equation of continuity}: \quad \Phi = \oint_{A_V} \rho\,\vec{v} \cdot d\vec{A} = \oint_{A_V} \rho\, v_n dA = 0.$$

This version of the equation of continuity is valid for all shapes and volumes and also in case the density is varying spatially.

If the density changes with time, i.e., the material is compressible, we will have to go even further. In that case, we can even treat the case that we have sources or sinks present.

For this purpose, we consider a small volume given by $dV = dxdydz$. The mass current flowing into this volume in the x-direction is then given by the following:

$$I_1 = \rho v_x(x)dydz.$$

Similarly, we can get the mass current flowing out of this same volume:

$$I_2 = \rho v_x(x + dx)dydz.$$

This gives the difference between in- and outflow in the x-direction dI_x:

$$dI_x = I_1 - I_2 = \rho \underbrace{(v_x(x) - v_x(x + dx))}_{\Delta v_x = \frac{\partial v_x}{\partial x} dx} dydz = \rho \frac{\partial v_x}{\partial x} dxdydz = \rho dV \frac{\partial v_x}{\partial x},$$

where dV is exactly the volume that we are actually considering. To get the total flux through the entire boundary of the volume dV, we have to look at the y- and z-directions as well, i.e., dI_y and dI_z and add up all three terms:

$$dI_{dV} = \rho \left(\frac{\partial v_x}{\partial x} + \frac{\partial v_y}{\partial y} + \frac{\partial v_z}{\partial z} \right) dV.$$

For an incompressible fluid without sources or sinks in the volume, this flux would be zero, or otherwise put:

$$\left(\frac{\partial v_x}{\partial x} + \frac{\partial v_y}{\partial y} + \frac{\partial v_z}{\partial z} \right) = div(\vec{v}) = 0.$$

Here, we have introduced the property of divergence $div(\vec{v}) = \frac{\partial v_x}{\partial x} + \frac{\partial v_y}{\partial y} + \frac{\partial v_z}{\partial z}$, which quantifies the geometry of the streamlines. If the density can change, this can lead to a change in the flux in the volume and $dI_{dV} \neq 0$. Because the volume we look at does not change, we can directly see that $dI_{dV} = -\frac{\partial \rho}{\partial t} dV$. Therefore, we get to the *general equation of continuity*:

$$\frac{\partial \rho}{\partial t} = -\rho div(\vec{v}).$$

Together with Fick's law, which tells us that a density gradient causes a diffusion current, the equation of continuity gives the diffusion equation that we will study more closely in Section 8.3. In general, every flow is described by an equation of continuity, and this already gives a lot of information about what different flows can do given the geometry of the containers they are flowing in. But we can also look at what makes a fluid flow at all, i.e., look at the equations of motion that describe the fluid, which will bring us back to Newton's laws of motion. We will now look at what these equations of motion can teach us about fluids. However, we will do this only for special cases, since in their most general form, known as the Navier–Stokes equations, the equations of motion for a fluid cannot actually be solved in many situations.

7.6.2 *Bernoulli's Principle

As we did in mechanics for pointlike objects, a fluid can be described by an equation of motion. We do not want to do this here, since this description is quite complicated. However, again like in mechanics, we can still make very useful statements by making arguments in terms of energies involved in changes between states. This often simplifies the solution of the problem and also allows us to formulate directly applicable laws.

We consider a laminar, stationary flow of an incompressible and friction-free fluid in a tube (or streamline bundle) of variable cross-section and variable elevation. The entrance surface A_1 and the exit surface A_2 are perpendicular to the flow, and the speed should not vary across the area of the surfaces (which is always the case as long as the surfaces are small enough). According to the equation of continuity, we obtain the fluid mass dm passing through the the tube during time interval dt:

$$dm = dm_1 = \rho v_1 A_1 dt = dm_2 = \rho v_2 A_2 dt.$$

Let's take account of all the energies involved in this process during the time dt (see Figure 7.29). The work performed by the pressure acting on the fluid dW (pressure × area × path length) must be equal to the change in kinetic and potential energy of the fluid: $dW = dT + dU$:

$$dW = p_1 A_1 v_1 dt - p_2 A_2 v_2 dt \quad dT = \frac{dm}{2}\left(v_2^2 - v_1^2\right) \quad dU = gdm(y_2 - y_1)$$

$$\Rightarrow \quad p_1 - p_2 = \rho\left(\frac{v_2^2}{2} - \frac{v_1^2}{2} + gy_2 - gy_1\right)$$

$$\Rightarrow p_1 + \frac{\rho}{2}v_1^2 + \rho gy_1 = p_2 + \frac{\rho}{2}v_2^2 + \rho gy_2$$

Since we have not specified points 1 and 2 in any special ways, this expression needs to be constant for the whole flow ($y \equiv h$):

$$p + \frac{\rho}{2}v^2 + \rho gh = \text{const.} \qquad \text{Bernoulli's principle}$$

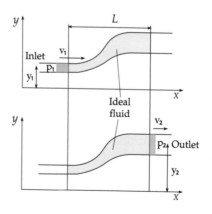

Figure 7.29 Sketch of the different contributions to the energy in fluid flow to obtain Benoulli's principle.

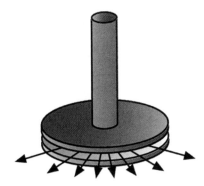

Figure 7.30 Hydrodynamic paradox.

Again, this is for an incompressible fluid without friction. If there is no change in gravity along the flow (i.e., no change in height), we can omit this term and obtain the following:

$$\frac{\rho v^2}{2} + p = \text{const.} \equiv p_0.$$

Here, p is the actual pressure that a manometer within this flow will measure. The second term, $\rho v^2/2$, is also a pressure (as one can check from its units) and is called *dynamic pressure*. p_0 is the total pressure. This can also be put in words:

Static pressure (p) plus dynamic pressure ($\rho v^2/2$) gives the total pressure (p_0).

There are several instruments and effects in daily life that can be explained by Bernoulli's principle. We will look at a few of them in more detail now.

Example: Hydrodynamic Paradox

If a gas flows from a pressurized tank against a movable plate, the plate is sucked in rather than blown away. Due to the high velocity of the gas between the two plates, the pressure is less than the outside air pressure. The two plates are pressed together by the external air pressure. You can use this to create your own hovercraft using a balloon and a CD (see Figure 7.30). If you glue the opening of a plastic bottle to one side af a CD and then attach an inflated balloon on that opening, the CD will glide on a table like a hovercraft held there by the pressure difference due to the outflowing air and Bernoulli's principle.

Example: Venturi Tube

Let's consider a stationary flow within a tube of variable cross-section (see Figure 7.31). The pressure p in the tube varies with the cross-section. The combination of the equation of continuity and Bernoulli's principle yields the following:

$$\frac{v_1}{v_2} = \frac{A_2}{A_1} \qquad \frac{\rho}{2}v_1^2 + p_1 = \frac{\rho}{2}v_2^2 + p_2 = \frac{\rho}{2}v_1^2\left(\frac{A_1}{A_2}\right)^2 + p_2$$

$$\Rightarrow \ p_2 = p_1 + \frac{\rho}{2}v_1^2\left(1 - \frac{A_1^2}{A_2^2}\right) < p_1$$

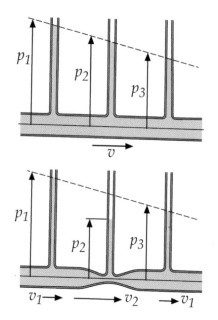

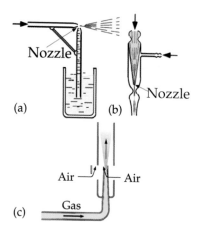

Figure 7.31 Venturi tube.

Figure 7.32 A vaporizer (a), water jet pump (b), and Bunsen burner (c) are variations of the Venturi tube.

If $A_1 = A_3$, we have $p_1 = p_3$. In an actual experiment corresponding to this situation, where water flows through a tube with constriction, the pressures p_2 and particularly p_3 are smaller than expected from the preceding. This is because in this case, there is also friction present, which is not taken into account in Bernoulli's principle, but is very strong in a narrow tube. This will be amended in the next section.

Variations of the Venturi tube used in everyday life are vaporizer (Figure 7.32a), water jet pump (Figure 7.32b), and Bunsen burner (Figure 7.32c). At the nozzle opening (small cross-section), the speed is large (equation of continuity!), thus the pressure is small (Bernoulli) so that the jet exerts a suction. In the case of the Bunsen burner, the lower pressure in the vicinity of the gas emerging at the nozzle helps to draw in the air necessary for maintaining the combustion.

7.6.3 Friction and Viscosity

In mechanics, we have already got to know of viscosity, and we have seen that a flow velocity gradient is generated when a shear stress is applied. Conversely, this also means that the maintenance of a velocity gradient requires a shear stress. This is the background of viscous friction. We will now look at how this affects flows.

In order to obtain the quantitative relationship between friction forces and the viscosity of a liquid, we idealize a shear deformation as we have discussed in Section 7.4 on elasticity. A fluid is put between two parallel plates and a shear stress is applied to one of the plates. Then this plate will be moving with a velocity v_0, and a linear velocity gradient $v(x) = ax$ is established between the two plates (x = horizontal coordinate in Figure 7.33). The velocity gradient, or otherwise put, the shear-deformation rate, is then proportional to the shear stress, and the constant of proportionality is the viscosity. This was first noted by Newton:

$$\tau = \eta \frac{dv}{dx} \quad \text{with} \quad \frac{dv}{dx} = a = \frac{v_0}{d}.$$

Typical values of the viscosity η are shown in Table 7.3. For gases, η increases with temperature, while for liquids it decreases. It is independent of pressure for gases.

So let's describe this quantitatively. We have a piece of the plate dA, with thickness dx and mass $dm = \rho \, dA \, dx$, which moves vertically (in the z-direction) with a velocity v_z. The equation of motion of the fluid then is as follows:

$$dm \frac{dv_z}{dt} = \tau(x + dx) \, dA - \tau(x) \, dA$$

if we now also take into account the friction forces present. On both sides, there is a volume dV, which can be divided to give the following:

$$\rho \frac{dv_z}{dt} = \frac{\partial \tau}{\partial x}.$$

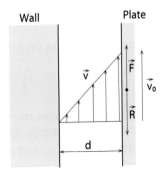

Figure 7.33 Demonstrating viscous friction.

Table 7.3 Viscosity of different materials.

Material	$\eta[\text{Pa} \cdot \text{s}]$			
	0°C	20°C	50°C	100°C
Air	0.000017	0.000017		0.000022
H_2O	0.00179	0.00101	0.00055	0.00028
Rhizinus-oil		0.95		
Glycerin		1.53		
	17°C	23°C	30°C	37°C
Blood		0.004		
Blood-plasma	0.0020	0.00173	0.0015	0.0013

Inserting Newton's law of friction yields the following:

$$\rho \frac{dv_z}{dt} = \eta \frac{\partial^2 v_z}{\partial x^2}.$$

This is basically the equation of motion for a fluid in the presence of friction, where an external force would have to be incorporated by a gradient in pressure. One thing that complicates this description is the fact that here we have to look at the temporal change of the velocity, which depends on both position and time, and the position again depends on time via the velocity. Therefore, the time derivative in the preceding description has to be evaluated at varying positions, which makes this equation (the Navier–Stokes equation) extremely difficult to solve. This is particularly important when trying to describe turbulent flows.

Transition to Turbulence

In turbulent flows, the inertial forces play a major role, i.e., the left-hand side of the preceding Navier–Stokes equation is dominant. The term quadratic in velocities can lead to turbulence, which results in a very complicated flow. The importance of the different parts can be obtained by transforming the Navier–Stokes equation by using normalized values. If we divide all lengths by a certain length scale L (given by the size of the obstacle in the flow), all velocities by a mean velocity u, and all times by a corresponding time L/u, then the Navier–Stokes equation becomes the following:

$$\frac{uL\rho}{\eta} \left(\frac{\partial \vec{v}}{\partial t} + \vec{v}\vec{\nabla} \cdot \vec{v} \right) = \left(\frac{\partial^2}{\partial x^2} + \frac{\partial^2}{\partial y^2} + \frac{\partial^2}{\partial z^2} \right) \vec{v},$$

where all times, lengths, and velocities are normalized. Which term is dominant, therefore, is directly given by the constant $Re = \frac{uL\rho}{\eta}$, which is a pure number and is called the Reynolds number.

If Re is very small, then we are looking at stationary and laminar flows. This can be seen from the preceding equation because for very small Re, the left-hand term effectively

A time sequence of mixing and demixing color patches in a rotating shear cell with a high-viscosity fluid. Left: initial condition with two colored patches; middle: after five rounds of shearing, the colors appear mixed in a band; right: after returning five rounds, the colors are present again in small patches as they were initially.

becomes zero and the flow is described by $\left(\frac{\partial^2}{\partial x^2} + \frac{\partial^2}{\partial y^2} + \frac{\partial^2}{\partial z^2} \right) \vec{v} = 0$, irrespective of time. This also means that for very small Re and therefore for flows around small objects, like bacteria, the flows will not depend on time. This means that flows in one direction can be exactly counteracted by a flow in the opposing direction. Therefore, in such a situation the symmetric beating of a wing would not lead to forward motion. This can be illustrated by shearing a highly viscous liquid very slowly in opposing directions. Initially mixed colors will demix on returning to the original position; see Figure 7.34. Bacteria are highly susceptible to this because of their small size and correspondingly small Reynolds number. Therefore, the movement of bacteria has to be achieved by asymmetrically beating flagella, where the two directions of flow are different because of the asymmetry in the beat, and therefore a total forward motion is obtained.

For very high Reynolds numbers, inertial friction is dominant and we can basically neglect the right-hand side of the previous Navier–Stokes equation. In this case, the friction will depend on the square of the velocity, as we have already seen in earlier chapters. Where this transition occurs depends on the geometry of the flow but is the same for all geometrically similar flows. Experimentally as well as through lengthy calculations, it can be shown that the transition to turbulence occurs at $Re_c \simeq 2300$ for the flow through smooth and straight tubes (see Figure 7.35). For a tube diameter of $L = 1$ cm, this leads to the critical speeds for water and air of respectively:

$$v_c = 0.23 \text{ m/s} \text{ for water} \qquad v_c = 3.2 \text{ m/s} \text{ for air}$$

7.7 Flow through Pipes and Blood Flow

7.7.1 Poiseuille Flow through a Pipe

In many applications, such as in the blood flow, which is what we finally want to be able to describe quantitatively, we have a situation where the flow of a fluid through a

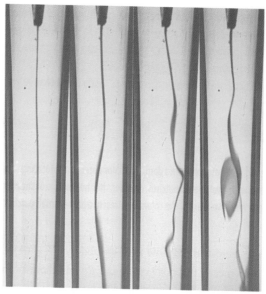

Figure 7.35 A single streamline in flows of different speeds. At low speeds (leftmost), the streamline is straight and does not move. At intermediate speeds, corresponding to Reynolds numbers around 100, the streamline still is compact, but undulates (second image from the left). Still increasing the Reynolds number, the flow becomes increasingly turbulent, until a fully turbulent stream is reached at the rightmost image.

cylindrical tube is maintained by a pressure difference between the two ends of the tube. In the circulation of blood, it is the pumping heart that creates this pressure difference, and the blood vessels are the tubes through which the blood flows. What one can easily check, e.g., with a garden hose, is that the flow rate, i.e., the amount of fluid transported, increases both with the pipe diameter and the pressure difference. A quantitative description of these relations is given by Poiseuille flow, which we will discuss in detail now. What we will also find in this is that the velocity distribution in the tube is actually inhomogeneous and the flow speed is greatest in the middle.

Poiseuille flow states that for a pipe of radius R and length L, where there is a pressure difference Δp that keeps up a (laminar) flow of a fluid with viscosity η, the velocity varies parabolically with distance from the center (see Figure 7.36)

$$v(r) = \frac{1}{4} \frac{\Delta p}{\eta L} \left(R^2 - r^2 \right)$$

and the total flux through the pipe is given by the following:

$$\dot{Q} = \frac{\pi \Delta p \, R^4}{8 \eta L} \quad [\mathrm{m^3 s^{-1}}]$$

This shows an important characteristic of Poiseuille flow in that the dependence of the flow on the radius of the pipe is very strong and goes with the fourth power of the radius. This means that even small changes in the diameter of the tube (or a blood vessel) can have

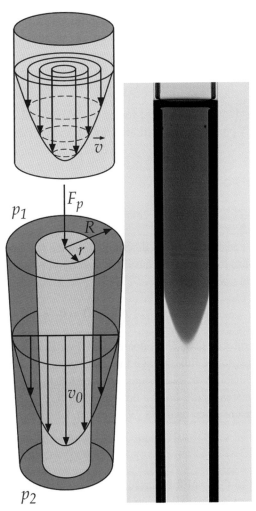

Figure 7.36 Construction for deriving the dependence of flow speeds in a cylindrical pipe as well as image of a viscous liquid colored on top flowing through a pipe, showing a parabolic distribution of speeds.

great consequences on the flow. For instance, if residues are left at the walls of arteries, the total flow is strongly influenced and in order to keep the same blood flow, the heart has to produce much higher pressures. This often leads to serious cardiovascular problems.

In order to see where this strong dependence on the tube diameter comes from, we will have to look at how one obtains the description of Poiseuille flow quantitatively. For this, we look at one streamline bundle within the fluid that because there is laminar flow will have the shape of a cylinder, just like the tube. If we take this cylinder to have radius r, the force on this volume because of the pressure difference Δp is given by the following:

$$F_p = \pi r^2 (p_1 - p_2) = \pi r^2 \Delta p.$$

There is also a viscous friction opposing this transport, which is given by the velocity gradient dv/dr and the outer surface area of the cylinder $2\pi rL$ sliding against the neighboring fluid:

$$F_\tau = \eta 2\pi rL \frac{dv}{dr}.$$

For a stationary flow, these forces cancel each other, which can then be used to obtain the velocity profile by integration of the equation for dv/dr we obtain:

$$\vec{F}_p + \vec{F}_\tau = 0 \;\; \Rightarrow \; \pi r^2 \Delta p + 2\pi rL\eta \frac{dv}{dr} = 0$$

$$\Rightarrow \quad \frac{dv}{dr} = -\frac{\Delta p}{2\eta L} r \;\; \text{Integration}: \; v(r) = -\frac{r^2 \Delta p}{4\eta L} + C$$

The constant C we can determine from the fact that at the boundary of the tube, the flow speed has to be zero, because it is directly in contact with the tube, which does not move. With this, we obtain

$$v(r = R) = 0 \;\; \Rightarrow \;\; C = \frac{R^2 \Delta}{4\eta L},$$

which directly gives the velocity profile we have previously stated.

To get to the flux through the tube, we first consider a hollow cylinder with a radius r and a thickness dr. The volume entering this hollow cylinder during time dt is limited by its surface area $2\pi r dr$ and the flow speed at the position of this hollow cylinder. The volume is therefore given by $dV = 2\pi r dr \, v(r)dt$, which leads to a flux:

$$\frac{dV}{dt} = v(r)2\pi r dr = \frac{\pi \Delta p (R^2 - r^2)}{2\eta L} r dr.$$

To get to the total flux through the tube, we have to sum up the contributions of all the hollow cylinders making up the tube, i.e., we integrate the preceding expression up the radius of the tube R, which yields the following:

$$\dot{Q} \equiv \frac{dV}{dt} = \frac{\pi \Delta p}{2\eta L} \int_0^R (R^2 - r^2) r dr = \frac{\pi R^4 \Delta p}{8\eta L},$$

which is the result quoted earlier. If we are interested in the mass of fluid transported rather than the volume, we have to multiply this result by the density ρ.

We can use this total flux to determine the average speed of flow inside the tube, $\langle v \rangle$. For a homogeneous speed over the whole tube of $\langle v \rangle$, the flux is simply given by speed times cross-sectional area, i.e., $\dot{Q} = \pi R^2 \langle v \rangle$. Putting in the flux due to Poiseuille flow from above, we can directly obtain the average speed as follows:

$$\langle v \rangle = \frac{R^2 \Delta p}{8\eta L}.$$

We can use this to determine the friction force of a cylinder pulled through a fluid at a constant speed. This will be the same as the total force acting on the tube in this case.

This force, R_v, acting on a tube of length L cross-section πR^2, is directly given by the pressure difference times the cross-section, as follows:

$$R_v = \pi R^2 \Delta p = 8\pi \eta L \langle v \rangle,$$

where we can insert an expression for Δp from the preceding relation for the average speed. As we have seen before, this gives a friction force that is proportional to the viscosity η and the speed v, but also to the length L, as we have used in the description of our simple model of electrophoresis. Now that we have done the full calculation, we even know the prefactor to be 8π.

The description of Poiseuille flow through a tube rests on the assumption that we are dealing with laminar flows. Therefore, we have to have average flow speeds that are below the critical speed for the transition to turbulence, i.e., $\langle v \rangle < v_c$. As we have noted previously, the critical Reynolds number for flow through a tube is roughly $Re_c \simeq 2300$, which gives a limit for the average flow speed of $\langle v \rangle < \frac{Re_c \eta}{\rho R} = \frac{2300\eta}{\rho R}$. For water ($\eta = 10^{-3}$ Pa s, $\rho = 10^3$ kg/m^3) flowing through an aorta of a radius of 0.5 cm, we obtain a critical speed of roughly 0.5 m/s, which just about corresponds to the highest speeds present in the aorta. Since blood actually has a somewhat higher viscosity than water (roughly three times higher), blood flow through all of our vascular system will be laminar and we can safely use Poiseuille flow to describe it, as we will do in the following subsections.

7.7.2 Resistance to Flow and Kirchhoff's Laws

We have just seen that a pressure drop is required to keep a flow in motion for fluids with friction. This means that no liquid flows without an external force or an applied stress. We will discuss something very similar in Section 9.4. in connection with electrical currents. There is also no electrical current flowing without the application of a voltage. This goes even further: what we can also see from Poiseuille flow is that the pressure drop (Δp) is proportional to the flux or current ($\dot{Q}$): $\Delta p = R \cdot \dot{Q}$. The proportionality constant R is also called the (flow) resistance. In the context of electrical currents, the voltage drop is proportional to the electrical current, $\Delta V = R \cdot I$, which is called Ohm's law and will be the subject of Section 9.4.

As we are working toward understanding the circulation of blood in our (or other animals') vascular system, we are looking for a description of the total flow resistance of a system of tubes forming a branched network. In particular, we want to now look at how the total resistance of a series of tubes or a number of parallel tubes depends on the properties of the individual parts of the network and their geometry. This is illustrated in Figure 7.37. So let us first ask, what happens when two pipes are branching off a single one?

In a closed system, in which the same liquid keeps being pumped around, liquid can be neither added nor lost. This means that the flow in pipe 1 must correspond to the sum of the flows in pipes 2 and 3, as follows:

$$\dot{Q}_1 = \dot{Q}_2 + \dot{Q}_3.$$

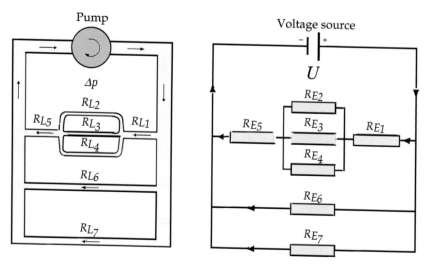

Schematic representation of a branched flow system and its corresponding network of resistors.

On the other hand, the pressure drop must be the same across the two adjacent tubes. This pressure drop must also correspond to that produced by a tube that replaces the parallel tubes. Thus, we have $\dot{Q}_2 = \Delta p/R_2$, $\dot{Q}_3 = \Delta p/R_3$ and $\dot{Q}_1 = \Delta p/R_{\mathit{eff}}$. This latter equation says that the replacement tube will of course have to take up all of the incoming flow. We can now insert all of these flows in the relationship between the three flows, which we have obtained from the conservation of flow to get the following:

$$\Delta p/R_{\mathit{eff}} = \Delta p/R_2 + \Delta p/R_3$$

and thus

$$1/R_{\mathit{eff}} = 1/R_2 + 1/R_3.$$

The total resistance is thus determined by the smaller of the two parallel resistances. This means that the current mainly flows through the large tube, which has a smaller resistance. Actually, for Poiseuille flow, the resistance is inversely proportional to the fourth power of the tube diameter, i.e., a larger tube has a much smaller resistance, and if we put a thick and a thin tube in parallel, most water will flow through the thick tube.

As a second example, we can consider what happens to the flow resistance if two tubes are put one after the other in series. In that case, the flux through both tubes must be the same, since there is nowhere for fluid to go or come from otherwise. Therefore, we must have $\dot{Q}_1 = \dot{Q}_2$. Furthermore, in both tubes there must be a corresponding pressure drop, $\Delta p_1 = R_1\dot{Q}_1$ and $\Delta p_2 = R_2\dot{Q}_2$, given by the respective flow resistances of the different tubes. For both tubes together, there is a total pressure drop, which is given by the sum of the pressure drops in the individual tubes, i.e., $\Delta p_{\mathit{eff}} = \Delta p_1 + \Delta p_2$. The current through the replacement tube again has to be the same, such that we obtain the following:

$$R_{\mathit{eff}} = R_1 + R_2.$$

In this case, the total resistance is determined by the larger of two the resistances. This also explains the behavior we have seen for the pressure reduction when flowing through a constriction, in the context of Bernoulli's principle. Behind the constriction, the pressure was much lower, which can be explained by the pressure drop due to the greatly increased flow resistance in the narrow tube (again, the resistance goes with one over the fourth power of the diameter).

In calculating the effective flow resistances, we applied Kirchhoff's laws, which also apply to electric currents. On the one hand, the junction rule, which describes branches, is the following: if we consider a branching from one tube to several tubes, then the current of the incident tube is given by the sum of the currents in all the outgoing tubes. As we have seen, this can be explained by the conservation of the flowing mass.

Second, we used the mesh rule, which describes the flow in a circuit. It states that the sum of the voltage drops in such a circuit corresponds to the total voltage applied from the outside.

These rules for the flow of a fluid with resistance are discussed in more detail in Section 9.2. in the context of electrical currents. However, as we have seen, they also apply to fluids such as, for example, in the blood circulation, which consists of a network of vessels of different sizes, as shown in Figure 7.38.

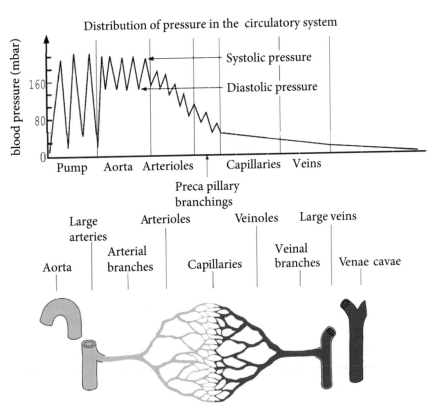

Figure 7.38 Schematic of the pressure distribution and the different types of blood vessels found in the human circulatory system.

7.7.3 *Blood Flow

We have now seen how fluids flow in tubes, what resistances we have to expect, and what the most important variables are that we need to take into account. Therefore, we now have all the tools together to describe how blood flows through the various vessels in the vascular system, which we are now going to do. In a closed circuit such as the bloodstream, the flux is certainly a very important variable. Since the entire blood flow must flow through the heart or through the aorta, we know from a relatively simple measurement of the flow through the aorta what the total flux through all the capillaries is, as well as the total flux through all other levels of the branched network of the vascular system. The proportions (size, surface, and volume) of the various vessels are schematically shown in Figure 7.38.

To get the flow through a single capillary, we only have to know the number of capillaries, since, according to the junction rule, the flow within the different sub-levels is evenly distributed. Since all capillaries are more or less equal, that is to say have the same flow resistance, we then get an even distribution of the flow through all capillaries. As we know the size of capillaries, we can estimate their number by considering how the size of the tubes in the vascular system depends on the level of branching and thus the number. Because of the conservation of the volume, this is the same question as how the flow must depend on the radius of the tube. To answer this, we consider what it costs an organism to make an extra tube of radius R.

On the one hand, there is resistance to flow. Increasing the number of tubes in parallel will decrease the total resistance, as we have seen. We can quantify this by the power needed to keep up the flow, given by the product of the pressure drop and the current, $P = \dot{Q}\Delta p$, which we can express as a function of only the current using Poiseuille flow: $P = \frac{8\dot{Q}^2 \eta L}{\pi R^4}$. Here, L is the length of the tube considered. On the other hand, there is an increased cost of an additional blood vessel in the additional volume of blood that has to be created. This is given by the volume of the vessel, i.e., $\pi R^2 L$. The total cost is then given by the sum of these two terms, or otherwise put $C = A\dot{Q}^2/R^4 + BR^2$, where we have used the constants A and B to describe the dependence on variables beyond our control and which remain constant, such as the tube length, the viscosity, the relative strength of the different costs, etc. For an optimally tuned flow system, these total costs for the vascular system C should be at a minimum as a functon of R. This is quantified by setting the derivative of $C(R)$ to zero, i.e., $\frac{\partial C}{\partial R} = 0$, which for the cost function described earlier gives $\frac{\partial C}{\partial R} = -4A\dot{Q}^2/R^5 + 2BR = 0$ or $2A\dot{Q}^2 = BR^6$. In other words, the most efficient flow through a network of branched tubes is obtained when the flow in the different levels depends on the radius as $\dot{Q} \propto R^3$. Thus, when we look at different branching levels, we know that the flow before branching corresponds to the sum of the flows after the split. If the daughter tubes are identical, then $N_1 \cdot R_1^3 = N_2 \cdot R_2^3 = const$, or otherwise put, $R_1 \propto N_1^{-1/3}$. This is a prediction that can be tested on anatomical data of the vascular system. For the example of a dog, this is done in Figure 7.39, where the dependence of the size of different levels of the vascular network on their number is shown on a double logarithmic plot. The straight line that indicates a power law dependence has a slope of $-1/3$, which is

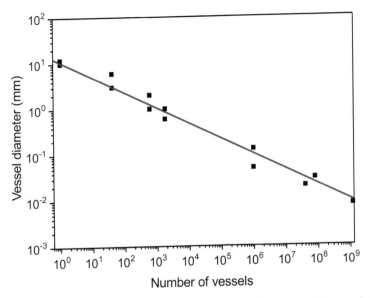

Figure 7.39 The number of blood vessels of different types plotted against their diameter. The data are plotted on a double logarithmic scale, and the power law corresponds to the distribution expected for optimal flow. The data are from the circulatory system of a dog (LaBarbera, 1990).

exactly the power law predicted for the branching of an optimal flow network. Thus, for the vascular system we find that the flux through a tube depends on the cube of its radius.

This seems to be in contradiction with the results for Poiseuille flow, which say that the flow in the tube is given by the fourth power of the radius. This should be valid for all laminar flows, and we have seen that the vascular system has laminar flows everywhere, so Poiseuille should be observed. What's wrong here? Well, Poiseuille flow is still observed for every single tube. The sizes and flow velocities in blood vessels correspond to a laminar flow, and therefore they have a velocity profile that is described by a parabola in each individual tube. However, the pressure drop that we have in the different levels in the bloodstream may also depend on its radius. If we require $\Delta p/L \propto 1/R$, then the flow is distributed optimally through the bloodstream with $\dot{Q} \propto R^3$ while still having Poiseuille flow in each tube. Is there a simple way for a biological regulation of the size in the vascular network to arrive at optimal flows? What other consequences does it have for the flow of blood when the flow in the system is proportional to R^3?

Let us look at the average flow speed: if $\dot{Q} \propto R^3$, then it must be that $\langle v \rangle = \dot{Q}/(\pi R^2) \propto R$. The flow speed is therefore directly proportional to the vessel size! This implies that the shear rate of the fluid acting on the vessel is the same for all sizes of vessels, $dv/dr = const.$ Since the shear stress is given by the shear rate, this means that all over the vascular system, the shear stress on the vessels is the same. This is the case as long as $\dot{Q} \propto R^3$ and the flow is optimal. Therefore, the optimal regulation of flow can be achieved by the mechanical control of tissue growth. If the shear stress is no longer equal to that in the rest of the circuit, the tube must either grow or shrink to accommodate it. In fact, the growth of tissue from blood vessels does indeed react to shear stresses.

7.8 *Cells and Tissues Are Neither Liquid nor Solid: Viscoelasticity

The ideal liquids that we have described so far have followed the friction law

$$\sigma = \eta \frac{d\epsilon}{dt} \tag{7.2}$$

with a constant viscosity η that did not depend on the time scale of the excitation or the velocity. For very many materials that we are dealing with in everyday life, this is, however, not quite true, and there are many exceptions to this, such as blood, saliva, paints, pastes, ointments, gels, etc. On the other hand, the elastic properties we have discussed earlier are also assuming a constant elastic modulus E in

$$\sigma = E\epsilon, \tag{7.3}$$

which again is not always a good approximation and typically starts to fail for any material at some point (level of strain or time scale). For instance, a glacier is made up of solid ice, but flows like a liquid on the time scale of centuries. This is basically done by the exchange of defects in the solid arrangement of the molecules, which allows for a very slow rearrangement of the entire structure in response to a stress, or otherwise put, a flowing of the material. Thus, while we can use ideal liquids and solids to obtain a description of several aspects of life, such as blood flow or the mechanics of bone, there are many instances in which we should go a little further in the description and take into account the dependence on the time scale as well. This is of particular importance in the description of the mechanical properties of cells and tissues, which typically are highly viscoelastic. This is what we will do in the following, and there are two distinct basic descriptions that correspond either to a solid that also has liquid properties or that of a liquid that can have solid, or elastic, properties.

7.8.1 Strange Liquids: Maxwell Materials

If the long-time behavior of a material is that of a liquid, this means that if we wait long enough or if we act on the material strongly enough, the material will start to flow, i.e., there will be viscous behavior that we have to describe. This corresponds to a Maxwell material, which describes a material that we can model as having a viscous part (denoted by ϵ_η) and an elastic part (denoted by ϵ_E) arranged in series as in Figure 7.40. This means that if we act on this material with a given stress, σ, this will act on both parts with the same strength. Therefore, for both these parts we can write down the characteristic behavior given by

$$\sigma = E\epsilon_E \tag{7.4}$$

Maxwell material   **Kelvin material**

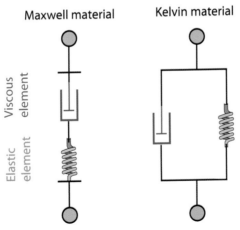

Figure 7.40 Two main models for viscoelastic materials. On the left, a spring and a dashpot are arranged in series giving rise to qualitatively fluid behavior in the Maxwell model. On the right, alternatively, the spring and dashpot are arranged in parallel, giving rise to qualitatively solid behavior in the Kelvin model.

for the elastic part and by

$$\sigma = \eta \frac{d\epsilon_\eta}{dt} \tag{7.5}$$

for the viscous part. The total deformation of the material will be given by the sum of the viscous and the elastic deformation, such that we have

$$\epsilon = \epsilon_\eta + \epsilon_E, \tag{7.6}$$

which implies that we have to obtain a description of these deformations from the applied stress. In order to do this, we solve the characteristic descriptions for the elastic and the viscous parts to obtain $\epsilon_E = \sigma/E$ and $\frac{d\epsilon_\eta}{dt} = \sigma/\eta$. If we consider the time derivative of the total deformation, we can insert both of these contributions to obtain the following:

$$\frac{d\epsilon}{dt} = \frac{d\epsilon_E}{dt} + \frac{d\epsilon_\eta}{dt} = \frac{d\sigma/E}{dt} + \sigma/\eta. \tag{7.7}$$

With this, we now have a description of the deformation as a response to a stress, where we see that the deformation does not only depend on the stress, but also on the total integrated stress during a period and the material is described by both an elastic modulus and a viscosity. If we consider a constant applied stress, we for instance see that the deformation initially is an elastic elongation, which then is constantly increasing in time described by the viscous term. On the other hand, if we stress a material by prescribing a constant deformation, we can use the preceding equation to investigate the relaxation of this stress. If the deformation is constant, we know that $\frac{d\epsilon}{dt} = 0$ and therefore that

$$\frac{d\sigma/E}{dt} + \sigma/\eta = 0. \tag{7.8}$$

This means that we have an equation describing the time dependence of the stress inside the material, which we can rearrange to the following:

$$\frac{d\sigma}{dt} = -\frac{E}{\eta}\sigma. \tag{7.9}$$

This is something that we have already seen many times, and we know that this means that the stress decreases exponentially on a given time scale as $\sigma = \sigma_0 \exp(-t/\tau)$, where $\tau = \frac{\eta}{E}$ is the characteristic time scale of the stress relaxation. The more viscous and the softer (the smaller E) the material, the longer it takes to relax the stress.

Thus we have a description of a material that shows characteristics of both a liquid and a solid, which, however, behaves as a liquid on very long time scales. This we can see from the previous result of the constant applied stress. We could, however, also obtain this directly from the fact that we modeled the material as a spring and a dashpot in series, i.e., we are always applying a force directly to a fluid, such that we have to obtain fluidlike behavior in the end. Also we see that this fluid can act as an elastic solid, i.e., store elastic energy on short time scales, where stresses do not fully relax. This time scale depends on the materials properties of the viscosity and the modulus and is longer for higher viscosities and lower moduli. You can basically visualize this by considering how honey can sustain a load for a certain time, while this time is a lot shorter for water. In very many instances, this is also the behavior of biological tissues, which react elastically on short time scales, but are plastic or liquid on longer time scales. For instance if you pull on your skin, it will return to its original form when released if you do this only for a short time; however, after longer stressing of the tissue, the deformation can stay in place.

7.8.2 Strange Solids: Kelvin Materials

A similar description can be given for materials that mostly behave as solids, but only react to stresses after a given time and thus do not incur stresses immediately. These are called Kelvin materials and are modeled in a way that distributes the stress between a viscous and an elastic part in parallel (see Figure 7.40). Therefore, there will always be a part of the force that acts on a solid, which means that after a long time (when all the dynamic effects have died down), the material will behave like a solid. Because the two parts are in parallel, the deformation of both parts is the same, $\epsilon = \epsilon_\eta = \epsilon_E$, and the total stress distributes itself between the two parts,

$$\sigma = \sigma_\eta + \sigma_E. \tag{7.10}$$

As before, we have the same relation between the stress and deformation for the individual parts, i.e.,

$$\sigma_E = E\epsilon \tag{7.11}$$

for the elastic part and

$$\sigma_\eta = \eta\frac{d\epsilon}{dt} \tag{7.12}$$

for the viscous part. This directly gives a relation between the applied stress and the deformation:

$$\sigma = E\epsilon + \eta\frac{d\epsilon}{dt}. \tag{7.13}$$

After a long time, when the deformation is constant, we can directly see from this that the material behaves as a solid, since it is described by Hooke's law. On the other hand, we can use this description to see on what time scale stresses and deformations can be incurred or relaxed in such a material. If we suddenly release a stress, then the deformation is described by $E\epsilon + \eta\frac{d\epsilon}{dt} = 0$, which we can rearrange to obtain the following:

$$\frac{d\epsilon}{dt} = -\frac{E}{\eta}\epsilon, \tag{7.14}$$

which again shows us that the deformation will go back to its original state of $\epsilon = 0$ exponentially on a time scale given by $\tau = \frac{\eta}{E}$. This means that such Kelvin materials will only deform on longer time scales, and this is, for instance, how bones behave.

As a matter of fact, on even longer time scales, bones will be remodeled in reaction to stresses, as we have seen in Wolff's law. This means that to fully describe bones, we would in principle need to add another viscous part corresponding to the time scale of weeks describing this process. This can in principle be done by the extension of the model to include more dashpots or springs in series to a Maxwell model. This more complicated Burgher's model is, however, beyond the scope of this text.

Exercises

7.1 Strain gauges

A strain gauge contains 20 paths of a meandering wire.

(a) How does the electric resistance of the strain gauge change is response to a strain of $\epsilon = \Delta l/l$? The electric resistance is given by $R = \rho \cdot l/A$ (see Section 9.4), where ρ is a geometry independent materials property called resistivity and l is the length and A the cross-sectional area of the wire. Poisson's ratio of the wire (aluminium) is $m = 0.34$.

(b) How accurately can you determine the strain if you can measure the resistance with an accuracy of 1%? For this purpose, invert the relation from (a) to obtain how the strain depends on resistance. Then assume that the strain is small (as is usual for metals; in fact, you will usually have values smaller than $\epsilon = 0.01$), such that you can Taylor-expand the obtained relation.

7.2 Freely hanging rope

(a) What is the maximum length of a freely hanging rope? The rope has a yield stress of $\sigma_Y = 0.50(1)$ GPa, a Young's modulus of $E = 200(2)$ GPa, and a density of $\rho = 1,000(10)$ kg/m^3.

(b) What is the uncertainty of this length?

(c) How much is the rope extended under its own weight in this case?

(d) What is the uncertainty of this extension?

7.3   Beam bending

(a) Assume that an actin strand is a beam with a square cross-section (side-length 8 nm) and having a length of 100 μm. The elastic modulus is roughly $E = 1.0(1)$ GPa and the yield strength is about $\sigma_Y = 500(50)$ MPa. The micro-tubulus is fixed at one end and bent with a force F at the other end. What is the maximum force allowed without breaking the microtubulus? Where would it break in this case?

(b) What is the uncertainty of this maximum force?

(c) Now compare this to a model of a microtubulus, where we assume everything to be the same except that we model it as a hollow beam with an external side length of 25 nm and a wall thickness of 4 nm. What is the maximum force in this case?

7.4   Elastic properties of materials

(a) The Young's modulus of a thread of spider silk is $E = 2.0(1) \times 10^9$ Pa. The thread snaps when extended by $50(5)\%$. What is the yield stress of spider silk? Compare to the value of steel ($\sigma_Y = 5 \times 10^8$ Pa).

(b) What is the uncertainty of this yield stress.

(c) How far can steel be extended? With the Young's modulus ($E = 2.0(2) \times 10^{11}$ Pa) and the yield stress determine the yield strain of steel. Assume that the steel is linear until it breaks and compare to the value for spider silk.

7.5   Interaction between molecules

Intermolecular forces are usually described by the so-called Lennard–Jones potential. This describes the potential energy of two molecules at a distance r, given by the following:

$$E_{pot} = -\frac{M}{r^6} + \frac{N}{r^{12}} \text{ mit } M, N > 0.$$

(a) What are the forces acting on the two molecules corresponding to the two different terms of this energy? Which one is attractive, and which one is repulsive?

(b) What is the minimum distance r_{min} between the molecules, given in terms of M and N in equilibrium? What potential energy doe the molecules have at position r_{min}, if the potential energy is zero for infinite distances?

(c) Sketch a graph of the potential energy normalized to the value of E_{pot} at the minimum position, E_{pot}/E_{pot}^{min}, as a function of r/r_{min}.

(d) What is the error of (r_{min} and $E_{pot}(r_{min})$) in case that M and N both have an uncertainty of 1%?

7.6 Laplace pressure

(a) An alveolus in the lungs is roughly spherical and has a radius of 50(2) μm. Assume that this is an air bubble surrounded by a thin water film (surface tension: 0.075(3) N/m). In that case, what pressure (including uncertainty) is needed to inflate an alveolus? For comparison, the maximum pressure difference when inhaling is about 2,400 Pa.

(b) Now assume that the alveoli can produce a surfactant that reduces the surface tension to about 0.03 N/m (which is actually the case in nature). How large does an alveolus become if it is still spherical and inflated with the usual pressure difference exerted when breathing, i.e., 1 kPa?

7.7 Laplace pressure 2

(a) A long cylindrical balloon is inflated inhomogeneously, as we have seen in this chapter. While the pressure inside is constant everywhere, in some places the balloon is inflated and taut, while in others it is deflated and limp. If we use this as a model for capillaries, what is the tension in this capillary tissue (in N/m) if the blood pressure is 4,000 Pa and the capillaries have a radius of 6 μm?

(b) If we consider an artery with a radius of 0.5 cm, having a yield tension in healthy humans of about 500 N/m. What pressure difference can the artery withstand? Compare this to systolic pressure of about 16 kPa.

7.8 Equation of continuity

A tube is split into two tubes having half of the original tube's diameter each. What is the speed of flow in each of the two smaller tubes (relative to the speed in the original tube)?

7.9 Equation of continuity and Bernoulli

(a) A syringe with a cross-sectional area of the main compartment of 80(4) mm^2 contains water. If there is no force applied to the syringe, there is a pressure of 1 bar (= 10^5 Pa), i.e., ambient pressure, everywhere in the syringe. Now you press with a force of 1.00(5) N onto the piston and water spurts out. What is the speed of the spurting water? Assume that at the open end there is still a pressure of 1 bar.

(b) What is the error of this speed?

7.10 Resistance to flow

You have two pipes of length L and radius r that are parallel (i.e., next to each other) in a network of tubes. What is the radius ρ that a pipe of the same length would need to have in order to replace the two tubes in the network without changing its flow resistance?

7.11 Resistance to flow 2

Now you have two pipes in series (i.e., one after the other), again having the same length L, but different radii, where the first pipe is twice as wide as the second one, i.e., $r_1 = 2r_2$. What is the radius of the replacement pipe? Take into account that the new pipe needs to replace both of the previous ones, i.e., its length has to be $2L$.

Quiz Questions

7.1   Elasticity
Which of these statements describes a stress–strain curve?

A The integral under the curve is the melting energy.
B Deviations from linear behavior indicate the occurrence of flows.
C The initial slope corresponds to the elastic modulus.
D The curvature of the graph indicates whether more or less deformation energy needs to be applied (relative to a linear material).
E Deviations from linear behavior indicate the yield tension.
F The yield tension is independent of material.

7.2   Stress and strain
Two beams of the same material are suspended from the ceiling and loaded with the same force. One of the two beams is twice as long, but has half the radius compared to the other. How much is the longer one extended relative to the shorter one?

A Eight times as much.
B Four times as much.
C Twice as much.
D They are both extended equally.

7.3   Elasticity 2
What needs to happen at the transition between a liquid and a solid (exactly at the melting point)?

A Atoms are more strongly bound.
B Atoms are ordered in a crystal.
C A shear modulus builds up.
D Latent heat of melting is released.
E Nothing.
F Atoms move less.

7.4   Laplace pressure
Two identical balloons are connected with a valve. The balloons are differently inflated and the valve is closed. What happens when you open the valve?

A Nothing.
B The smaller balloon inflates the larger one.
C The larger balloon inflates the smaller one.
D This depends on how much the balloons are inflated.

7.5   Buoyancy
A wooded block is swimming in water, such that part of the block is above the water level. Now you are pushing the block under water, such that it is completely submerged. What happens to the buoyant force acting on the block?

A It decreases.

B It stays the same.

C It increases.

D This depends on how deeply it is submerged.

7.6 Resistance to flow

What happens to the resistance to flow of a network of tubes if you add one additional tube in parallel?

A It increases.

B It decreases.

C It stays the same.

D This cannot be determined.

7.7 Equation of continuity and Bernoulli

Water is flowing though a constriction in a tube. What happens to air bubbles solved in the water at the point of the constriction?

A They become larger.

B They become smaller.

C They stay the same.

8 Heat, Temperature, and Entropy

8.1 How the Interplay of Many Particles Leads to the Whole Being More than the Sum of Its Parts

While we have seen that conservation of energy can give a simplified description of many systems in Section 6.4, we have also seen that there definitely are instances where conservation of energy in the mechanical sense breaks down. Examples of this are everywhere as soon as damping or friction starts to play a role, as we have for instance seen in Section 7.17. For instance, if we pump water through a pipe, the motion would eventually come to a halt if we did not keep pumping. In the language of energies, this would correspond to a case where initially, the pump creates potential energy, which is then converted to kinetic energy in the form of a fluid flow. However, when the flow comes to a standstill due to friction, there is no longer any kinetic energy, but also no potential energy, and we would think that energy is not conserved. However, if we consider an additional form of energy, namely the heat, we can solve this conundrum and reintroduce a more general conservation of energy. Friction leads to the fact that the individual molecules of the water impart some kinetic energy onto the walls of the tube, but this does not lead to an additional total movement of the tube. In other words, the motion of all particles, or molecules of the tube (their velocities), cancel out on average, such that the total tube is at rest. However, if we consider their kinetic energies (given by the square of their velocities), this does not cancel out because there are only positive contributions. It is this disordered energy that corresponds to heat and that we have to describe in order to get back conservation of energy. Since all processes that we actually encounter in daily life or in biological systems are influenced by friction, it is necessary to understand how we can describe heat and corresponding processes. In this chapter, we will find out that there are basically two different versions of description, one being based on macroscopic properties and one based on microscopic properties:

Macroscopic: In this "classical" thermodynamics, the physical parameters of a whole system and their relationships are considered: There are variables describing the state of the system, such as temperature, pressure, and volume, that can be physically defined. These different state variables depend on each other, and the connection is described by state equations. Finally, there are thermodynamic potentials, such as internal energy, entropy, or free energy, that are represented as functions of state variables and whose development describes the state of equilibrium.

Microscopic: Here a system is considered as a multiparticle system consisting of molecules or atoms, a so-called ensemble. With the help of the mechanical laws for the particle system, i.e., the conservation of energy and momentum, as well as a statistical treatment of possible movements, conclusions can be drawn about the mean values and fluctuations of physical variables. This works well because atoms are so small that any piece of macroscopic material contains an extremely large number of particles. These mean values and fluctuations are found as state variables and can therefore justify the above macroscopic description. The area is also called statistical mechanics.

Using the considerations of statistical mechanics, macroscopic variables can be identified with corresponding microscopic mechanical variables. For example, the absolute temperature of a system corresponds to the mean kinetic energy of the molecules. The internal energy corresponds to the total potential and kinetic microscopic energy stored in the system. The pressure is attributed to the mean momentum changes that the molecules make when they make contact with the wall of the vessel. We will confine ourselves almost exclusively to this microscopic point of view in order to understand heat, since the new concepts can be obtained from what we already know, but also because this will give us a much more detailed understanding of how the motion of molecules influences properties that we can observe and hence describe many phenomena based on molecular processes in biology, such as diffusion and mechanical properties of the cytoskeleton.

Due to the statistical nature of this description, we will only be able to make proper statements for cases where we can take some form of average. These particularly include the so-called *equilibrium states*. This is understood as a state of the (isolated) system in which the macroscopic state variables (i.e., the averages of the microscopic ones) do not change in time. Another instance of a case we can describe statistically are the stationary states. While the macroscopic variables do not change in these states either, they are not to be confused with equilibrium, as there is still an energy flow in this case, i.e., these systems are not isolated. An example of this would be a room with a thermostat where we heat something to keep it at a constant elevated temperature, but any state of homeostasis in a living system that is driven by the metabolic rate also falls in this category of states. Since the macroscopic variables (i.e., the averages) do not change, such nonequilibrium steady states can be treated, and we will also look at a few examples in more detail. Such states correspond much more closely to biological systems than the standard equilibrium thermodynamics, since a living system always contains flows of energy. If we have the case that we have a constant flow of energy and thereby change the state variables, we are concerned with fully out-of-equilibrium thermodynamics, which is still an active field of research and where there are no general fundamental laws that have (yet) been found. However, in some special cases, such as transport due to random processes, this is possible, and we will see how this influences the behavior of biological systems from the microscopic to the macroscopic level.

This distinction between different types of systems can be done more formally, where we distinguish among the following:

Isolated systems do not exchange anything with their environment.

Closed systems can exchange energy (heat, work) with the surroundings, but do not exchange matter.

Open systems can exchange energy as well as matter with the environment.

The first two are well described by thermodynamics, as they will quickly reach either equilibrium or stationary states, but the latter case, which comprises most of the interesting systems biologically, are hard and such systems will often have to do with nonequilibrium states. The prime example of an open system is a living cell that is in contact with its external world in every respect. The exchange of matter also has the consequence that, for example, no atom that is currently making up your body was a part of you 10 years ago. In fact, it is the property to be an open system, which allows the cell to be alive at all.

To illustrate this, let's take an example of a process in which conservation of mechanical energy is violated, even taking into account friction. What happens when we slowly take a book from a table and slowly place it on the floor? Then we have no additional heat energy in the ground or in the book, but nevertheless the system has changed from a state with potential energy and without kinetic energy to one without either. What happened here? To do this, we must also look at the person who moved the book. Thus, the system becomes an open system in which the exchange of matter of the person with the environment must be taken into account as well. In particular, we must consider the energy that the muscles of the person consume, or what happens in the muscle when we take the book from the table. Although we gain energy when we place the book on the ground, the muscles have had to produce work continuously. The power stroke of the myosin heads is pointing in the direction of the muscle contraction. However, the thermal movement of the molecules takes up some of the energy, and another part is that the consumption is slightly less than in the opposite direction. This consumption is compensated by the supply of ATP; after all, we are dealing with an open system.

8.1.1 Irreversible Processes

Let us generalize these observations on processes that violate the conservation of mechanical energy. For this purpose, we consider the damped pendulum (Figure 8.1). The pendulum is raised and then released. In the course of time, it is decelerated by the impacts with the

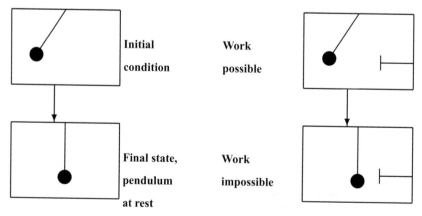

Figure 8.1 The potential energy of the initial pendulum is converted into kinetic energy of the gas atoms by collisions. The pendulum rests in the equilibrium state. No work is possible anymore.

gas atoms. In the end, it is at rest, but the gas atoms on average have gained a little bit of kinetic energy. The pendulum energy is initially present in the form of potential energy, and at the end evenly distributed as kinetic energy over many atoms. This has drastic consequences. For example, while we could break a window using the pendulum, i.e., the kinetic energy can be converted into work, this does not work with a gas at rest. While is not absolutely impossible, it is extremely unlikely because a huge fluctuation is necessary to create a current of gas directed toward the window that has enough momentum to break it. Such fluctuations do not occur in any conceivable case. The distribution of the energy to a large number of particles – in this case, the gas atoms – is called *thermalization*. The share of energy that is transferred and that can no longer be used for work is called *heat*. The essential point is that the macroscopic state of the system, i.e., its motion or its position, is caused by the averaging over all its constituting particles, i.e., of many vectors of position or momentum. In the case of a disordered movement or arrangement, this average cannot be distinguished from zero. However, the average energy is not equal to zero, since the squares of the velocity (kinetic energy) or the position (binding energy of a spring) enter into this. This means that when considering heat, we always have to deal with an energy in which the macroscopically relevant quantity enters the microscopic (i.e., for each individual particle) quantity as a square. The number of such possible quantities over which the potential energy can be distributed as heat is called the number of thermodynamic degrees of freedom. We will discuss this in more detail later.

The fact that a system generated out of equilibrium evolves toward the equilibrium state if left on its own accord can be demonstrated in a variety of ways:

Vortex formation: If a vortex is produced by an external action in a homogeneous liquid or gas, it vanishes after a short time. A homogeneous state is established by the collisions of the molecules.

Heat conduction: If a metal rod is heated strongly at one point, it shows an uneven temperature distribution with a maximum at the heated point. If the heating is switched off and the rod is left to itself, a constant, elevated temperature distribution is established again.

Diffusion: Two superimposed liquids with an initially sharp boundary mix again into a homogeneous solution – sometimes with an extremely long relaxation time of weeks or even longer.

These examples present only a small selection of so-called *irreversible processes*. This we call processes that run such that the temporally reversed process, which might be observed, for example, in a film running backward, is extremely unlikely, and therefore almost never occurs.

8.1.2 Brownian Motion and the Statistics of Many Particle Systems

The difference between the movement of the average and the mean square displacement can be illustrated very well by Brownian motion. Looking at a drop of milk under the microscope, one clearly sees a disordered movement of small fat droplet in the water, which is obviously directed randomly. This so-called Brownian motion comes from the collisions

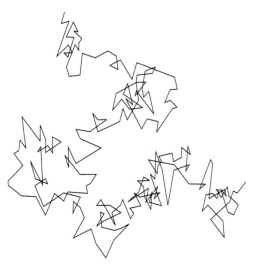

Figure 8.2 The random walk of a droplet in solution, also called Brownian motion.

of the droplets with the molecules of the water. It was discovered by the botanist Robert Brown in the pollen of a species of night candles and initially he thought this was an active movement akin to that observed in animal sperm. However, he could show the same effect with soot particles, showing that this is an inanimate motion present everywhere.

The individual velocities of the droplets can be very different. Statistical mechanics provides information about the probabilities with which different velocities (and other mechanical properties) of the individual particles occur. This will be discussed in Section 8.3.7 (on Boltzmann distribution).

The shaking movement of Brownian motion consists of translations and rotations in any direction. This can be observed by the naked eye in smoke or dust particles that float in the air when illuminated by obliquely incident sunlight. If one records the distance traveled, for example, a fat droplet, which can nowadays be done easily by connecting a video camera to a microscope, an image similar to Figure 8.2 could result. The short line segments connect points at which the observed object is located every 30 s.

Of course, we cannot calculate the complete motion for each individual molecule. But we can take the average of the motion of many particles, or the mean value of a particle over a long period of time. If we only look at the position, that is, the distance x, which a particle moves in a given time, then this mean is zero; on average, the particles do not move. If we look at a single particle, we may have to wait a long time for this to happen, but when we look at a lot of particles, those who have gone far to the left compensate for those that have gone far to the right. The average position of the particles is thus relatively simple in equilibrium. However, if we look at the mean of the squared distance, $\langle x^2 \rangle$, the different sides no longer cancel out (the square always gives a positive contribution). The longer we wait, the farther away a particle may have come. That is, the mean square distance $\langle x^2 \rangle$ increases with time. It can also be said that the mean value of the distribution is constant, but its width increases. Think back to Section 2.1, when we looked at statistical errors, where many measurements (particles) have yielded a very precisely determined mean value, even if the distribution was quite wide. We can formulate this fact precisely and thus get a description of diffuse transport. We will do this in Section 8.3. This diffusivity depends on the nature of the particle, the fluid, and the temperature. The exact relation goes back to Einstein, and we shall derive it shortly.

If our ear were somewhat more sensitive, we would perceive the Brownian motion of the air molecules, which leads to a constant bombardment of the eardrum, as a constant noise. We have described in Chapter 4 that the sensitivity of our ear does not miss much for this to be an issue. Even with an electric current in a resistor or a transistor, the moving charges that bring about this current cause jumps caused by impacts. With great amplification, the voltage fluctuations at the output of the transistor can be visualized on an oscilloscope and can also be converted into audible signals by a loudspeaker because the distances between the jumps are in the range of 10^{-3}s at 1 kHz.

The Brownian motion, the *noise* of the transistor current, as well as the folding of polymers (see Section 8.5), are examples of thermally induced motion caused by collisions between the particles of the system. The kinetic energy of the particles is expressed in the absolute temperature. We will now look at this in more detail.

8.1.3 Temperature

The concept of temperature will play an important role in the following. But it is physically and objectively not so easy to grasp. Subjectively, the answer is simple. What feels warmer has a higher temperature than what feels colder. That is, our nerve stimuli give us a sense of temperature. This can be quite deceptive, for example, when you look at an aluminum plate and a piece of wood of the same temperature. For the heat felt, heat conduction is at least as important as the temperature (see Section 8.3.6).

In general, it is not the temperature of the thermometer that interests us, but that of the other body. The thermometer must therefore be brought to the temperature of the body to be measured, that is, we must bring it into thermal contact with it. With this contact, the two systems exchange thermometers and body heat. When they are in contact with each other for a sufficiently long time, they reach a state of the same temperature.

This condition is called *thermal equilibrium*.

Instead of the Celsius scale customary in daily life, the *absolute temperature scale* with the unit Kelvin is used in the International System of Units. A temperature difference of 1K corresponds to just 1°C, but the zero point of the Kelvin scale is at −273.15°C, at the absolute zero point of the temperature.

That there is such an absolute zero is not trivial when considering macroscopic thermodynamics. If, however, we consider a microscopic interpretation of the temperature, we find that the temperature corresponds precisely to the average kinetic energy of the particles occurring in the disordered motion. Thus we also see the connection of the temperature with heat, since this is the energy of the disordered motion. It should be understood, however, that it is the fluctuation of the speeds that is important for the temperature, not necessarily the magnitude of the speed itself. If you throw a ball at high speed, all the particles in the ball have a high speed, but the temperature does not change. The contribution of the disordered movement has therefore remained the same, and all particles have experienced an additional velocity in a particular direction. Mathematically, this is the main difference between $\langle v^2 \rangle$ and $\langle v \rangle^2$. In the case of gas, there is a large difference due to the disordered motion. In this case, $\langle v \rangle$ is equal to zero with good accuracy. When the ball is fast, this is no longer the case. Since all particles move essentially in the same direction with the same velocity, $\langle v^2 \rangle \simeq \langle v \rangle^2$.

This definition of the temperature naturally gives an absolute zero, namely where all particles have the same kinetic energy (zero). At the top of the scale, there are no limits to the temperature. For instance, inside the sun the temperature can be about 15 million degrees.

8.2 Temperature and the Ideal Gas

Let us now look at a quantitative, microscopic interpretation of temperature and other thermodynamic variables. To this end, we investigate the model system of the ideal gas, in which this approach can be well illustrated. An ideal gas can be seen as a collection of small spheres, all of which are colliding completely elastically with each other and are so small that they only meet when they are directly flying toward each other. The results will then be generalized to other states that are related to thermal motion. Let us first look briefly at which parameters can characterize an (ideal) gas and how these are connected before we derive a microscopic interpretation.

8.2.1 Ideal Gases

The macroscopic parameters necessary to describe the equilibrium state of a gas quantity are *pressure*, *volume*, and *temperature*. Temperature we still understand as the property indicated by a thermometer, which is in thermal contact with our gas. For the pressure exerted by the gas on the walls of a vessel, we think of the force per unit of area (as discussed in hydrostatics) measured in the following:

$$\text{Pascal} \ = \ \frac{\text{Newton}}{\text{m}^2}$$

A gas is considered ideal if one can neglect the interaction between the individual molecules, which is appropriate when the distances between the individual molecules are sufficiently large on average that they do not notice their mutual attraction and the reciprocal repulsion. Very small distances for a short time make an elastic impact possible.

For ideal gases, there is a relationship between the pressure p, the volume of a mole (containing $N_0 = 6.023 \times 10^{23}$ atoms) V, and the absolute temperature T (in Kelvin), the *state equation*. This was found empirically:

$$pV = RT.$$

If the volume V contains ν moles instead of one, we correspondingly have the following:

$$pV = \nu RT.$$

For the temperature T, we must use the absolute temperature scale, measured in Kelvin. The constant R is called the *universal gas constant*. Its value is as follows:

$$R = 8.31 \frac{\text{Joule}}{\text{mol K}}.$$

It is highly noteworthy that R has the same value for all gases, which is why it is called a universal gas constant. The value is not material dependent. Instead of reckoning with moles, one can also solely consider the number of particles. Then, the state equation becomes the following:

$$pV = Nk_B T, \qquad\qquad (8.1)$$

where N is the particle number and k_B is the Boltzmann constant. The Boltzmann constant is given by the ratio of the ideal gas constant and the Avogadro number: $k_B = R/N_A = 1.38 \times 10^{-23}$ J/K. For the microscopic interpretation, k_B is important, since in that case we are interested in individual molecules. In the future, we will only use this form of the ideal gas law.

The state equation $pV = Nk_B T$ applies to all sufficiently dilute gases. The nature of the molecules does not matter. For a mixture of ideal gases, therefore, *Dalton's law* of the partial pressures applies. If the volume V contains $N_1, N_2, \ldots N_n$ molecules of different gases $1, 2, \ldots n$, the total pressure is given by the sum of all the partial pressures $p = p_1 + p_2 + \ldots + p_n$ and we have the following:

$$pV = (N_1 + N_2 + \ldots + N_n)k_B T \qquad \text{or} \qquad p_i V = N_i k_B T.$$

8.2.2 Microscopic Description: Pressure

We have now cited an empirical law that links the macroscopic parameters of our system. We now go back to the microscopic system and study how statistical methods and the consideration of individual atoms can help us understand this relation.

The macroscopic pressure, that is, the force on a piston, or the walls, is the result of many collisions of individual atoms with the respective wall. Let us lock the N molecules of the gas in a cubic container with a volume V, which has the length L in x-direction (see Figure 8.3). If a molecule m collides with the wall in the x-direction, where the wall with area $A = V/L = L^2$, the x-component of the momentum changes its sign. Therefore, the corresponding change in momentum between the incoming (i) and outgoing (o) molecule is as follows:

$$\Delta p_x = p_{ix} - p_{ox} = -mv_x - (+mv_x) = -2mv_x.$$

The momentum transmitted to the wall is therefore $+2mv_x$. The same molecule hits the wall again after it has been reflected on the opposite side, i.e., after the time $\Delta t = 2L/v_x$. Thus, in one second, the force (= momentum change per unit of time, Newton's principle of action) on the molecule is as follows:

$$F = \frac{\Delta p}{\Delta t} = \frac{2mv_x}{\Delta t} = 2mv_x \frac{v_x}{2L} = \frac{mv_x^2}{L}.$$

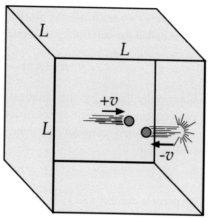

Figure 8.3 Illustration of the change of momentum of a particle hitting a wall as the microscopic origin of pressure.

For N molecules, therefore, we have a total force of

$$F_{tot} = \sum_{i=1}^{N} \frac{mv_{xi}^2}{L} = \frac{m}{L} \sum_{i=1}^{N} v_{xi}^2 = \frac{m}{L} N \langle v_x^2 \rangle.$$

In the last step, we have used the average of v_x^2, defined as follows:

$$\langle v_x^2 \rangle = \frac{1}{N} \sum v_{xi}^2.$$

This is because we cannot assume that all molecules have the same speed. For the (average) pressure on the wall, we therefore find the following:

$$\langle p \rangle \equiv p = \frac{F_{tot}}{A} = \frac{mN \langle v_x^2 \rangle}{LA} = \frac{mN \langle v_x^2 \rangle}{V}.$$

8.2.3 Microscopic Description: Kinetic Energy

In the following, we will try to relate the aforementioned mean value of the squared velocity $\langle v_x^2 \rangle$ to the kinetic energy.

The kinetic energy for a particular molecule is as follows:

$$E = \frac{m}{2} v^2.$$

Its average therefore is as follows:

$$\langle E \rangle = \frac{m}{2} \langle v^2 \rangle.$$

If we have a large number of molecules N that are independent and all have different speeds, the total mean kinetic energy is just N, the average of a single particle:

$$E_{kin} = N \cdot \langle E \rangle = N \frac{m}{2} \langle v^2 \rangle.$$

Here E_{kin} is the total kinetic energy of all molecules in the system and $\langle E \rangle$ is the average kinetic energy of a single molecule. If there is no overall motion, i.e., the molecules all move about randomly in all directions, we have the following:

$$\langle v_x^2 \rangle = \langle v_y^2 \rangle = \langle v_z^2 \rangle \quad \Rightarrow \quad \langle v^2 \rangle = \langle v_x^2 \rangle + \langle v_y^2 \rangle + \langle v_z^2 \rangle = 3\langle v_x^2 \rangle$$

$$\langle v^2 \rangle = \frac{2E_{kin}}{Nm} = \frac{2\langle E \rangle}{m} \qquad \langle v_x^2 \rangle = \frac{2}{3}\frac{E_{kin}}{Nm} = \frac{2\langle E \rangle}{3m}$$

If we insert this in the equation we derived in the last section for the microscopic derivation of the pressure, we obtain the following:

$$pV = \frac{2}{3}E_{kin} = \frac{2}{3}N\langle E \rangle.$$

This looks very similar to the ideal gas law. If we compare our expression for the energies with the state equation for the ideal gas, we obtain the following:

$$pV = \frac{2}{3}E_{kin} \quad \text{compared to} \quad pV = Nk_BT$$

$$\Rightarrow E_{kin} = \frac{3}{2}Nk_BT \qquad \langle E \rangle = \frac{3}{2}k_BT$$

The absolute temperature thus corresponds to the mean kinetic energy of the molecules.

From these microscopic considerations, therefore, one can understand the ideal gas equation for macroscopic systems by comparing the mean kinetic energy of a molecule with the absolute temperature according to the following relation:

$$\langle E \rangle = \frac{m}{2}\langle v^2 \rangle = \frac{3}{2}k_BT.$$

The relation between the average total energy, here present only as kinetic energy, and the temperature is of far more general validity, as our simple derivation in the bead model suggests. It also applies to atoms in liquids and solid bodies. The temperature is proportional to the mean kinetic energy of the molecules! We know that the instantaneous kinetic energies differ from the mean kinetic energies due to fluctuations. *Temperature is a typical macroscopic term associated with the mean value of a system. For a single atom, for the microscopic system, we cannot define temperature.* Furthermore, the assumption is that the motion is completely disordered, so if $\langle v \rangle = 0$ is the same, just like for the fast ball we have discussed in Section 8.1. The temperature of a gas corresponds to the total kinetic energy involved in the disordered motion of the molecules divided by a suitable constant $(3k_B/2)$. Where this factor comes from we will see when we discuss entropy. For a directed beam of atoms, one cannot define a temperature.

The result that the total energy (and the average energy of the individual atom) is independent of the container size is plausible. Since we have neglected the interaction between the atoms, except occasional collisions between them, the kinetic energy of the translation motion does not depend on the distance between the individual atoms. Therefore, the container size (at a fixed temperature) does not play any role. If the gas were not ideal, these conclusions would no longer hold. For if the interactions between atoms are not negligible with a sufficient density or sufficiently small distances, the potential energy changes, and this also contributes to the total energy. The total energy will then depend on the size of the container.

The average energy is the same for all types of molecules at the same temperature. It follows that the speeds are mass dependent. Heavy molecules move at the same temperature slower than light ones. Therefore, the atmosphere of earth does not contain hydrogen, since the hydrogen molecules in the atmosphere have such high speeds that they disappear from the gravity range of the earth.

A transformation of the ideal gas law is also provided by Dalton's law, which describes particle density well:

$$p = \frac{Nk_BT}{V} = nk_BT, \qquad n = \text{ particle density } \equiv \frac{N}{V}.$$

In a mixture of gases, the kinetic energies of all the individual components must be added, but the temperature is the same for all components in thermodynamic equilibrium. If n_i is the particle density of component i, and p_i the corresponding partial pressure, Dalton's law states the following:

$$p_{tot} = \sum_i p_i \qquad \text{with} \qquad p_i = n_i\,k_BT.$$

The total pressure is equal to the sum of the partial pressures of the constituents of the mixture.

8.2.4 *The Equipartition Theorem

Up to now, we have only talked about translatory movement of a particle, because we regarded the molecules as monatomic. Biatomic molecules, such as H_2, N_2, and O_2, may also have rotational motions about the molecular axis (moment of inertia $J_\parallel$, angular frequency $\omega_\parallel$) and about the axis perpendicular to it ($J_\perp$, $\omega_\perp$) (see Figure 8.4). The total energy of a molecule is then given by the following:

$$E = \frac{m}{2}(v_x^2 + v_y^2 + v_z^2) + \frac{1}{2}(J_\parallel\omega_\parallel^2 + J_{\perp 1}\omega_{\perp 1}^2 + J_{\perp 2}\omega_{\perp 2}^2).$$

We have six terms in the energy. They are reduced to five because $J_\parallel \approx 0$, because the moment of inertia is the nuclear radius, not the molecular distance, as in $J_\perp$. At high

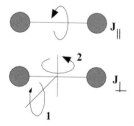

Figure 8.4   Biatomic molecules may have rotational motions about the molecular axis.

temperatures, a further degree of freedom from the vibration is added. For each term, the thermal disordered motion gives the same contribution, in fact, $(1/2)k_BT$, as follows:

$$\langle E \rangle = \frac{5}{2}k_BT = \text{ average energy of a two } - \text{ atomic molecule.}$$

This is the statement of the *equipartition theorem* of statistical mechanics. We get a contribution of $(1/2)k_BT$ to the mean energy of the molecule (or atom) per thermodynamic degree of freedom f_{th}. Thermodynamic degrees of freedom mean something slightly different from degrees of freedom in classical mechanics. f_{th} is the number of spatial and momentum coordinates that appear squared in the expression for the total energy of a particle. As we have discussed, it is precisely these positions or movements where an average energy results even when we do not have a macroscopic movement. This average energy of the fluctuations is precisely that caused by thermal motion. Quantitatively speaking, the equipartition theorem states that for the average energy of a single particle,

$$\langle E \rangle = \frac{f_{th}}{2}k_BT,$$

where f_{th} is the number of thermodynamic degrees of freedom.

For single atomic gases, $E = \frac{m}{2}(v_x^2 + v_y^2 + v_z^2) \Rightarrow f_{th} = 3, \quad \langle E \rangle = \frac{3}{2}k_BT.$

If there are also binding energies which we can describe by means of a spring law, e.g., in a solid state, these must also be considered.

For solids, $E = \frac{m}{2}(v_x^2 + v_y^2 + v_z^2) + \frac{k}{2}(x^2 + y^2 + z^2) \Rightarrow f_{th} = 6, \quad \langle E \rangle = 3k_BT.$

8.3 Transport via Random Processes

8.3.1 Random Walk and the Gaussian Distribution

Random movements and random collisions that have no correlation with their own past lead to probability distributions for location or velocity components. They have the form of the so-called *Gaussian normal distribution*, which we have already encountered in Chapter 2. It has the form

$$P(x) = \frac{1}{\sqrt{2\pi}\sigma} \exp\left(-\frac{(x-x_0)^2}{2\sigma^2}\right)$$

and is also called the bell curve due to its shape.

The Gaussian normal distribution is symmetrical about its center at $x = \langle x \rangle = x_0$. For the variation $\langle x^2 \rangle - \langle x \rangle^2$, we get σ^2, which is also called the *standard deviation*. At half the maximum height, the Gaussian bell curve has a width $2\sqrt{2\ln 2}\sigma = 2.355\sigma$.

The normal distribution occurs when x is a so-called *random variable* and its value is not the result of a single, but many successive, random steps. Let us now look at how the Gaussian distribution comes about in such a random process. We consider a random process in which a particle is continually scattered from one location to another by collisions. The probability distribution for the location of this particle should then be the Gaussian distribution. For the sake of simplicity, we restrict the movement of the particle to one direction, i.e., forward and backward. We divide the distance into equidistant steps δx, where a particle at the location x is always pushed to the right or left with the probability of 1/2. So if a particle is at position x at time t, it has a 50% chance of being at position $x - \delta x$, a time step δt later, and again a 50% change of being at position $x + \delta x$. Therefore, the probability of finding a particle at position x at time $t + \delta t$, $p(x, t + \delta t)$ is given by half the probabilities at $x + \delta x$ and $x - \delta x$ a time step before, since half the particles of the neighboring positions are scattered into x. We can put this this in an equation:

$$p(x, t + \delta t) = \frac{1}{2}(p(x + \delta x, t) + p(x - \delta x, t)).$$ (8.2)

For the temporal change of $p(x, t)$, we can then obtain the following:

$$\frac{\partial p(x, t)}{\partial t} = \frac{p(x, t + \delta t) - p(x, t)}{\delta t} = \frac{1}{2\delta t}(p(x + \delta x, t) + p(x - \delta x, t) - 2p(x, t)).$$

The term on the right-hand side can be translated into the second derivative by the location, whereby we obtain the following equation for the probability of finding the particle at the time t at the location x:

$$\frac{\partial p(x, t)}{\partial t} = \frac{\delta x^2}{2\delta t}\frac{\partial^2 p(x, t)}{\partial x^2} = D\frac{\partial^2 p(x, t)}{\partial x^2}.$$ (8.3)

This equation describes how the spatial distribution of the probability determines its temporal change. Since the probability and its changes depend on two different parameters (space and time), we have to take the partial derivatives here. Such an equation is also called a partial differential equation. If we solve this equation, we get a time-dependent distribution in space. The longer the random process has taken place, the more uniform (i.e., broader) will be the distribution. If we start with all particles at a given point at a given time, then the diffusion equation 8.3 is solved by a Gaussian distribution, where the variance σ^2 depends on the time, $\sigma^2 = \sigma^2(t)$. This means that the solution of the equation has to be $p(x, t) = \frac{1}{\sqrt{2\pi}\sigma(t)}\exp\left(-\frac{x^2}{2\sigma^2(t)}\right)$. From this ansatz, we can determine the time dependence of σ by inserting it into the equation. The time derivative then is as follows:

$$\frac{\partial p(x, t)}{\partial t} = \left(-\frac{1}{\sqrt{2\pi}\sigma^2}\frac{d\sigma}{dt} + \frac{x^2}{\sigma^3} \cdot \frac{1}{\sqrt{2\pi}\sigma}\frac{d\sigma}{dt}\right) \cdot \exp\left(-\frac{x^2}{2\sigma^2}\right)$$

and the following is the second spatial derivative:

$$\frac{\partial^2 p(x, t)}{\partial x^2} = \frac{1}{\sqrt{2\pi}\sigma^3}\exp\left(-\frac{x^2}{2\sigma^2}\right) \cdot \left(\frac{x^2}{\sigma^2} - 1\right).$$

If the Gaussian distribution is to solve the diffusion equation, these two terms must applicable to the diffusion equation, and we obtain the following:

$$\left(\frac{x^2}{\sigma^2} - 1\right) \cdot \frac{1}{\sqrt{2\pi}\sigma^2}\frac{d\sigma}{dt} \cdot \exp\left(-\frac{x^2}{2\sigma^2}\right) = D\frac{1}{\sqrt{2\pi}\sigma^3}\exp\left(-\frac{x^2}{2\sigma^2}\right) \cdot \left(\frac{x^2}{\sigma^2} - 1\right).$$

On both sides of the equation, we have a term $p(x,t)\left(\frac{x^2}{\sigma^2}-1\right)\frac{1}{\sigma}$, which we can eliminate to obtain the following:

$$\frac{d\sigma}{dt}=D\frac{1}{\sigma}$$

or in other words

$$\sigma\,d\sigma=D\,dt,\rightarrow\sigma^2=2Dt.$$

On the one hand, this shows the time dependence of the width of the distribution, but on the other hand it also shows that the Gaussian distribution indeed solves the diffusion equation.

The mean square displacement of the particles (i.e., the variance σ^2 of the distribution) increases linearly with time. The fact that the quadratic distance increases linearly with time tells us directly that the diffusion becomes more inefficient at long times because of slower (remember that $v=\Delta L/\Delta t$, so for diffusion $v\propto 1/\sqrt{t}$). This has important implications, as we will see in the later course of the chapter.

As we have seen, the molecules of a gas have the tendency to fill the whole space available to them evenly as a result of the thermal molecular motion. Similar behavior is shown by ions or atoms in a solvent. The consequence of this is that in the absence of external volume forces, the average particle density n or concentration c is uniform in equilibrium. Density refers to gases (number of particles/volume), concentration in solutions (number of particles/volume of solution). If, for example, the partition between two vessels is removed with different gases or liquids, they are irreversibly mixed until the densities n_i of the particle type i, and thus the partial pressures $p_i=n_ik_BT$ are the same everywhere. This applies to a solution in which the concentration of the solute is not the same everywhere. By the diffusion current (time change of mass per flow-through area), concentration differences are compensated, as outlined previously. The *Fick diffusion law* for the one-dimensional case is as follows:

$$j_x=-D\frac{dc}{dx}.$$

The diffusion constant D is the same as in the previous diffusion equation. We do know from experience that in dilute gases diffusion is much faster than in liquids, i.e., D is bigger in gases than in liquids. We will quantify this further later in this chapter. There we will also describe the dependence of the diffusion constant on the properties of the particles.

If the diffusing particles are charged (ions), then a charge transport is associated with diffusion. If they have very different masses, their different velocity causes a charge separation as long as a concentration gradient exists. If, as a result of the diffusion, the concentration c reaches the same value everywhere, $c=$ const. The diffusion current j vanishes according to Fick's law. The gas mixture or the solution is in equilibrium. The different partial pressures are independent of position. Diffusion as the transport of matter plays a very important role in the organism (e.g., O_2 supply). Since it runs very slowly in liquids, the distances must be small and the area for the diffusion current should be as small as possible (pulmonary vesicles, blood capillaries, kidneys). From the properties of the diffusion of oxygen in the tissue, one can also calculate the ideal size of capillaries, as discussed in Section 8.3.3.

8.3.2 Diffusion in Cells and Embryos and Morphogenesis

Another example to show the limit of diffuse transport efficiency is the diffusion of morphogens in an embryo. Morphogens are substances that lead to the differentiation of cells in the development of an organism and are thus of great importance for the shaping of the form of the organism. The transport of morphogens is often described by diffusion, which can also be efficient from steady to a limit of the embryo size of about 500 μm. A well-studied example of such a morphogen is the protein bicoid in the embryo of the fruit fly *Drosophila*. This protein is deposited with the mother at a point of the egg and diffuses from there through the embryo. The concentration of the protein in the later stage determines the front and back of the fly (high concentration – front; low concentration – rear). So we have the diffusion with a point source at one end. In addition, the protein is degraded in the embryo, resulting in a reaction-diffusion equation for the protein. If we want to describe the temporal and local dependence of the concentration c, we get the following equation:

$$\frac{\partial c}{\partial t} = D\frac{\partial^2 c}{\partial x^2} - kc.$$

Here D is the diffusion constant of bicoid and k is its degradation rate in the embryo. This equation has a solution that does not change with time that is reached after a short time. This steady concentration profile is also the one used by the embryo to make the distinction between the front and the rear. If the concentration does not change in time, $\frac{\partial c}{\partial t} = 0$, so for the stationary case we have the following:

$$D\frac{\partial^2 c}{\partial x^2} - kc = 0$$

or

$$\frac{\partial^2 c}{\partial x^2} = \frac{k}{D}c.$$

This equation describes an exponential concentration profile (also called a concentration gradient) where the (normalized) concentration has dropped to the value $1/e$ on a certain length λ: $c(x) = c(0)\exp(-x/\lambda)$. The value of λ is determined directly by the diffusion constant and the degradation rate: $\lambda = \sqrt{D/k}$. The concentration profile of bicoid in the embryo can be determined experimentally, and the concentration decreases exponentially to a good approximation.

The transport over long distances needs another mechanism, *convection*, that is, the transport of matter with a flow (collective movement). We have seen that diffusion on long scales becomes less and less efficient. Convection will be treated in the context of nonequilibrium phenomena (Section 8.6.1).

8.3.3 The Size of Capillaries

We can now also consider the size of capillaries. Capillaries indicate the size at which the transport becomes more efficient by diffusion than by a macroscopic flow. This is not unlike the instability that leads to convection. On small scales, diffusion is very efficient in material transport. On large scales, it is not efficient at all. For a capillary, the flow velocity of the capillary, v_{cap} must be less than the velocity of the transport by diffusion: $6D/R_{cap}$. The condition becomes $v_{cap} = 6D/R_{cap}$. Here, we have to take into account the speed from Section 7.7.3 on the blood circulation system. We have found that the number of capillaries is given by $N = (R_{Aorta}/R_{cap})^3$ and the total flow is given by $\dot{Q}_{tot} = N\dot{Q}_{Kap} = Nv_{Kap}\pi R_{Kap}^2 = \pi R_{Kap}^2(R_{Aorta}/R_{Kap})^3 6D/R_{Kap} = 6\pi DR_{Aorta}^3/R_{Kap}^2$. This gives us the size of the capillaries: $R_{cap} = \sqrt{6\pi DR_{Aorta}^3/\dot{Q}_{tot}}$. If we use typical diffusivities for oxygen in blood, we obtain for the size of the capillaries $R_{cap} \simeq 6\ \mu m$, which corresponds exactly to the size one finds in almost all animals. The size also cannot depend on the animal, since the flow increases as the third power of the size, that is to say, the dependence on size in the radius of the aorta cancels out and we are left with a condition that only contains general constants, which are not dependent on the size or type of the animal.

8.3.4 The Coefficient of Diffusion and the Einstein Relation

When we look at the diffusion constant, there is a direct way to interpret it, namely as a product of the particle velocity and the average distance between two collisions as schematically represented in Figure 8.5. The velocity of the particles between the collisions is $v = \delta x/\delta t$, and δx is the average distance between two collisions, resulting directly in $D = v\delta x/2$. The mean distance that a particle can travel between two collisions can also be calculated using statistical methods. This distance is called *mean free path* and is often called ℓ.

The mean free path decreases with the surface area of a molecule, which determines the collision probability and also decreases with the particle density in the gas:

$$\ell \propto \frac{1}{area}\frac{1}{density} = \frac{1}{\sqrt{2}\pi d^2(N/V)}.$$

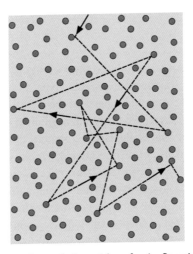

Figure 8.5   Illustration of the mean free path in a random walk of a particle performing Brownian motion.

Here, d is the molecular diameter, and N/V is the number of particles N in volume V. For oxygen, we find $d = 2.9 \times 10^{-10}$ m, and the particle density at room temperature can be calculated from the state equation (1 mole at 0°C takes in a volume of 22.4 l) $N/V = 2.7 \times 10^{25}$ m^{-3} to get $\ell = 0.1$ μm. At an average speed of $\langle v \rangle = 445$ m/s (Table 8.1), the average collision rate (the inverse of the time between two collisions) $\langle v \rangle / \ell = 4.5 \times 10^9$ s^{-1}. This means that diffusion leads to very long and tortuous paths because of the many impacts to overcome a relatively short distance. This happens, for example, also in the sun, where the light that is emitted in the nuclear reactions is scattered multiple times in the interior of the sun, such that it takes more than a million years until it reaches the surface of the sun!

Another consideration of the nature of the diffusion constants is given by Einstein. Let us take a particle on a random walk, which should not take place in a vacuum but in a substance. Then the particle feels a frictional force, so at every step of the path, work has to be done. The friction must be described by the viscous friction, since the particles are very small, and we have seen that in such small particles the Reynolds number is very small, which implies that viscous friction dominates, regardless of the material and velocity of the particles. Then the work done per step of the random walk:

$$\Delta W = f v \delta x.$$

Here f is the friction coefficient, and we do not need to worry about the vector, since we know that friction is always opposed to the motion. If we use the definition of the diffusion constant $D = v \delta x / 2$, we get a relationship between the work done per step and the diffusion constant $\Delta W = 2 f D$. This work is supplied by thermal motion. We thus must take into account that the driving force is given by the momentum change per time, i.e., $\Delta W = F \delta x = \delta p \delta x / \delta t = 2 m v^2$, since $\delta p = 2 m v$. The theorem of equipartition now implies $m \langle v^2 \rangle = k_B T = \Delta W / 2$. So diffusion is directly driven by thermal motion. In other words, $\Delta W = 2 k_B T = 2 f D$, which is the Einstein relation describing how temperature, friction, and diffusion are coupled:

$$D = \frac{k_B T}{f} = \mu k_B T \tag{8.4}$$

Here we have also introduced the mobility $\mu = 1/f$, which describes how fast a particle subjected to friction can become given at a certain force. This allows us to directly understand the essential aspects of the diffusion of different particles. If we consider the particle as a sphere, we know the coefficient of friction $f = 6 \pi \eta r$, where r is the radius of the sphere. That is, the larger a particle, the smaller the diffusion constant. A steel ball on the table does not diffuse around! On the other hand, the higher the temperature, the higher the diffusion constant. This makes sense, since we have seen that the temperature is the "driving force" of diffusion. It should be noted, however, that the viscosity of the material in which the substance diffuses will also depend on temperature. This is included in the coefficient of friction and will also influence the diffusion constant. We will discuss the temperature dependence of the viscosity in Section 8.3.5 and then will also return to the temperature dependence of the diffusion constant.

8.3.5 The Boltzmann Distribution

Let us now apply the preceding results to diffusion, or to the description of random processes, on the ideal gas, and then make a generalized consideration of the probability distributions of processes with thermal motion. We have just seen that thermal motion is the driving force of these random processes.

Let us consider a velocity component, e.g., v_x, of a molecule in an ideal gas as a random variable. For each collision with other molecules, v_x changes *randomly*, so it should be described by a Gaussian distribution. We therefore expect a normal distribution for the velocities v_x

$$P(v_x)dv_x \propto \exp(-v_x^2/2\sigma^2),$$

which is also found experimentally. The average $\langle v_x \rangle$ is zero, since the gas as a whole does not move. Because all directions are equal for symmetry reasons, also the components v_y nd v_z are following a Gaussian distribution and also have the average $\langle v_y \rangle = \langle v_z \rangle = 0$. In addition, we know the value of $\sigma^2 = \langle v_x^2 \rangle$ from our previous considerations regarding the temperature of an ideal gas. Here the mean kinetic energy was proportional to temperature, such that we find $\langle v_x^2 \rangle = k_B T/m$. If we insert this into the probability distribution, we find the following:

$$P(v_x)dv_x \propto \exp\left(-\frac{mv_x^2}{2k_B T}\right).$$

Again, all directions are equal because of symmetry, and we can have the same description in v_x, v_y, and v_z. This thus gives the probability distribution for the velocities:

$$P(\vec{v})d\vec{v} \propto \exp\left(-\frac{mv^2}{2k_B T}\right) = \exp\left(-\frac{E_{kin}}{k_B T}\right)$$

The mean kinetic energy of a gas molecule was found in our statistical derivation of pressure:

$$\langle E_{kin}\rangle = \frac{m}{2}\langle v^2\rangle = \frac{3}{2}k_B T.$$

At room temperature ($T \equiv 300$ K), this corresponds to an energy of roughly 6×10^{-21} Joule. This means that one mole of gas (i.e., 30 g) has, because of thermal motion, a total energy of more than 3,000 Joule. This corresponds to a potential energy of lifting the same amount of gas by 100 km!

Similarly, the corresponding speed $v_{rms} \equiv \sqrt{\langle v^2 \rangle} = \sqrt{3k_B T/2m}$ is considerable. Table 8.1 compiles the values for a few gases and compares them with the speed of sound in the corresponding gas. The speed of sound is actually smaller than the average speed but strongly correlated with it. This is understandable since the disturbance of the local pressure and density ratios, which corresponds to the sound wave, also spreads through the collisions of the individual molecules. The wave cannot propagate faster than the average velocity of the molecules, and it propagates about a factor of 1.5 more slowly because not all molecules move in the same direction as the wave.

Table 8.1 The molecular speed obtained from the average kinetic energy v_{rms} compared to the speed of sound v_{sound} (in [m/s]).

Gas	Molar mass [g]	v_{rms}	v_{sound}
Hydrogen	2.0	1920	1284
Helium (0°C)	4.0	1370	965
H_2O (134°C)	18.0	645	494
Nitrogen (0°C)	28.0	517	334
Oxygen (0°C)	32.0	483	316
CO_2 (0°C)	44.0	412	259

The Boltzmann distribution tells us that the probability of finding a kinetic energy of a particle higher than the mean kinetic energy is an exponentially decreasing function, where $k_B T$ is just the mean kinetic energy. This is also valid for more general processes where thermal motions are responsible for deviations from the average energy. Such deviations can have far-reaching consequences because they are responsible for, e.g., water evaporating. It is also the faster nuclei in the sun, which can overcome electrical repulsion and thereby lead to the energy production of the sun by means of nuclear fusion.

As a further example, we consider the decrease of the air pressure with height. According to the ideal gas law, the pressure is given directly by the particle density at a constant temperature, so we consider only the density (or concentration of the particles). The air molecules have a certain mass, i.e., they are also subject to gravity. The particles are small, so they fall with viscous friction with the terminal velocity $v = mg/f$, where f is the friction coefficient, m the mass of the particles, and g the acceleration due to gravity. This results in a particle current density downward, $j_g = c \cdot v = c \cdot mg/f$. This means the concentration near the earth's surface is greater. This concentration gradient dc/dh then leads to a diffusion current density upward $j_D = -Ddc/dh = -k_B T/f \cdot dc/dh$ when we use the Einstein relation. If the pressure, i.e., the concentration, does not change with time continually, these two particle current densities have to cancel each other, i.e., $j_g = j_D$ or $-k_B T/f \cdot dc/dh = c \cdot mg/f$, which yields the following:

$$dc/dh = -c\frac{mg}{k_B T}. \tag{8.5}$$

Solving this equation, we obtain an exponential decrease of the concentration with height $c(h) \propto \exp\left(-\frac{mgh}{k_B T}\right)$. Similarly, the pressure decreases and reaches about half the value it has at sea level at a height of 5 km.

This concentration dependence is also equal to the probability of finding a particle at the height h, which is $p(h) \propto \exp\left(-\frac{mgh}{k_B T}\right)$. If we consider that a molecule of the mass m has to get an additional energy $E_{pot} = mgh$ in order not to be on the earth's surface, but at height h, we find a dependence for the probability of finding a particle with this additional energy to be $p(E_{pot}) \propto \exp\left(-\frac{E_{pot}}{k_B T}\right)$.

This is very similar to the relationship for the ideal gas, and we can generally state that the probability of an energetically unfavorable process is described by the ratio of the energy barrier ΔE to the thermal energy $k_B T$. The probability is then given by the following:

$$p(\Delta E) \propto \exp\left(-\frac{\Delta E}{k_B T}\right). \tag{8.6}$$

This is also called the Boltzmann distribution, which is very fundamental in describing the properties of thermal motion.

As a further example, we can look at the temperature dependence of the diffusion constant. We have said that an essential point will be how the viscosity of a material depends on the temperature. When we established the viscosity microscopically, we had considered the overcoming of an energy barrier W_η of the binding of the respective atoms in neighboring layers. The probability that this energy barrier is overcome by thermal fluctuations is given by the Boltzmann distribution by $p(W_\eta) \propto \exp(-W_\eta/(k_B T))$. The higher this probability is, the lower the viscosity of the substance. We expect a dependence of the viscosity of $\eta \propto \exp(W_\eta/(k_B T))$, and because the diffusion constant is inversely proportional to the viscosity, we obtain a temperature dependence of the diffusion constant of

$$D(T) \propto k_B T \exp\left(-\frac{W_\eta}{k_B T}\right). \tag{8.7}$$

Here we took into account the direct proportionality of the diffusion constant with temperature from Section 8.3.4. However, this direct dependence is, in most cases, many times slower than the indirect one via the viscosity and the Boltzmann distribution, such that we have in good approximation $D(T) \propto \exp(-W_\eta/(k_B T))$.

8.3.6 *Heat Conduction and the Age of the Earth

Heat conduction is the process in which a system exchanges energy with its surroundings without performing work. A large part of the calories we consume every day in order to be able to exchange energy and matter with the environment is to compensate for losses caused by heat conduction. After all, we also have to eat if we do not do (mechanical) work.

In collisions of molecules, for example, with the wall of the vessel, kinetic energy, which is in the disordered molecular motion, is exchanged until the mean energies of the molecules of the system and those of the wall are equal, i.e., until thermal equilibrium between the wall and the system is reached. This transmitted or conducted energy is the heat that is exchanged. The proper physical variable is the *heat current density* j_W [energy/m^{2}s], and the relevant equation is the heat conduction equation:

$$j_{Wx} = -\lambda_W \frac{dT}{dx}.$$

The thermal conductivity λ_W, just like the diffusion constant, is proportional to the mean free path ℓ and the mean velocity $\langle v \rangle$, but also contains the density n of the molecules as well as the Boltzmann constant k_B (since an energy $\approx k_B T$ is transported):

Table 8.2 Thermal conductivities [J / msK] of some materials at $T = 300$ K.

Diamond	2320	Glass	12
Ag	407	Concrete, stone	12
Cu	384	Dry earth	12
Al	220	Wood	≈ 0.1
Iron	50–75	Wool, cotton	≈ 0.04
Stainless steel	≈ 15	Glass wool	≈ 0.04
H_2O	0.6	Air	0.026

$$\lambda_W \propto nk_B \langle v \rangle \ell \quad \left(\lambda_W = \frac{1}{3} \ell \langle v \rangle n \left(\frac{3}{2} k_B \right) \text{ for ideal gases} \right) \quad \left[\frac{\text{Joule}}{\text{msK}} \right]$$

Because of the factor n, λ_W is much smaller in gases than in liquids and solids. Since both the mean free path and the average velocity increase with the temperature, the heat conduction coefficient also increases with temperature. Values of λ_W for some substances are compiled in Table 8.2.

The total heat quantity ΔQ flowing through a cross-section A of a wall per time unit is given by the heat current density in the stationary (time-independent) case as follows:

$$\frac{\Delta Q}{\Delta t} = A j_W = -A \lambda_W \frac{\Delta T}{\Delta x} = \frac{T_i - T_a}{d} A \lambda_W.$$

In the final term, it is assumed that the temperature gradient between the internal temperature T_i and the outside temperature T_a over the wall (thickness d) is constant. This amount of heat is then the undesired heat loss in a building. The quantity $k_W = \lambda_W/d$ [Joule/m²sK] is the heat transfer coefficient or the k value of the wall. A typical value for well-insulated houses is about $k_W \approx 0.3$. Since the heat flow is proportional to the surface, heaters and heat sinks, and heat exchangers in general, must have large surfaces. Also this directly gives the size dependence of the metabolic rate we discussed in the introduction.

From this law of heat conduction, Lord Kelvin calculated the age of the earth at the end of the nineteenth century by assuming that the earth was initially a ball of molten stone, which had cooled down to an outside temperature of solid stone of a thickness of about 100 km. This could be extrapolated at that time from the temperature increase in the interior of the earth by 3 K every 100 m from the measured temperatures of mines. With a heat conduction coefficient of $\lambda = 1.7 W/(\text{m K})$, a density of $\rho = 3,000$ kg/m³, and a specific heat of $c = 1,500$ J/(kg K), we can recapitulate this calculation. For a temperature gradient of a decrease of 3 K every 100 m depth, we get a heat flow of due to the heat conduction equation:

$$\frac{dQ}{dt} = A \lambda \frac{dT}{dx}. \tag{8.8}$$

Using the preceding numbers and $A = 4\pi R_E^2$ this gives a heat flux of $\dot{Q} = 2.5 \times 10^{13} W$. This heat flux is supplied by the heat of the earth, thus cooling it down during all of its existence, leading to the solid crust of a thickness of about 25 km. If the earth initially had a temperature of $T = 3,300$ K (the current temperature in the center), the lost heat is therefore given by the following:

$$Q = \rho c \Delta T \cdot V = \rho c \Delta T 4\pi R_E^2 L/2. \tag{8.9}$$

The factor of one-half in the second term comes from the fact that the temperature decreases linearly with depth, such that we effectively only have to cool down half the volume. Numerically, we obtain a total heat loss of $\Delta Q \simeq 8 \times 10^{28}$ J. Dividing this lost heat by the rate of heat loss (i.e., the heat flux), we directly obtain the age of the earth. This yields $T_E = \Delta Q/\dot{Q} = 10^8$ years. This is much shorter (by more than a factor of 10) than we currently think. At the time, this was also the most important scientific argument against Darwin's theory of evolution, since the development of complex life would take more than a hundred million years. Darwin himself, therefore, no longer categorically excluded the possibility of inheriting acquired traits. It turned out, however, that Kelvin had not taken into account something important in his consideration (which he could not possibly have known): the heat losses are counteracted by a heat source. This heat source is a radioactive decay of atomic nuclei (see Section 5.7) inside the earth, which almost completely overcome the massive heat loss. Rutherford was the first to recognize this at the beginning of the twentieth century and then immediately determined the still valid age of the earth to about 4.5 billion years.

In analogy to the Ohm's law in the electrical current and the electrical resistance occurring in it ($I_{el} = U_{el}/R$, electric current $\equiv$ transported charge per unit of time $=$ voltage difference divided by the resistance; see Section 9.2), heat resistance is defined by the thermal resistance R_W ($R_W = 1/Ak_W$):

$$I_W = \frac{\Delta Q}{\Delta t} = \frac{T_1 - T_2}{R_W} \quad \Leftrightarrow \quad I_{el} = \frac{dQ}{dt} = \frac{U_{1el} - U_{2el}}{R}.$$

Analogously, heat resistors of several layers must be connected in series to the total resistance:

$$R_{tot} = \sum_i R_i = \frac{1}{A}\left(\frac{d_1}{\lambda_{W1}} + \frac{d_2}{\lambda_{W2}} + \cdots \frac{d_N}{\lambda_{WN}}\right).$$

And the temperature differences across the individual layers can be calculated as the voltage drops in a voltage divider. We have already discussed these so-called Kirchhoff's laws in the context of viscous friction in Section 7.7.2 and will discuss them again in Section 9.4 in connection with electrical currents.

Heat can be transmitted not only by means of impacts, that is, by direct contact of two systems, but also by radiation. Any system whose temperature T is not equal to zero emits and absorbs electromagnetic radiation, so-called heat radiation, which also propagates in vacuum and transmits energy (e.g., sun $\rightarrow$ earth). If a system is in thermal equilibrium with its environment ($T_S = T_U$), it absorbs exactly as much radiation energy as it emits.

The collisions between the molecules of a gas or a liquid can be used to exchange energy and momentum and thus also to transport the molecules involved. We have already encountered the transport of momentum in Newton's law of friction, which applies to the internal friction of liquids and gases and defines the viscosity. The *transport of energy* is called *heat conduction*, the *transport of mass diffusion*. If the objects jiggling around are

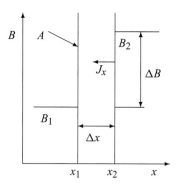

Figure 8.6 The flux density of quantity B through a layer of thickness Delta x and cross-sectional area A.

charged, then *electric charge* can be transported as well as mass, and we observe an *electric current*. This is covered in greater detail in Section 9.4.

The initial situation is approximately the same in all cases. For an interval Δx (thought or really existing as in a membrane), there is a difference ΔB in a physical quantity B, i.e., the gradient $\Delta B/\Delta x$ (or the derivative dB/dx at very short intervals) is different from zero. This results in a stream J_x (that is, a time change in size) from the page where B is larger to the page where B is smaller. The current is greater the larger the difference ΔB is, and the larger the flow through area A is (see Figure 8.6):

$$J_x \propto -A\Delta B \quad \Rightarrow \quad \text{current density} \equiv \frac{\text{current}}{\text{volume}} = j_x \equiv \frac{J_x}{A\Delta x} \propto -\frac{\Delta B}{\Delta x}$$

The basic relation for the current density in transport phenomena is therefore obtained by taking the derivative rather than the differences. In addition, we generally deal with a three-dimensional current distribution, and the current density and the gradient are vectors $(\vec{j})$. The derivatives then become partial derivatives:

$$j_x = -K \cdot \frac{dB}{dx} \quad \text{or in three dimensions}: \quad \vec{j} = -K \cdot \text{grad } B$$

where K is a constant, which depends on the process, for example, the diffusivity D in case of diffusion.

Before we identify the variable B and the associated constant for the processes mentioned, we must remember the constraints under which we have derived the transport equation. We have assumed a stationary, i.e., time-independent current. This is not necessarily always the case; time-dependent terms can also occur in the transport equation. Furthermore, as with all processes mediated by randomly distributed collisions, we observe fluctuations. Remember the noise in electrical conduction and Brownian motion in diffusion.

Diffusion: The *concentration c* (number of particles per unit volume) differs on both sides of the layer. The proportionality constant is called *diffusion constant D*.

Heat conduction: On both sides of the layer, the *temperature T* is different. The proportionality constant is the *thermal conductivity* λ_W.

Electrical conductivity: On both sides of the layer, the *voltage* U_{el} is different. The proportionality constant is the *electrical conductivity* σ.

Viscous friction: On both sides of the layer, the *speed* v is different. The proportionality constant is the *viscosity* η.

8.4 Entropy and the Laws of Thermodynamics

Now that we have advanced the microscopic description of thermal motion in detail, let us return to the macroscopic properties. The considerations that we have made so far, together with the conservation of energy, yield the two basic laws of thermodynamics on the macroscopic level. We will also learn about the concept of entropy, which we will again mostly look at in the microscopic and statistical interpretation.

8.4.1 Work, Internal Energy and Heat

The first law of thermodynamics is the general, macroscopic form of conservation of energy. It links the various forms of energy or their change. To do this, the heat Q is introduced as a macroscopic form of energy, which will appear in processes with friction. More specifically, heat is the energy transfer that is not associated with work performed on the macroscopic scale. Q is assumed to be positive when energy is supplied to the system considered.

Thus, even macroscopically, energy can neither be generated nor destroyed, but can only be converted from one form to another.

We denote the total (internal) energy with U, its change with ΔU. It can contain different components, e.g., the potential energy of oscillators, electrical or magnetic energy, as well as kinetic energy. For this kinetic energy, two parts have to be distinguished: first, the kinetic energy of ordered collective motion when the system moves as a whole (= macroscopic motion), i.e., $m\langle v\rangle^2/2$; and second, the kinetic energy of the *disordered* thermal molecular motion, which is $m(\langle v^2\rangle - \langle v\rangle^2)/2$, which corresponds to the temperature T.

The system can also absorb or release energy by performing mechanical work. For example, a gas can be compressed. We denote the work done as in mechanics by W, where positive values mean that energy is applied to the system in question. In the general case, we would also include chemical or electrical work. Here, we subsume these under the same variable W.

With these definitions of the different types of energies present in a given situation, conservation of energy and hence the *first law of thermodynamics* is given by the following:

$$\Delta U = W + Q.$$

The increase in internal energy of a system for any given process is equal to the sum of the work done during that process plus the heat supplied. Here's an example.

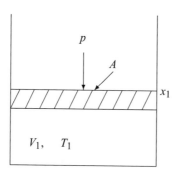
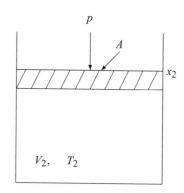

Figure 8.7   Expanding a compressed gas.

Let us consider a process in which the volume of one mole of an ideal (monatomic) gas underneath a movable piston is increased by heating it, while leaving the pressure constant (see Figure 8.7). The heat is provided by contact with a body at higher temperature.

Since the volume changes and the piston moves, mechanical work is performed ($|\vec{F}| = $const.$= pA$, $-\vec{F} \parallel d\vec{r}$):

$$W = W_{1\rightarrow 2} = \int_1^2 \vec{F} \cdot d\vec{r} = -pA\Delta x \qquad W = -p\Delta V < 0.$$

Here W is the work being done to the system. Thus W is negative when the system expands against an external force, as is the case here. W would be positive when the system would be compressed by an external force.

As the temperature increases, the internal energy also changes. For an ideal monatomic gas, we had found $\langle E \rangle = \frac{3}{2}k_B T$ for the average energy of an atom, and therefore the total internal energy of N times $\langle E \rangle$ is given by

$$U = \frac{3}{2} Nk_B\, T.$$

Therefore, the change in internal energy is given in our example by the following:

$$\Delta U = \frac{3}{2}Nk_B\Delta T > 0.$$

Knowing this, we can obtain the supplied heat from the first law of thermodynamics as follows:

$$Q = \Delta U - W = \frac{3}{2}Nk_B\Delta T + p\Delta V.$$

There are several different formulations of the first law of thermodynamics, all of which have the same physical content and are thus interchangeable:

(i) Energy can neither be generated nor destroyed; it is always conserved.
(ii) Heat is a form of energy.
(iii) If a system is to produce work, the internal energy has to be changed or heat has to be supplied.

8.4.2 Specific Heat

The amount of heat necessary for a specific temperature increase is referred to as specific heat. We distinguish two situations, depicted in Figure 8.8.

The amount of heat required to increase the temperature of a mole by 1 K is called molar heat. Depending on which of the two situations occurs, C_V, the specific heat at constant volume, and C_p, the specific heat at constant pressure, are defined as follows:

$$C_V = \left(\frac{\delta Q}{dT}\right)_{V=\text{const.}} \equiv \frac{\delta Q_V}{dT} \qquad C_p = \left(\frac{\delta Q}{dT}\right)_{p=\text{const.}} \equiv \frac{\delta Q_p}{dT}$$

And with the first law of thermodynamics, we find the following:

$$\delta Q = dU \;\Rightarrow\; C_V = \left(\frac{\partial U}{\partial T}\right)_V \qquad \delta Q_p = dH \;\Rightarrow\; C_p = \left(\frac{\partial H}{\partial T}\right)_p = \left(\frac{\partial U}{\partial T}\right)_p + p\left(\frac{\partial V}{\partial T}\right)_p$$

C_p and C_V depend on each other. Their difference can be determined from the thermal and caloric state equations.

For an ideal gas, the internal energy U is independent of pressure and volume:

$$U = E = N\frac{f}{2}kT.$$

With the state equation $V = RT/p$, one finds the following for the molar specific heats:

$$C_V = \left(\frac{\partial U}{\partial T}\right)_V = \frac{dU}{dT} = \frac{f}{2}Nk_B$$

$$C_p = \left(\frac{\partial U}{\partial T}\right)_p + p\left(\frac{\partial V}{\partial T}\right)_p = \frac{dU}{dT} + p\frac{dV}{dT} = \frac{f}{2}Nk_B + Nk_B = C_V + Nk_B$$

In solids, the atoms or ions are bound to resting positions around which they vibrate. They behave in the same way as three-dimensional oscillators. We therefore found the number of thermodynamic degrees of freedom as $f = 6$. The internal energy of a mole of solid therefore is $U = 3RT$, and because of the small compressibility of solids, $C_p \approx C_V = 3R = 25\text{J/(mol K)}$. This is the rule of *Dulong and Petit*, which is also found experimentally. However, at low temperatures, this breaks down, and there is no reason

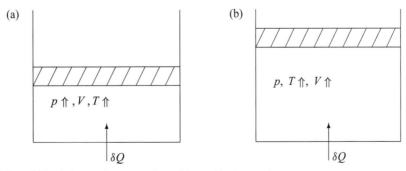

Figure 8.8   Specific heat: (a) fixed piston and constant volume; (b) movable piston and constant pressure.

within thermodynamics to be found for this. These effects were explained by Einstein on the basis of quantum mechanical considerations. The behavior of specific heat at low temperatures is a direct, macroscopic demonstration for microscopic, quantum mechanical effects.

8.4.3 Entropy

As we have already seen, irreversible processes occur in thermodynamics, i.e., processes in which a sequence of event follows as prescribed order. These are typically associated with the dissipation of energy, i.e., the generation of heat. In order to quantify the description of such irreversible processes, Rudolph Clausius has introduced the concept of entropy. Originally, this was treated as an empirical, macroscopic quantity that is closely linked to the heat supplied or extracted in a process.

Today, like the heat (and temperature) itself, it is also connected to statistical properties of a system of many particles. More specifically, the entropy is a measure of the number of possible microscopic states of the system having the same macroscopic state. The central statement of the second law of thermodynamics, then, is that a system of very many particles takes its most probable state with time. We will make this quantitative in the next subsection.

Since the second law describes irreversible processes, it also implies a direction to the flow of time. We have already seen such a course of time in random processes for diffusion, where the probability distribution became ever broader, and an original difference in concentration was slowly compensated. A comparable demixing does not take place. This is very different from mechanics, where there actually is no direction to the flow of time, and backward and forward are both on the same footing.

In order to describe the irreversibility, the entropy should describe the proportion of energy that can no longer be converted into work – therefore, it will also be directly linked to the heat.

The entropy change is defined as follows:

$$dS := \frac{\delta Q_{rev}}{T} \qquad S_2 - S_1 = \int_1^2 \frac{\delta Q_{rev}}{T}$$

In particular, in an isolated, adiabatic, and reversible process ($\delta Q = 0$), the entropy does not change.

From the first law of thermodynamics, we can obtain a general expression for the entropy change in the case that only volume work is performed ($\delta W = pdV$), namely

$$dS \equiv \frac{\delta Q}{T} = \frac{dU}{T} + \frac{pdV}{T} \qquad S_2 - S_1 = \int_1^2 \left(\frac{dU}{T} + \frac{pdV}{T} \right).$$

For an ideal gas, we can rewrite this using $U = C_V T$ and $p/T = Nk_B/V$. We then find the following:

$$dS = C_V \frac{dT}{T} + R \frac{dV}{V} \qquad S_2 - S_1 = \int_1^2 \left(C_V \frac{dT}{T} + R \frac{dV}{V} \right) = C_V \ln \frac{T_2}{T_1} + Nk_B \ln \frac{V_2}{V_1}$$

Thus, both an increase in volume as well as a rise in temperature lead to an increase in entropy.

8.4.4 Microscopic Description: Entropy and Probability

If we have a gas enclosed in an isolated system and let it expand on its own accord from the initial volume V_1 into the final volume $V_1 + V_2$, this is an irreversible process for which we can determine the entropy change due to macroscopic considerations (see Figure 8.9). In an ideal gas (a good approximation for real gases at room temperature) the internal energy is independent of the size of the container. Since we simply let the gas expand on its own accord within a closed container, there is no work being performed nor is there heat supplied to the isolated system. Therefore, the total internal energy and thus the temperature must stay the same.

We therefore consider an isothermal expansion from V_1 to $V_1 + V_2$.

The entropy change is then, according to the above treatment, as follows:

$$\Delta S = S_2 - S_1 = Nk_B \ln \frac{V_1 + V_2}{V_1} > 0.$$

As we have guessed, this irreversible process leads to an increase in entropy. What does this mean microscopically?

We can calculate the probability P for all atoms of the gas to return to the original volume. The probability of finding a certain atom in the volume V_1 is proportional to the proportion of this volume in the total volume, i.e., proportional to $V_1/(V_1 + V_2)$. The probability that N atoms are in the volume V_1 is given by the product of the probabilities for all the single particles, as follows:

$$P = \left(\frac{V_1}{V_1 + V_2} \right)^N$$

This very unlikely process would macroscopically lead to the inverse entropy change as previously discussed. The entropy would be decreasing by the following amount:

$$\Delta S = Nk_B \cdot \ln \frac{V_1}{V_1 + V_2} = k_B \cdot \ln \left(\frac{V_1}{V_1 + V_2} \right)^N.$$

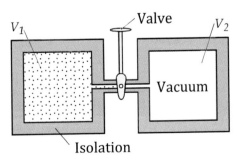

Figure 8.9   Expanding a gas into a vacuum by opening a valve.

If we compare the expression of the probability of the state with the entropy decrease, we see the following:

$$\Delta S = k_B \cdot \ln P$$

This is what Boltzmann has noted: the entropy is proportional to the logarithm of the microscopic probability! This realization is so important that it was engraved on his tombstone in Vienna.

If we consider just one mole and $V_1 = V_2$, the result for the probability is as follows:

$$N = 6 \times 10^{23} \qquad P = \frac{1}{2^N} = 10^{-1.8 \times 10^{23}},$$

which is an insanely small number. We can get a better intuitive grasp of this probability argument when we consider only a few particles, as is done in Figure 8.10. Even with only four particles, the uniform distribution is much more probable, since there are six possibilities to distribute the particles uniformly, but only one each has all the particles in one half.

Left	Right	L	R	Total	Fraction
● ● ● ●		4	0	1	6.25%
● ● ●	●	3	1		
● ● ●	●	3	1		
● ● ●	●	3	1		
● ● ● ●		3	1	4	25.00%
● ●	● ●	2	2		
● ●	● ●	2	2		
● ●	● ●	2	2		
● ● ●	●	2	2		
● ● ●	●	2	2		
● ● ● ●		2	2	6	37.50%
●	● ● ●	1	3		
●	● ● ●	1	3		
●	● ● ●	1	3		
●	● ● ● ●	1	3	4	25.00%
	● ● ● ●	0	4	1	6.25%
Total number of possibilities				16	100.00%

Figure 8.10 Illustration of probability interpretation of entropy. Randomly distributed particles in a container with two halves are distributed with greater frequency equally. For many particles, the likelihood that all the particles in a half are extremely strong decreases.

So we can reinterpret the second law of thermodynamics, i.e., that entropy always increases with irreversible processes, to a statement that such processes always go in the direction of a more probable state.

From this consideration of the probability of a macroscopic state that can arise from many random processes, we can also make plausible the Boltzmann distribution, which we found earlier. The probability of a state is given exactly by e to the power of the entropy divided by k_B: $P \propto \exp(\Delta S/k_B)$. This directly results in the Boltzmann distribution $P \propto \exp(-\Delta E/(k_B T))$ if we remember the link between entropy and heat as well as the first law of thermodynamics.

If we want to change the entropy, that is, to change the probabilities of the microscopic states, we must do work on the system. The Boltzmann distribution tells us how these probabilities and the work are related. That is, processes in which the entropy is changed (or the entropy multiplied by the temperature) give a force countering the macroscopic force that performs this macroscopic work. If we consider the previous example of a gas enclosed in a vessel and decrease its volume using a piston, the microscopic probabilities change. The (infinitesimal) change in entropy then is $dS = Nk_B ln((V+dV)/V)$, where N is the number of particles and V is the volume, which changes by the small amount dV. For small changes, we can Taylor-expand the logarithm and obtain the work that is to be done: $dW = TdS = Nk_B TdV/V$. For a piston with area A, the change in volume is $dV = Adx$, thus the piston moves by dx. So we obtain a work done, which is directly proportional to the displacement $dW = TdS = Nk_B TA/Vdx$, which means we can read off a constant force against which the work has to be done:

$$F = dW/dx = Nk_B TA/V.$$

Dividing this by the area of the piston, we obtain the pressure we have to work against and obtain $p = Nk_B T/V$, which is just the ideal gas law! The force of a gas that counteracts compression is thus given only by the narrowing of the possibilities of the movement of the gas molecules and an example of an entropic force. These play an important role in polymer chemistry and also in biology.

The elasticity of a rubber band is such a case. Here, the restoring force is determined by the lowering of the entropy associated with the lengthening of the rubber molecules. Thus, the spring constant of rubber is due to the thermal movement of the molecules, which implies that this restoring force must become stronger when we increase the temperature. You can see this yourself by heating a rubber band with a blow dryer while a weight is hanging on it. The increased restoring force will pull up the weight. This process is described quantitatively later in this chapter, once we have considered how thermal motion changed the properties of long molecules.

A further example of such an entropic force is that which occurs between two colloids (particles of about 1 μm in size) in solution. The disordered collisions of the water molecules lead to a Brownian motion of the colloids, as we have already seen. If the colloids are very close, there are fewer collisions with molecules in the gap between the colloids than outside of the gap. This means that the more probable impacts from outside push the colloids closer together. Thus there is an attraction of the colloidal particles purely based on the disordered motion of water molecules and therefore of the maximization of entropy.

When high-density colloids are put in solution, this interaction can even lead to counter intuitive effects. Such an attraction due to entropic forces can also lead to crystallization of the colloids, meaning that the ordered crystal is the state of maximum entropy in this case! Basically, one can understand this when one considers that the number of microscopic states of colloids moving via Brownian motion is greater when they are on regular lattice sites than when they form a static, disordered distribution. In this case, the maximization of entropy as exemplified by Boltzmann actually dictates that the ordered arrangement of the colloids, a colloidal crystal, is formed preferentially. This example is intended to show you that entropy is *not* a measure of the disorder in the system, as is usually stated, but rather, as we have considered, a measure of the probability of the macroscopic state. When we look at random processes, this is usually the uniform state associated with disorder, but it does not have to be this way!

8.4.5 Entropy and Information

We have just discussed, entropy is not a measure of disorder but is somehow micro-scopically related to the probability of different states. There is another way to view this interpretation of entropy in the sense of information. When considering the information content of a message, be it in a DNA sequence or in a piece of computer code, it is important to know what else could be transported. For instance, consider a long sequence of alternating zeros and ones that a computer needs to transmit (or a long sequence of alternating As and Gs in a piece of DNA). What is the information content of that message? In fact, it is very low, since in fact all that would in principle need to transfer is the length of the sequence and the fact that ones and zeros alternate. This is similar to the case when we consider the entropy of a state. Typically, the ordered state has a much lower entropy, since in a crystal, all you need to specify is how a unit cell is formed and how many of these there are. The previous example of the colloidal crystal does not fulfill this because the colloids in the crystal undergo thermal motion as well, such that the actual positions are not as well specified.

Another way to look at this is to consider what compression algorithms do in a computer. If they are well developed, the information in the compressed file should not be lost. Thus the compressed file should have the same amount of information as the uncompressed one, even though many fewer zeros and ones are needed to transmit it. If you try and replay a compressed video file without decompressing it, it will look like white noise. The compactified information in the file looks random to us. This is closely connected to the fact that the information content of a sequence is directly given by its entropy. As we have seen, the Boltzmann version of entropy for a distribution of microscopic states is given by $S = k_B \ln p$, where p is the probability distribution of the microscopic states. If we look at the average of this entropy, given by $\langle S \rangle = k_B \sum_i p_i \ln p_i$, where i number all possible states that we are considering, we obtain a measure of the macroscopic state that we are considering. Now let's apply this to the information content in a sequence. Here, the same concept of entropy will be important, but given that we will not want to connect the quantity of entropy to heat and temperature, we will not need to take into account the Boltzmann constant k_B. Therefore, the average entropy (or information) will be given by

$\langle S \rangle = \sum_i p_i \ln p_i$. This is almost the same as the definition of information used in computer science to determine the information in a sequence. Because computer codes are usually written in a series of zeros and ones, i.e., in bits with two possible states, it is simpler from the point of view of calculations to take the logarithm base two rather than the natural log. With this replacement, the preceding entropy is exactly equal to the information content in the sequence, i.e., the minimum length a sequence can be compressed to. Again, without a decompression, this will usually not make much sense, as you can experience when you try to play a compressed video. But also, the genetic information stored in the sequence of DNA in all of our cells is highly compressed and contains a lot of information, i.e., entropy. This is also why physicists often speak of DNA as a random polymer sequence, because a random sequence is the one with the largest information as defined earlier. So what the physicists mean in that case is that the genetic information is perfectly compressed in DNA. While this is not quite true, it is a good first approximation to describe overall properties of DNA, which are independent of a specific function, which would be dependent on the "decompression-algorithm" that transcriptional and translational control are forming.

Considering this equivalence between entropy and information, we can have another look at the second law of thermodynamics, which states that the entropy in the universe tends to increase. This implies that the information contained within the universe tends to increase. In other words, unpredictable events have to happen in the universe, since otherwise the information would remain the same all the time. This can either be due to the fact that there are nonlinear interactions taking place, which tend to be unpredictable, or on the fundamental level due to quantum mechanical uncertainties encapsulated, e.g., by Heisenberg's uncertainty principle that we have met in the context of waves (Section 4.8).

8.5 *The Influence of Thermal Motion on Materials Properties

8.5.1 Elastic Properties of Single Molecules

As we have already seen in the introduction of forces and their origin in muscles, it is technically possible nowadays to apply and measure very small forces acting directly on individual molecules. With this, it is possible to determine mechanical and elastic properties of individual biomolecules. An example of this is a single molecule of DNA. For this purpose, a DNA strand (usually from a λ phage) is connected to the functionalized surface of a microbead. This microbead is then held in an optical tweezer, whereby the force of the optical tweezers can be measured directly by the intensity of the laser. As shown in Figure 8.11, the DNA molecule is joined at both ends to a microbead, so that there is a location at both ends where a force can be applied. The force can also be applied with a micropipette, sucking the microbead to the opening of the pipette and then determining the force from the bending of the micropipette. Once the situation has been reached in which both ends of the DNA molecule are attached to a microbead, and these two microbeads are held somewhere, one can determine the elastic properties of the DNA by measuring the

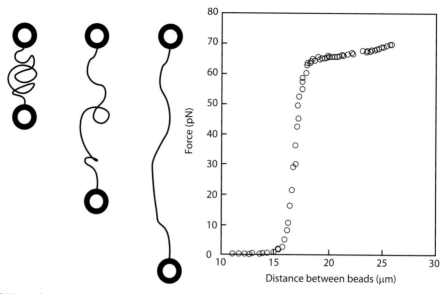

Figure 8.11 A DNA stretching experiment. The DNA between the two microbeads is shaped into a coil by thermal impacts. By means of an optical tweezer or a micropipette, the microbeads are pulled apart and the force required for this is measured. This gives a force-extension curve for the single molecule (right). Here, the first part of the curve is connected with the pulling out of the curled-up molecule. The actual Young's modulus of DNA comes into play only at larger extensions, where the molecule (the coil) is stretched. Upon still further stretching, a structural change occurs in the DNA molecule, which is associated with the disruption of the base pairing. Data from Smith et al., 1996.

force needed to pull apart the DNA a certain distance. The first thing to notice, however, is that the two microbeads have a distance that is much shorter than expected from the length of the DNA strand. The plasmid of a λ phage typically used in such DNA stretching experiments comprises about 48,000 base pairs, and thus has a length of about 16 μm. The initial expansion is typically only 6–8 μm, which is only about half that length. Thus, the long DNA molecule has collapsed into a coil. We will see later in this chapter that this collapse occurs due to the thermal impacts of the water with the DNA molecule. To do this, however, we first have to deal with how such a molecule is bent, which we will do in the next section. We will also be able to calculate how much DNA is coiled up because of this. The force that must now be exerted on the microbead is therefore only needed to pull the coil straight. Thus the entropy of the DNA is lowered, and this lowering requires the measured force. It is only when the DNA is almost completely extended that the actual Young's modulus of the DNA is applied and a much greater force is needed for a much smaller extension. We see here an example of a curved stress–strain relation of a biomaterial, as discussed previously. When the DNA is further expanded, a transition occurs in the DNA at a certain force, during which the conformation changes. This leads to a strong expansion with a very small force. What happens here is that the hydrogen bonds between the base pairs are separated by the applied force and effectively we go from double-stranded DNA to a single-stranded DNA.

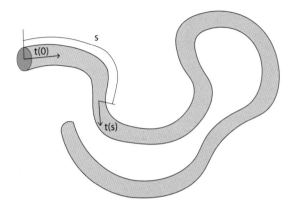

Figure 8.12 The shape of a polymer in solution is strongly bent by the thermal movement of the water. The space curve follows a probability distribution, which is described by a Gaussian function.

8.5.2 Polymers

We have seen what energy must be used to bend a beam. The bending energy was given by $E_B = \frac{ELI_z}{2R^2}$. In order to bend a rod into a loop, an energy of $E_B = \frac{\pi EI_z}{R}$ is necessary, since $L = 2\pi R$. In a solution, however, the molecules are always traveling with an average kinetic energy $k_B T$, as we have seen in heat theory. That is, collisions of a long-chained molecule with the molecules of the solution always lead to a disordered bending of the energy $k_B T$ of the long molecule. If $k_B T$ corresponds approximately to the bending energy of a loop, the long-chain molecule is compressed to form a coil, as we have already seen in the DNA single-molecule experiments and is illustrated in Figure 8.12.

But since we now know the bending energy, or on what it depends, we can estimate on which length scale a polymer is bent. To this end, we assume that the bending energy is just the average kinetic energy for an angle of about 180 degrees. This means that, on average, the thermal energy is sufficient to bend the polymer by 180 degrees, i.e., to completely change its direction. This is only a mean value, and the direction will fluctuate in all directions, since all kinetic energies can occur between 0 and 2 $k_B T$. Mathematically, we use the bending energy that $L = \pi R$ and for the energy we set to $E_B = 3/2k_B T$ (three thermodynamic degrees of freedom). This yields the following:

$$\frac{3}{2}k_B T = \frac{\pi EI_z}{2R}.$$

To a good approximation, the radius of curvature of a polymer in solution is always $R = \frac{EI_z}{k_B T}$. As already mentioned, the thermal motion is disordered, which is why the bending with this radius of curvature will always go in different directions. This is the curling up of the molecule we want to describe. The direction of the molecule is thus described by a random process, which is driven by thermal motion. This means that the probability distribution of the polymer is described by a diffusion equation, as described in Section 8.3. The step width of this random process is then given by the radius of curvature, which we have derived previously for the polymer. This length is also called the persistence length:

$$\xi_P = \frac{EI_z}{k_B T}.$$

Later we will describe at which length-scale a molecule behaves like a solid rod. If we describe the molecule as a cylindrical rod with radius r, we can also insert the area moment of inertia, and we obtain the following:

$$\xi_P = \frac{E\pi r^4}{4 k_B T}.$$

For the further description, we thus assume that the polymer consists of rods of length ξ_P, which point in random directions. If we move along the length of the polymer, we make a random walk with the step length of a persistence length. These steps along the polymer represent the time in the diffusion equation, and thus we can directly specify the probability distribution of the polymer. The probability of finding a part of the polymer at $\vec{x}$ is given by a Gaussian distribution $p(x)$:

$$p(x) = \frac{1}{\sqrt{2\pi R_G^2}} \exp\left(-\frac{x^2}{2R_G^2}\right), \tag{8.10}$$

where $R_G^2 = \xi_P L/6$ is the width of this Gaussian distribution, where L is the length of the polymer molecule. For DNA, the persistence length can be determined to be $\xi_P^{DNA} = 50$ nm. So for a DNA molecule in solution, the width of the local distribution, i.e., the region with a high probability of finding a DNA molecule, is approximately given by $R_G = \sqrt{\xi_P L/6}$. For DNA from a λ phage with a length of $L = 15$ μm, this yields $R_G = \sqrt{50\ \text{nm} \times 15\ \text{μm}/6} = \sqrt{750 \times 10^{-3}/6}\ \text{μm} = \sqrt{1/8}\ \text{μm} \simeq 0.4$ μm. In other words, a 15 μm long DNA molecule in solution is coiled up to less than a micron in size. If we now draw on such a molecule, the force that we have to expend, for a certain length change is given by changing the probability distribution $p(x)$ and thus the entropy over the Boltzmann relation. As a function of the distance, the entropy is given by $S = k_B ln(p) = -k_B \frac{x^2}{2R_G^2}$. This corresponds to a thermal energy of $E_{th} = TS = -k_B T \frac{x^2}{2R_G^2}$. The change of this energy with position immediately gives the force that has to be used for such an extension:

$$F = \frac{\partial E_{th}}{\partial x} = -k_B T \frac{x}{R_G^2}. \tag{8.11}$$

This describes a spring with a spring constant $c = \frac{k_B T}{R_G^2}$, which explains the elastic properties of the DNA molecule at small expansions.

This applies in general to polymers, in particular, in rubber bands, where the long-chain molecules are cross-linked and the elastic restoring force comes through the curling up of the molecules. This means that at higher temperatures, the restoring force of a piece of rubber is greater, which can also be seen in Figure 8.13.

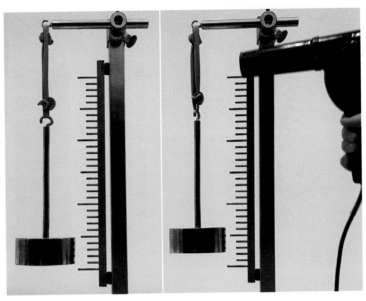

Figure 8.13 Due to the elasticity of rubber being governed by the enropy of conformation of the long molecules, rubber gets stiffer when heated. This can be shown by suspending a weight from a rubber band (left). When heating the rubber band, the weight is lifted, due to the increased spring constant (right).

8.5.3 The Cytoskeleton

We have seen that the thermal motion can have a great influence on the structure of polymer molecules. In the description of this process, the length of persistence, especially, depended on the size and geometry of the molecule. This means that differently sized molecules naturally show very different behavior in a cell. Remember the definition of the persistence length: $\xi_P = \frac{\pi E r^4}{4k_B T}$.

Let us look at two examples in the cytoskeleton. Tubulin is responsible for the construction of the microtubules, which form the backbone of a cell. As shown in Figure 8.14, the tubulin molecules form a tube with a diameter of about 25 nm. With a Young's modulus of a protein of about $E \simeq 250$ MPa, we obtain a persistence length for microtubules of $\xi_P^{MT} \simeq 1$ mm, which is many times greater than a cell. Therefore, it is also understandable that the microtubules form the backbone of the cell, because they are actually effective bracing. This is quite different for actin. These are also formed by polymerizing individual proteins, as shown in Figure 8.14. However, this strand has only a diameter of about 8 nm, i.e., about three times smaller than a microtubule. Because of the r^4 dependence of the persistence length on diameter, this means an approximately 100-fold reduced persistence length of about $\xi_P^{Ac} \simeq 12$ μm. Thus we can see that the actin in the cytoskeleton can be used for the more flexible structures in which the elastic deformations of the cell are given by those of the actin network. Finally, a DNA molecule (which is no longer part of the cytoskeleton) with the diameter of 2 nm will, as we have already seen, become even more flexible, namely about 250 times, which also means the aforementioned value of the peristance length of $\xi_P^{DNA} = 50$ nm.

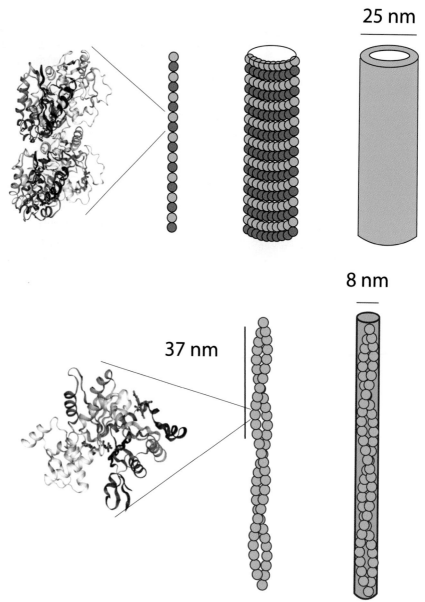

Figure 8.14 Two important biopolymers in the cytoskeleton, tubulin (top) and actin (bottom). Protein structure data from Nettles et al., 2004, and Graceffa and Dominguez, 2003, respectively. The filaments formed by these proteins can be approximated well by tubes of different diameters, giving rise to their very different mechanical properties.

8.5.4 Packing DNA within Cells

DNA is therefore a very flexible molecule because of its size and length. This means that it is not very difficult to pack the genetic information into a cell, since the thermal movement within the cell will lead to a great reduction in size. If we look at the genome length of different species, it is noticeable that in the prokaryotes that do not have a nucleus, the genome length is much shorter than in the eukaryotes and that the length for more complex animals increases more or less linearly. In the case of the eukaryotes, this is no longer the case at all. This can be understood when we consider that up to a certain genome length, the DNA can be packed without further measures within the cell due to the thermal motion. This length is given by the fact that the radius of gyration of the DNA must not be larger than the radius of the cell. A better criterion is half the radius of the cell, since within two radii of gyration we find 95% of the molecule (Gaussian distribution). So we have the condition that $R_{cell} = 2R_G = 2\sqrt{\xi_P L/6}$. We solve this for L to obtain $L = \frac{3R_{Zelle}^2}{2\xi_P}$. For a typical size of a cell of $R_{cell} = 5$ μm, this gives $L = \frac{3 \times 5000\,\text{nm}\,5\,\mu\text{m}}{2 \times 50\,\text{nm}} = 150 \times 5$ μm $= 0.75$ mm. A length of 1 mm is about 3 million base pairs of genetic information (0.34 nm per base pair), which is just about the genome length of the most complex prokaryotes. Longer genoms could not be packed into a cell, unless the cell becomes larger, but this is not practical due to the efficient transport of nutrients, etc. In other words, for more complex organisms, the DNA must be packed more efficiently, which also happens in eukaryotes in the nucleus. There, the DNA is wrapped around histone proteins, which fold in a three-dimensional structure, thus allowing the DNA to be packaged much more efficiently. The length of a DNA molecule that can be packaged with the help of histones is approximately given by $R \propto (\ell^2 L)^{1/3}$, where ℓ is the size of a histone (cf. Figure 8.15). If we use the size of the histones $\ell = 5$ nm, we see that DNA strands with a length of 1 m (as in humans) can be packaged into a nucleus of size $R \simeq 3$ μm.

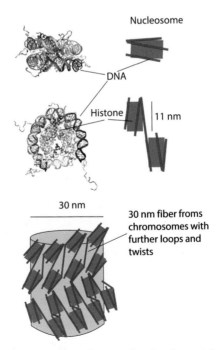

Figure 8.15   The packaging of DNA in the cell nucleus by means of histones. Data from Davey et al., 2003.

A large force is required for the strong bending of the DNA molecule to a bending radius of 5 nm. With thermal impacts, one obtains exactly at the persistence length, which is 10 times larger. For this purpose, electrostatic forces are necessary, which are described in Chapter 9 in more detail. We will also look at how the DNA is wrapped around a histone.

This compactification is, however, bought at the price of a more complicated reading of the genetic code. In the end, the complex has to be deliberately disentangled in order to read the DNA double strand with an RNA polymerase molecule. This means that the regulation of gene expression becomes much more complicated in such creatures, which then also leads to the fact that the complexity of the organism can no longer be read directly from the length of the genome.

8.5.5 Electrophoresis

Now that we have dealt with the packing of long-chained molecules, let us again deal with electrophoresis. In a gel, a long-chained molecule can only perform very limited motions. This is illustrated in Figure 8.16, where a (fluorescently labeled) actin strand is moving through a gel. Here, the actin strand can only move along its own contour, since the movements are limited laterally by the gel. This movement is called reptation. This

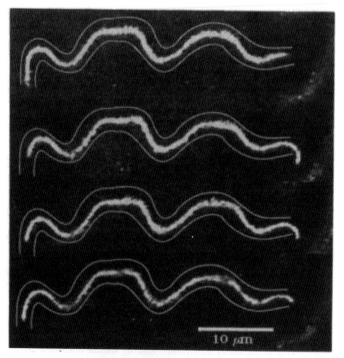

Figure 8.16 Illustration of the reptation of an actin strand in a gel. Time snapshots of the movement every 20 s are shown. Due to the restriction of the possibilities of movement by the network of the gel, the movement of the tangled structure of the molecule follows its own path. Figure from Käs et al., 1994, used with permission.

has a great influence on the mobility of a long molecule in a gel. Instead of a diffusive movement in three dimensions, therefore, only one-dimensional movement takes place. The time to move around your own length is thus given by $\tau_{rep} = L^2/D_1$, where D_1 is the diffusivity in one dimension. This is given by the viscosity of the solvent η and the length of the molecule using the Einstein relation: $D_1 \propto \frac{k_B T}{\eta L}$. Therefore, we obtain a reptation time of $\tau_{rep} \propto L^3 \eta/(k_B T)$. If you think back to the description of viscoelastic materials in Section 7.8, this corresponds to the time scale on which a Maxwell material behaves as a liquid, i.e., where stresses are relaxed, $\tau = \frac{\eta}{E}$.

If we now wish to treat the mobility of such a long molecule in the gel, we know that the diffusivity in three dimensions is proportional to the mobility we are looking for. This diffusivity is, however, precisely given to the fact that in the reptation time the radius of gyration is swept, that is, $\mu \propto D_3 = R_G^2/\tau_{rep}$. With the relation for the radius of gyration from the last section $R_G^2 \propto L$ and the dependence of the reptation time on the length $\tau_{rep} \propto L^3$, we obtain a mobility $\mu \propto L^{-2}$. If all the important parameters are taken into account, one obtains $\mu \propto \frac{\xi P}{\eta L^2}$, irrespective of temperature. As we have seen in mechanics, the electrophoretic mobility is just the mobility multiplied by the charge Q. In DNA, this charge increases in proportion with its length, $Q = \sigma L$, where σ is the number of charges per length (in the case of DNA, a negative elementary charge per base pair). Thus, the electrophoretic mobility $\mu_E = Q\mu \propto L^{-1}$. As we have also discussed in mechanics, the velocity of a molecule is just proportional to electrophoretic mobility, which means that shorter sections will move faster than longer ones by a factor of $1/L$. So within the same time, the shorter ones travel farther. If the DNA pieces are shorter than the persistence length, this argument is no longer valid. Then the pieces all move at approximately the same speed. DNA fragments smaller than 50 nm can therefore no longer be well separated using gel electrophoresis.

8.6 *Nonequilibrium Processes

Many processes we have looked at so far have been equilibrium processes. In these, there is no addition of matter or energy to the system and they evolve by themselves. In this case, the second law of thermodynamics applies, such that the system always strives to the most probable state (which is usually homogeneous). This, however, is no longer the case when there is a constant source of energy from the outside of the system. One striking example of such a nonequilibrium process is life itself, where in order to survive any living being must constantly be fed. As soon as thermal equilibrium is reached, or actually long before that, any living thing is dead. This means that living systems are not subject to the second law of thermodynamics. The frequently made statement that random processes cannot lead to a sufficiently complex system, which could be considered to be alive, because of maximization of entropy, is therefore invalid. Let us look at some nonequilibrium processes as an illustration of how a continuous supply energy can produce patterns that correspond to a lower entropy than a homogenously mixed system.

8.6.1 Convection

If a thin layer of a liquid is heated from below, a temperature gradient will occur, between the warm and cold above. If this difference in a given layer exceeds a certain size, flow patterns are formed in the liquid that form a regular pattern. With a layer of oil in a frying pan, it is easy to observe in the kitchen. This pattern forms cellular shapes, which are also called Rayleigh–Benard cells (see Figure 8.17). The currents that occur are called convection. These currents transport the molecules over macroscopic distances. For example, if you open a bottle of perfume at one end of the room and after a few minutes smell the perfume at the other end, the etheric molecules have been transported via convection currents in the air (and not by diffusion, as is often claimed).

We shall examine this instability more closely and derive a quantitative criterion for its occurrence. To do this, we consider a small volume of liquid near the heating plate. Since this is warmer than the environment, it expands, so it becomes lighter. Specifically, the density ρ of the material is described by the thermal expansion coefficient α: $\rho(T + \Delta T) = \rho(T)(1 - \alpha \cdot \Delta T)$. Therefore, a buoyant force acts on this bit of liquid given by $F_A = \Delta \rho V g$, where V is the volume we consider (let's make this a sphere, so we have $V = 4\pi R^3/3$). The density difference $\Delta \rho$ is given by the thermal expansion, i.e., $\Delta \rho = \alpha \rho \Delta T/d \cdot dz$. Here ΔT is the temperature difference across the whole layer thickness d, and dz is a small change in position along the height of the layer. Putting all of this together, we obtain a buoyant force on the small volume we consider of $F_A = \frac{\alpha \rho \Delta T}{d} \frac{4\pi R^3}{3} \cdot g \cdot dz$, which would accelerate it upward into cooler layers. However, the volume is also braked by viscous friction. This is given by (we consider a spherical volume) $F_R = 6\pi \eta R v$, where v is the speed of the rising volume. If both forces are equal, we can obtain the time it takes for the volume to travel its

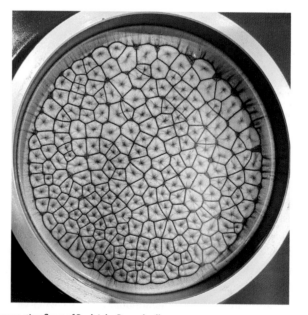

Figure 8.17 Pattern formation in convection flows of Rayleigh–Benard cells.

own diameter. Force balance says $\frac{\alpha\rho\Delta T}{d}\frac{4\pi R^3}{3}\cdot g\cdot dz = 6\pi\eta Rdz/\tau$, where τ is the rise time we are looking for. Solving this yields the following:

$$\tau = \frac{9\eta d}{2\alpha\rho\Delta TgR^2}.$$

So the thicker the layer, the longer the rise time, and the greater the temperature difference, the shorter the rise time. If this rise time is longer than the time it takes for the volume to equilibrate to the temperature of the surroundings by means of heat conduction, the flow will automatically come to a standstill. That is, in order to obtain a criterion for the occurrence of the convection currents, we must compare the rise time with the time of the heat conduction. The heat-conduction equation has already been discussed, and it has the form of a diffusion equation: $\frac{\partial T}{\partial t} = \kappa\frac{\partial^2 T}{\partial x^2}$, where κ is the thermal conductivity coefficient. The time of the heat conduction is obtained from the consideration of the volume element: to diffuse over a distance R, the heat takes a time $\tau_W = \frac{R^2}{6\kappa}$. Thus we have an instability when the heat exchange time τ_W is longer than the rise time τ: $\tau_W > \tau$ or $\frac{R^2}{6\kappa} > \frac{9\eta d}{2\alpha\rho\Delta TgR^2}$. This gives a condition for a dimensionless combination of all the relevant parameters as follows:

$$\frac{\alpha\rho\Delta TgR^4}{\eta d\kappa} > 27.$$

If we further consider that our arguments should be independent of the size of the arbitrary volume R, we must express this as a function of the layer thickness. We have seen that the rise time is shorter for larger volumes. For the instability, we want to consider the minimum rise time, that is, as large a volume as possible. The maximum volume would fill the entire layer, $R = d/2$. However, such a volume can no longer rise, and we must consider a somewhat smaller one. We will take a volume of $R = d/3$ as an example. In that case, the criterion for the occurrence of instability becomes as follows:

$$\frac{\alpha\rho\Delta Tgd^3}{\eta\kappa} = Ra > 2{,}200.$$

The dimensionless combination of the essential parameters is also called the Rayleigh number. A much more complicated, but more accurate calculation, determines the volume to be considered and becomes an instability point $Ra > 1{,}700$.

8.6.2 Granular Demixing and Instabilities

As a second example, we make a model of our thought experiment of the distribution of many particles on two halves of a container. So we make a cylinder, which is separated in the middle by a wall, and in this wall we make an opening. If we give small particles (sand grains or something similar) into this container and shake, something very strange happens. The shaking causes the particles to collect in one half! At the beginning, it is not clear which half it will be, but ultimately the particles will land in one of the two halves (see Figure 8.18). What is the reason? With this thought experiment, we had justified the increase of the entropy, since the mixed state is much more probable than the unmixed!

Figure 8.18 Snapshots of a time sequence of shaking an initially mixed collection of beads. Due to an instability caused by inelastic collisions, the particles separate, until at the end (right) all particles are in a single compartment.

The point is that the entropy does not enter the problem at all. Without the shaking, that is, the constant supply of energy, nothing happens here. So we have a non-equilibrium phenomenon. The reason why nothing happens is that the shocks between these real particles always dissipate some energy (i.e., convert to heat). This was not the case in the thought experiment. That is, if the particles are left to themselves, they will quickly come to rest. The more particles are in a container, the more bumps take place. The mean free path is smaller. As more impacts occur, more energy is dissipated. So if there are not the same number of particles in the two halves, then more energy dissipates with more particles. Thus, the particles in this half also have slightly less mean kinetic energy and thus a lesser probability to come to the height h of the hole. Thus, more particles of half with fewer particles therein are transported into that with more particles therein, which enhances the effect. This means that the emptying of the less-filled half is always faster when it has begun. A small fluctuation at the beginning is thus strengthened in this case to a maximum difference.

Again, let us look at this more quantitatively. To this end, we first consider the flow of particles between the different halves. This particle flow is given by the particle density at the height of the hole $n(h)$ times the average speed of the particles u, so the flow from 1 to 2 is given by $j_{12} = n_1(h)u_1$ and conversely from 2 to 1 by $j_{21} = n_2(h)u_2$. In a stationary state, these flows are the same, such that the same number of particles move from right to left and vice versa. In order to characterize this, we must consider what the mean speed of the particles is. This is supplied to the particles by collisions with the wall moved by the shaking. The total momentum has to be distributed over all of the particles, and the speed of the wall is determined by the shaking frequency and amplitude. Therefore, we can assume $(af)^2 \propto u^2 N$, where a is the amplitude and f the frequency. This will give the average speed: $u \propto af/\sqrt{N}$. The exact proportionality will depend on the properties of the particles and how elastic the collisions are, but we will not need these details and simply write $u = Aaf/\sqrt{N}$. Finally, we need the particle density at height h. However, this will be given by a Boltzmann distribution, where instead of the real temperature we must have the average kinetic energy of the particles. So we get $n(h) = N\exp\left(-gh/u^2\right) = N\exp\left(-\frac{ghN}{Aaf}\right)$. The flux of the particles from one half to the other is then given by $j = n(h)u = Nu\exp\left(-\frac{ghN}{Aaf}\right) = NAaf/\sqrt{N}\exp\left(-\frac{ghN}{Aaf}\right) = \sqrt{N}Aaf\exp\left(-\frac{ghN}{Aaf}\right)$. If we now consider the stationary case, the fluxes in both compartments must be equal. We then obtain the following:

$$\sqrt{N_1}\exp\left(-\frac{ghN_1}{Aaf}\right) = \sqrt{N_2}\exp\left(-\frac{ghN_2}{Aaf}\right).$$

This equation is true if $N_1 = N_2$, but also $N_1 = 0$ and N_2 large or vice versa.

What really interests us, however, is which of these two states is stable, i.e., in which state the original state is restored when the particle numbers change somewhat. For this, we go into the state, where N_1 and N_2 are both roughly $N/2$, to be exact: $N_1 = N/2(1 + \epsilon)$ and $N_2 = N/2(1 - \epsilon)$, and ϵ is some small number. Then the change of the particle numbers to the right is described by the difference of the two currents. More specifically,

$$\frac{dN_1}{dt} = \frac{Nd\epsilon}{2dt} = -j_{12} + j_{21} = \sqrt{N_2} \exp\left(-\frac{ghN_2}{Aaf}\right) - \sqrt{N_1} \exp\left(-\frac{ghN_1}{Aaf}\right)$$
$$= \left(\sqrt{(1-\epsilon)} \exp\left(\frac{ghN}{2Aaf}\epsilon\right) - \sqrt{(1+\epsilon)} \exp\left(-\frac{ghN}{2Aaf}\epsilon\right)\right) \exp\left(-\frac{ghN}{2Aaf}\right).$$

Since ϵ is small, we can Taylor-expand the square root to obtain the following:

$$\frac{d\epsilon}{dt} = \left((1 - \epsilon/2) \exp\left(\frac{ghN}{2Aaf}\epsilon\right) - (1 + \epsilon/2) \exp\left(-\frac{ghN}{2Aaf}\epsilon\right)\right) \frac{2}{N} \exp\left(-\frac{ghN}{2Aaf}\right).$$

Similarly, we can expand the exponential, which yields the following:

$$\frac{d\epsilon}{dt} = \left((1 - \epsilon/2)\left(1 + \frac{ghN}{2Aaf}\epsilon\right)\right) - (1 + \epsilon/2)\left(1 - \frac{ghN}{2Aaf}\epsilon\right)\right) \frac{2}{N} \exp\left(-\frac{ghN}{2Aaf}\right).$$

or

$$\frac{d\epsilon}{dt} = \left(\left(-\epsilon/2 + \frac{ghN}{2Aaf}\epsilon\right) - \left(\epsilon/2 - \frac{ghN}{2Aaf}\epsilon\right)\right) \frac{2}{N} \exp\left(-\frac{ghN}{2Aaf}\right)$$
$$= \left(-1 + \frac{ghN}{Aaf}\right) \frac{2}{N} \exp\left(-\frac{ghN}{2Aaf}\right) \cdot \epsilon.$$

This means that the fluctuation increases if $\left(-1 + \frac{ghN}{Aaf}\right)$ is positive. In this case, the state in which the number of particles on both sides is equal is unstable and the solution with the majority (or all) of the particles in one of the two halves is realized. This also corresponds to a stationary state since then the currents on the respective sides are both zero and thus equal.

8.6.3 Pattern Formation in Living and Nonliving Systems

Instabilities as the ones described in the preceding section lead to a separation of qualitatively different states, such as the upward and downward flow of liquid in convection or the density of grains in granular demixing. This separation basically constitutes the most basic form of a pattern, which for instance in convection is periodically repeated giving rise to the convection cells that one can observe in Figure 8.17 or in rice that has cooked long enough or also in the sun, where these are the origin of sunspots. Thus it has been interesting to study the formation of patterns due to such instabilities, which might give rise to patterns also in living organisms. The first study of this has been done by Alan Turing (Turing, 1952), who tried to model the pattern formation in morphogenesis by studying the reaction-diffusion system of two antagonistic morphogens. These consist of an activator (a) and an inhibitor (h) of a certain trait, which can diffuse and which can also interact. The corresponding model then is given by Gierer and Meinhardt (1972):

$$\frac{\partial a}{\partial t} = D_a \frac{\partial^2 a}{\partial x^2} + \rho_a \left(\frac{a^2}{h} - a \right) \tag{8.12}$$

and

$$\frac{\partial h}{\partial t} = D_h \frac{\partial^2 h}{\partial x^2} + \rho_h (a^2 - h). \tag{8.13}$$

Here, the interactions of the activator and inhibitor are described by the last terms in the equations. The inhibitor is activated by the activator (therefore the term with a^2 in the second equation) and the activator is inhibited by the inhibitor (therefore the term with $1/h$ in the first equation). In order for such a system to give a pattern, the activation has to be short range, whereas the inhibition has to be long range. Intuitively, this means that if there is locally a fluctuation increasing the activator, this gets increased, but due to the long-range inhibition, the increase remains localized. This therefore gives rise to a pattern of spots. In order to do this quantitatively, what one does is similar to the situation we looked at with granular demixing, i.e., one looks at what happens to a steady state when there is a fluctuation. A steady and homogeneous state is characterized by no variation in x and t, i.e., the preceding model reverts to the following:

$$0 = \left(\frac{a_0^2}{h_0} - a_0 \right) \tag{8.14}$$

and

$$0 = (a_0^2 - h_0) \tag{8.15}$$

This means that we quantitatively know the steady and homogeneous state to be $h_0 = a_0^2$ and $a_0 = 1$. A small fluctuation from this steady state is then given by δa and δh, and we have $a(x, t) = a_0 + \delta a(x, t)$ and $h(x, t) = h_0 + \delta h(x, t)$. The main trick now is that we are only interested in what happens for small fluctuations, i.e., in the case where both δa and δh are small compared to a_0 and h_0. This means that we can Taylor-expand the nonlinear reaction terms in the preceding equations, as follows:

$$\frac{a^2}{h} - a = \frac{(a_0 + \delta a)^2}{(h_0 + \delta h)} - a_0 - \delta a \simeq \frac{a_0^2}{h_0}(1 + 2\delta a/a_0)(1 - \delta h/h_0) - a_0 - \delta a.$$

If we now insert the values of $h_0 = 1$ and $a_0 = 1$, we obtain the following:

$$(1 + 2\delta a)(1 - \delta h) - 1 - \delta a \simeq (1 + 2\delta a - \delta h) - 1 - \delta a = \delta a - \delta h.$$

Similarly, we can expand the following:

$$(a^2 - h) \simeq 1 + 2\delta a - 1 - \delta h = 2\delta a - \delta h$$

With these approximations, we obtain from the equations for a and h two equations for δa and δh, which are much simpler and can be solved in order to see whether or not the steady and homogeneous state is stable or unstable. The equations we obtain are as follows:

$$\frac{\partial \delta a}{\partial t} = D_a \frac{\partial^2 \delta a}{\partial x^2} + \rho_a (\delta a - \delta h) \tag{8.16}$$

and

$$\frac{\partial \delta h}{\partial t} = D_h \frac{\partial^2 \delta h}{\partial x^2} + \rho_h (2\delta a - \delta h). \qquad (8.17)$$

For this purpose, we have to assume a specific shape of the fluctuation in the form of $\delta a(x,t) = \delta a_0 e^{\lambda t} \cos(kx)$, such that the space and time derivatives are given by $\frac{\partial \delta a}{\partial t} = \lambda \delta a$ and $\frac{\partial^2 \delta a}{\partial x^2} = -k^2 \delta a$. Inserting this ansatz therefore gives us a system of two linear equations in δa and δh given by the following:

$$\lambda \delta a = -D_a k^2 \delta a + \rho_a (\delta a - \delta h)$$

and

$$\lambda \delta h = -D_h k^2 \delta h + \rho_h (2\delta a - \delta h)$$

or separating the different variables,

$$(\lambda + D_a k^2 - \rho_a)\delta a = -\rho_a \delta h$$

and

$$(\lambda + D_h k^2 + \rho_h)\delta h = 2\rho_h \delta a.$$

We can insert the first equation into the second one and obtain the following:

$$(\lambda + D_h k^2 + \rho_h) = -\frac{2\rho_h \rho_a}{\lambda + D_a k^2 - \rho_a}.$$

This can be rewritten as follows:

$$(\lambda + D_h k^2 + \rho_h)(\lambda + D_a k^2 - \rho_a) + 2\rho_h \rho_a = 0$$

or

$$\left(\lambda^2 + \lambda \left(\rho_h - \rho_a + k^2(D_h + D_a)\right)\right) + D_h D_a k^4 + \rho_h \rho_a + k^2(D_a \rho_h - D_h \rho_a) = 0.$$

This gives a condition for the parameters of diffusivity and reaction such that solutions exist, which we can obtain by solving this quadratic equation for λ (we assume $\rho_a = \rho_h = \rho$ for simplicity):

$$2\lambda = -k^2(D_h + D_a) + \sqrt{(k^2(D_h + D_a))^2 - 4(D_h D_a k^4 + \rho^2 + k^2(D_a - D_h)\rho)}.$$

We can rearrange this somewhat to simplify the term in the square root to get the following:

$$2\lambda = -k^2(D_h + D_a) + \sqrt{(k^2(D_a - D_h))^2 - 4(\rho^2 + k^2(D_a - D_h)\rho)}$$

Whether such a λ that one obtains can be positive or negative decides whether a fluctuation grows or is damped out, i.e., whether the homogeneous and steady state is stable or not. The preceding equation actually can only have positive values for λ if the diffusivity of the inhibitor is much larger than the diffusivity of the activator (otherwise, the negative first term always wins). This corresponds to the case described previously, where the activation is short range and the inhibition is long range, such that a pattern can form. The calculation, however, yields more information, such as how the diffusivity has to relate to the reaction rates, as well as a length scale ($L = \sqrt{D_h/\rho}$) of the corresponding pattern below which a pattern does not form. If we neglect the diffusivity of the activator above and introduce the length scale, we find a solution for λ of

$$2\lambda/\rho = -k^2 L^2 + \sqrt{(k^4 L^4 - 4 + 4k^2 L^2)},$$

which is positive as long as kL is larger than 1. Actually, close to $kL = 1$, we can solve this approximately to obtain the following:

$$\lambda \simeq \rho(kL - 1).$$

What this lengthy calculation shows is that instabilities of reaction-diffusion systems can lead to patterns that were not encoded anywhere simply due to the presence (and enhancement) of fluctuations. Moreover, the scale of the pattern depends on a few parameters such as diffusivities and reaction rates. If the morphogens in such a system act as regulators of gene expression, patterns in, e.g., developmental biology can be formed, thus controlling morphogenesis. The dependence on the critical length scale can, for instance, be used to describe regeneration experiments, where amputated parts of an organism may be too small to form a pattern and hence give rise to a change in the observed morphogenesis. However, we have not specified in any of this that the reaction system has to be inside a living organism, and these patterns also appear in direct chemical reactions.

In fact, living systems typically have many more redundant systems of controlling morphogenesis, such that often a description as the preceding one does not give the full picture. For instance, if we consider the formation of the anterior–posterior axis in the *Drosophila* embryo by the maternal bicoid protein gradient we have described in Section 8.3.2, many more aspects can be important. For instance, downstream of bicoid is the gene hunchback, which corresponds to a specification of the center of the embryo. This center is specified by the concentration profile of hunchback more reliably than would be expected from the fluctuations of the concentrations of bicoid, and the control of this requires a more complicated reaction-diffusion system than the Turing system previously described, but can still be achieved (Aegerter-Wilmsen et al., 2005).

Exercises

8.1   Diffusion of molecules

(a) What is the diffusivity of a long molecule? Use the Einstein relation $D = \mu k_B T$. Assume the molecule is a cylinder with length L, such that the mobility is given by $\mu = 1/f = 1/(8\pi \eta L)$, where η is the viscosity of the solvent (take water at room temperature, 20°C, where one finds $\eta = 1.00(1) \times 10^{-3} Pas$) and use $L = 31.5(3)$ nm for the length of the molecule.

(b) What is the uncertainty of this diffusivity.

8.2   Diffusion of molecules 2

(a) It is often said that diffusion is responsible when we quickly smell an opened bottle of perfume. Test this statement quantitatively. Usually it takes less than a minute until one can smell the perfume in another corner of the room, i.e., a distance of maybe 10 m away. How long does it take to transport a substance this far by diffusion? The diffusivity of gases (and for gases in other gases, i.e., perfume in air) is roughly 16(1) mm²/s. This is the actual value of methane in air. To do this estimate, determine the time it takes for a diffusive cloud to reach a width $\langle x^2 \rangle = 6Dt$ of 10 m.

(b) How long does it take for the same type of molecule to pass a distance of 500 μm? The diffusivity is again 16(1) mm²/s.

(c) Determine the uncertainties of the values in (a) and (b).

8.3   Diffusion and Brownian motion

You have studied the Brownian motion of fat droplets in milk using a microscope. In the course of this study, you have followed one individual droplet with a radius of $r = 0.40(4)$ μm during 10–15 minutes. Every 2.0(1) s, you have determined its position for 400 consecutive time points. From these, you have determined a histogram of the changes in position between consecutive time steps, from which you have observed and obtained the following result:

Δx (μm)	N
−0.5 to 0.5	117
0.5–1.5	84
−0.5 to −1.5	92
1.5–2.5	46
−1.5 to −2.5	32
2.5–3.5	8
−2.5 to −3.5	15
3.5–4.5	3
−3.5 to −4.5	2
4.5–5.5	0
−4.5 to −5.5	1

(a) Plot a histogram of these data and determine $\langle \Delta x \rangle$, i.e., the mean distance traveled in 2 s, as well as $\langle \Delta x^2 \rangle$, i.e., the mean square displacement. Determine the diffusivity of the droplet from these quantities.

(b) Use the Einstein relation to determine the Botzmann constant from your measurements. Since the fat droplet essentially diffuses in water, use the viscosity of water, $\eta = 1.00(5) \times 10^{-3}$ Pa s. Also, the experiment was done at room temperature, i.e., $T = 300(3)$ K.

(c) What is the uncertainty of the Botzmann constant you have thus determined? Compare the value and uncertainty to the literature value.

8.4   Air pressure

(a) If you have a properly sealing fridge in which you are cooling down air from 27°C to −3°C, what is the force you have to apply to open up the fridge door again once it has been closed for a while? For simplicity's sake, consider that the handle is in the middle. The outside air pressure is 1,000 hPa and the door 1 by 2 m^2.

(b) If the fridge is not properly sealing, by how much does the air in the fridge change (measured in moles)? The fridge is 50 cm deep and otherwise the same as in (a).

8.5 Heating a room

You want to increase the temperature of your living room (volume $V = 50(1)$ m^3) from 290 K to 293 K. The air pressure is 1,000(20) hPa.

(a) How much does the internal energy of the air in the room change?
(b) How much heat do you have to generate?
(c) What is the error of this amount of heat?

8.6 Thermal equilibrium

Consider a box made up of two compartment with a dividing wall, and a gas on the left as well as on the right. Both parts, with volumes V_1 and V_2 respectively, contain an ideal gas, where the values of p, V, and T are different on both sides. After this, the system is allowed to evolve on its own accord, i.e., reach thermal equilibrium. At the end of this, there are values p', V', and T'. Determine how these values relate for the two parts respectively, i.e., whether or not the properties are equal in equilibrium.

The dividing wall puts the two parts in thermal contact, but otherwise the system is isolated. Distinguish the following three different cases and determine which of the properties p, V, and T are equal in the two parts in equilibrium. Argue qualitatively.

(a) The dividing wall is fixed.
(b) The dividing wall is mobile.
(c) The dividing wall is fixed but permissive of gas transport.

8.7 Thermal expansion

Show that the thermal expansion coefficient is given by $\alpha = 1/T$. The coefficient is defined as the relative change in volume with temperature, i.e., $\alpha = \frac{1}{V}\frac{\partial V}{\partial T}$, while keeping the pressure constant. Use the ideal gas law $pV = Nk_B T$.

8.8 Maxwell–Boltzmann distribution

The Maxwell–Boltzmann distribution describes the probability $P(v)$ of finding an atom (of mass m) in a gas (at temperature T) to have a speed v and is given by
$P(v) = const \times v^2 exp\left(-\frac{mv^2}{2k_B T}\right)$.

(a) Determine the maximum of this function $v_{max}(T)$ by differentiating it and setting the derivative to zero. What is the physical meaning of this speed?
(b) Determine the second moment (or variance) of this distribution, i.e., $\langle v^2 \rangle = (\int v^2 P(v)dv)/(\int P(v)dv)$. What is the physical meaning of this property? Use the following integrals: $\int x^2 exp(-x^2/2)dx = \sqrt{\pi/2}$, and $\int x^4 exp(-x^2/2)dx = 3\sqrt{\pi/2}$.

8.9 Maxwell–Boltzmann distribution

(a) The reaction rate of a chemical reaction (with an activation energy ΔE at temperature T) is described by a Boltzmann-type distribution, i.e.,

$k = k_0 \exp(-\Delta E/(k_B T))$. If you have measured a certain reaction rate k at room temperature ($T = 293(3)K$) with an activation energy $\Delta E = 4.0(4) \times 10^{-20} J$, at what temperature do you have to carry out the reaction to observe a doubling of the reaction rate?

(b) Determine the errors of the value determined in (a).

8.10 Specific heat

(a) What is $C_V = \frac{\partial Q}{\partial T}$ for an ideal two-atomic gas? Consider the different possibilities of motion in order to determine the number of thermodynamic degrees of freedom.

(b) What is the specific heat at constant pressure? It is defined as $C_p = \frac{\partial Q}{\partial T}$; however, the heat is for the case that the pressure remains constant. To determine this quantitatively, determine the part of the internal energy given by work due to the pressure, i.e., $W = p\Delta V$, using the ideal gas law. Give a numerical value for a one-atomic gas such as helium.

8.11 Elastic properties of DNA

Consider the force extension curve of a single DNA molecule shown in the text. A DNA molecule 16 μm long has been stretched in this case.

(a) Determine the Young's modulus of DNA. For this purpose, assume that DNA is a cylinder with a diameter of 2.0(1) nm and concentrate on the part of the curve where the molecule is stretched out.

(b) What is the error of the modulus thus determined?

(c) At the beginning of the curve, the slope is due to a disentanglement of the curled-up molecule. For such an entropic spring, we had determined a spring constant of $\frac{3 \times 6 \times k_B T}{L \xi_P}$ in the text, where ξ_P is the persistence length, T the temperature (in K), and k_B the Boltzmann constant. Assuming the experiment was carried out at room temperature (i.e., 20°C), determine the persistence length of DNA from the slope of the curve.

(d) Also determine the error in (c).

(e) For a cylindrical molecule, we have seen that the persistence length is given by $\xi_P = \pi E R^4/(16 k_B T)$, where R is the radius and E the elastic modulus of the molecule. Given the persistence length of DNA (you should have found about 50 nm in (c)), determine the elastic modulus.

8.12 Packing of long-chained molecules

(a) How long can a piece of DNA be to still fit inside a (spherical) cell with a radius of $R_c = 5(1)$ μm? Assume that DNA is curled up by thermal fluctuations, leading to a random arrangement corresponding to a Gassian distribution with a radius of gyration $R_g^2 = \xi_P \cdot L/6$. The persistence length of DNA is roughly $\xi_P = 50(5)$ nm. In order for the entire molecule to fit inside the cell, the radius of the cell has to be about twice the radius of gyration. How many base pairs does this correspond to?

(b) What's the uncertainty of the preceding estimate?

(c) What energy do you have to spend to bend DNA to a radius of curvature of $r = 5.0(5)$ nm, i.e., the size of a histone? Use that the bending energy is given by $E_{Bieg} = \pi ER^4 L/(8r^2)$, as well as the definition of the persistence length: $\xi_P = \pi ER^4/(4k_B T) = 50(5)$ nm, where $R = 1.0(3)$ nm is the radius and $E = 2.50(3) \times 10^8 Pa$ the elastic modulus of DNA. All of this happens at room temperature (293 K). For a turn with radius r, you need a length of $L = 2\pi \cdot r$ of the molecule.

(d) What's the error in c)?

8.13 Convection

We want to determine the onset of convective flows in a gas (this actually explains the distribution of perfume in a room on fast time scales). For this purpose, we first need to determine the different properties of gases needed in the Rayleigh criterion described in the text. In another question, we have already determined that the thermal expansion coefficient of a gas is $\alpha = 1/T$.

(a) What is the mean free path, ℓ in a gas at a temperature of $T = 293$ K and a pressure of $p = 10^5$ Pa. We have determined in the text that the mean free path is given by $\ell = \frac{1}{2\pi d^2 N/V}$. Assume a size of the gas molecules of $d = 2\text{Å}$.

(b) What is the viscosity of a gas? Consider that in gases, the viscosity is due to random collisions between gas molecules, giving a kinematic viscosity of $\nu = \eta/\rho = \langle v \rangle \ell/3$ in three dimensions. Use the standard deviation $\sqrt{\langle v^2 \rangle} = \sqrt{3k_B T/m}$, from the Maxwell–Boltzmann distribution you have determined earlier as a measure of the average speed. For air, the mass is about $m \simeq 30$Da, such that $\langle v \rangle \simeq 500$ m/s. The density of air is about $\rho = 1$ kg/m^3.

(c) The thermal conductivity can be determined analogously to the viscosity, since it is also determined by random collisions of the gas molecules. Thermal diffusivity will therefore be very close to kinematic viscosity, i.e., $\kappa \simeq \nu \simeq 2.0(2) \times 10^{-5}$ m^2/s. With the thermal expansion coefficient $\alpha = 1/T$, we can now fully calculate the Rayleigh number $Ra = \frac{g\Delta T d^3 \alpha}{\kappa \nu}$ for a gas, if we know the temperature difference and the layer thickness. The onset of convection takes place when the Rayleigh number exceeds its critical value of $Ra_c \simeq 2.0(2) \times 10^3$. So if we have a temperature difference of $\Delta T = 0.010(1)$ K within a room at room temperature $T = 293(3)$ K, how high does this room have to be such that convection will take place and leads to appreciable currents of air in the room?

(d) What is the uncertainty of the value obtained in (c)?

Quiz Questions

8.1 Energy

Which of the following statements is correct?

A Processes with friction destroy energy.

B The potential energy of a compressed spring is $kx^2/2$.

C The kinetic energy of translational motion is $mv/2$.

D Heat is a form of energy.

E In a closed system, the sum of potential and kinetic energy is constant except if there is friction or other nonconservative forces at play.

F A spatial integral of an energy gives rise to a force.

8.2   Thermal expansion

If you heat up a metal ring, how does the inner diameter of the ring change because of thermal expansion?

A It increases.

B It decreases.

C It remains the same.

8.3   Boltzmann distribution

In a container filled with a mixture of hydrogen, oxygen, and nitrogen at a given temperature, which molecules in the mixture have the largest kinetic energy?

A Those of hydrogen.

B Those of oxygen.

C Those of nitrogen.

D All are equal.

8.4   Entropy

What happens to the entropy of an ideal gas if you double the number of particles (while keeping temperature and volume the same)?

A It doubles.

B It is halved.

C It remains constant.

D It increases by $k_B ln(2)$.

E It decreases by $k_B ln(2)$.

8.5   Maxwell–Boltzmann distribution

What is the width (variance) of the velocity distribution in an ideal gas?

A $k_B T/m$

B $m/(k_B T)$

C $(k_B T)^2$

D $\sqrt{k_B T}$

E $1/(k_B T)^2$

F $1/\sqrt{k_B T}$

8.6   Ideal gas

For an ideal gas (for instance, helium is a good approximation of an ideal gas), the energy content is directly proportional to the temperature according to the ideal gas law.

If a helium balloon is heated from 2°C until it contains twice the energy, what temperature do you have to heat it to?

A 4°C

B 4 K

C 278 K

D 277°C

E 138 K

9 Electrical Charges and Currents

9.1 Electric Charges, Fields, and Potentials

As we have already seen in the discussion of the various forces in mechanics, almost all forces we know from everyday life are actually due to electromagnetic interactions. This is due to the fact that atoms consist of electrically charged particles that are separated into the atomic shell (electrons) and in a very small, dense nucleus (protons). For macroscopic bodies, the nuclei are shielded by the electrons and thus the interactions of the charged electrons determine friction and normal force, but also cohesive or adhesive forces. This is why we are now looking to get a deeper understanding of this interaction with a few basic examples, which for an understanding of molecular biology are of particular importance, since it is these molecular-scale forces that determine the function of the individual biomolecules. In the course of this, we will also revisit the Bohr model of the atom and see how electric interactions of the electrons actually explain how molecules form and what the different properties of the elements are.

9.1.1 Electric Charge and Coulomb's Law

The basic physical phenomenon at the heart of Coulomb's law is that there are electric charges. What we mean by this can be summarized by a few observations one can demonstrate. This does not mean that we understand why there should be something like electrical charges or what they actually are. But we can work with them and have a quantitative definition from which to work further. Let's state the basic tenants of electrostatics in words rather than an equation:

1. There are positive and negative electric charges. Microscopically, the negative charges correspond to excess electrons, the positive ones to excess protons. This can be demonstrated by means of the frictional electricity: charge generated by rubbing cat fur on a plexiglas rod attracts the opposite charge generated by rubbing leather on a glass rod. On the other hand, like charges will repel.
2. Electrical charges only appear in integer multiples of the elementary charge $e = 1.602189(5) \times 10^{-19}$ C. Why the electric charge only appears in well-defined packages (is quantized), we do not know.
3. Units: $[q] = 1$ C $= 1$ Coulomb $= 1$ ampere $\times$ second.
4. The total electric charge (i.e., the sum of all electrical charges present) is conserved. Electrical charge cannot be created or destroyed. The charge we obtain in frictional

electricity is obtained by macroscopically separating positive and negative charges (i.e., electrons and nuclei).

In terms of an equation, we can give the same substance by describing Coulomb's law, which tells us what the force is between two electric charges. For two charged point particles, the Coulomb force is formally written as follows:

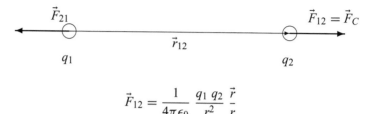

$$\vec{F}_{12} = \frac{1}{4\pi\epsilon_0} \frac{q_1\, q_2}{r^2} \frac{\vec{r}}{r}$$

This is the force that a body 1 with the electrical charge q_1 exerts on a body 2 with the electrical charge q_2. Here $\vec{r}$ is the distance vector that points from body 1 to body 2. If q_1 and q_2 have the same sign, the force is repulsive, as we have stated previously based on experiments. For different signs of the charges, the force is attractive, again as we have demanded earlier. The natural constant appearing in the equation has the value of $\epsilon_0 = 8.85 \times 10^{-12}\ \frac{(\text{As})}{\text{Vm}}$.

9.1.2 The Electric Field

Compared to the forces we have encountered so far, the Coulomb force is somewhat strange. It can act at a distance, i.e., the two bodies interacting do not actually have to touch. In order to make sense of this, we introduce a new concept, namely that of the electric field. Then the Coulomb force is a contact force again, where one of the charges (the test charge q) is in contact with the electric field $\vec{E}_Q$ produced by the other charge, Q. When we are looking at temporally changing fields in electrodynamics, this will become of great importance, since a change in the charge Q will not instantaneously change the electric field everywhere and we will see that the electric force propagates at the speed of light as well as that an electric field can be created by a temporally varying magnetic field. Basically, this fact gives rise to all of optics as we have alluded to in Chapter 5. For now, let us just define the electric field via the force $\vec{F}$ felt by a test charge q:

$$\vec{E}_Q = \frac{\vec{F}_{Qq}}{q}.$$

$\vec{E}_Q$ then is the *electric field* generated by the charge Q: in this parlance, Q generates $\vec{E}_Q$, and subsequently $\vec{E}_Q$ interacts with q (see Figure 9.1). The field is, so to speak, the transmitter of the force from Q to q. If you remember Newton's reaction principle, we can of course just as well say that the charge q generates a field $\vec{E}_q$, which interacts with Q, and these two interactions are equal and opposite. Also in accordance with Newton's reaction principle, the field generated by Q does not exert a force on Q itself. We have said in Section 6.1.3 that forces always act between two different bodies! So while we could always see the

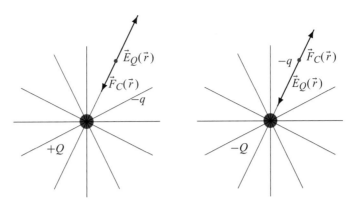

Figure 9.1 The electric field of a spherically symmetric charge distribution. For a positive charge, the field lines point outward radially, whereas for a negative charge the field lines point inward toward the point charge. Therefore, the force is directed toward the charge if the test charge q has the opposite sign as the field generating charge Q (i.e., the interaction is attractive) or radially outward from the center when the two charges have the same sign (i.e., it is repulsive).

fields from two charges as on the same footing, we will for clarity in the following always call the charge that generates the field Q and the test charge q.

Electric fields are of great importance in many biological settings, and we will encounter several of them in this chapter, for instance when we want to describe the properties of DNA, membranes, or how charged particles interact in aqueous solutions. This will be done shortly in Section 9.1.5. Taking all of these points together, we will then look at nerve conduction in Section 9.5, where we will look, for instance, at what timing neurons can have based on the physics of charging a membrane.

9.1.3 The Electrostatic Potential

As we have seen in mechanics, it is often easier to treat a problem in terms of energies rather than forces, since we then typically only have to consider the starting and end points of a process rather than its whole temporal development. This is also the case in electric problems, which is why we introduce a quantity corresponding to the work exerted in an electric field $\vec{E}$. For electrostatic fields, this can be done well and given conservation of energy, the work performed to move a charge q between two points 1 and 2 corresponds to a potential energy U. Just like we got from force to the electric field, we go from potential energy to potential difference or *voltage V* in electric phenomena, such that the voltage multiplied by the charge yields the potential energy difference and thus the work performed:

$$\int_1^2 \vec{F}_C(\vec{r})d\vec{r} = q \int_1^2 \vec{E}(\vec{r})d\vec{r} = -(U(2) - U(1)) \equiv -q(V(2) - V(1)).$$

The unit of a voltage is the volt (1 V = 1 nm/As). We obtain a voltage from a given electric field configuration in the same way as we obtain a potential energy difference in mechanics for a given force:

$$\int_1^2 \vec{E}\, d\vec{r} = -(V(2) - V(1)) = \text{Potential difference} \equiv \text{Voltage} \equiv V.$$

Just as only differences in the potential energy can be measured, but not the absolute value, only differences of the electrostatic potential can be measured, that is, voltages. The electrostatic potential, however, is usually set to be zero far from the field-generating charges; that is, with $V(1) = V(\infty) \equiv 0$ and $V(2) = V(r)$, we obtain the following:

$$V(r) = -\int_\infty^r \vec{E}\, d\vec{r}.$$

If we use the field corresponding to Coulomb's law in this definition, we obtain the electrostatic potential for a point charge:

$$V(r) = \frac{Q}{4\pi\epsilon_0}\frac{1}{r}.$$

If we are given a potential energy U or the electric potential V, we can obtain the forces acting or the electric field respectively. This is done by virtue of taking a gradient of the potential, i.e., looking at the spatial variation of the potential. Given that we are usually looking at situations, where this variation can happen in all three spatial dimensions, we obtain a vector (the electric field) from a scalar (the potential) by taking the derivatives in the three different spatial dimensions. This vector is then called the gradient of the potential:

$$\vec{E} = -\text{grad}\ V \qquad\qquad \vec{F}_C = -\text{grad}\ U$$

$$\text{grad}\ U = \left(\frac{\partial U}{\partial x}, \frac{\partial U}{\partial y}, \frac{\partial U}{\partial z}\right)$$

Again, this can also be done in mechanics, where we can obtain the forces acting from the potential energies involved. Consider, for instance, a situation where we have only gravitational potential energy, $E_{pot} = mgz$, then we directly obtain the force acting from $\vec{F} = -\text{grad}E_{pot} = (0, 0, -mg)$. Since E_{pot} only depends on z, the derivatives with respect to x and y are zero and the derivative with respect to z is mg. Thus we get that gravitational potential energy $E_{pot} = mgz$ corresponds to a constant force of mg acting in the negative z-direction (i.e., downward), as it should be.

9.1.4 Gauss' Law of Electrostatics

We have encountered the flow or flux of a vector field $\vec{S}$ through a surface A (with $d\vec{A} \equiv \hat{n}dA$) in fluid dynamics (Section 7.6), where the vector field we had considered was the flow speed of the fluid. In that case, the flux directly corresponded to the amount of transported material by the velocity field of the flow we consider. In more general terms, we can define a flux of any vector field $\vec{S}$ as follows:

$$\Phi = \int_A \vec{S}\cdot d\vec{A} \equiv \int_A (\vec{S}\cdot\hat{n})dA = \int_A S_n dA = \int_A S\cos\alpha\, dA$$

The unit vector $\hat{n}$ is perpendicular to the surface element dA and thus indicates the direction of that surface. S_n then is the component of the vector field pointing into that direction. For

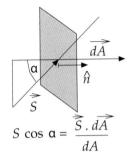

$$S \cos \alpha = \frac{\vec{S} \cdot \vec{dA}}{dA}$$

Figure 9.2   Flux of a vector field through a surface.

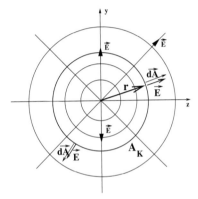

Figure 9.3   Choosing a spherical surface around a point source ensures that the field is always perpendicular to the surface, making calculations easier.

$A \parallel \vec{S}$ ($\alpha = \pi/2$), the flux is virtually zero, whereas for $A \perp \vec{S}$ ($\alpha = 0$), the flux is maximal (see Figure 9.2).

If we choose a *closed surface* as a boundary for which we determine the flux, then the incoming flux must be equal to the outflow, and thus the total flux is equal to zero, at least as long as there is no *source* in the interior of the volume V bounded by the surface A_V:

$$\text{Source free vector field}: \quad \Phi = \oint_{A_V} \vec{S} \cdot \vec{dA} = \oint_{A_V} S_n dA = 0$$

Let us now look at what happens if we have a source of electric field corresponding to the charge Q. Then we will choose the surface of integration A_V such that it represents the surface ($A_S = \oint_S dA = 4\pi r^2$) of a sphere with the charge Q at its center (see Figure 9.3). This will simplify the calculations, since we know in that case that the field will always be perpendicular to the surface, i.e., we have the following:

$$\vec{E} \parallel \vec{dA} \qquad \Rightarrow \qquad \vec{E} \cdot \vec{dA} = E \, dA$$

Since nothing enters this volume, but there are only field lines flowing out, the flux will not be zero. For a point source, we can calculate the flux integral using Coulomb's law:

$$\vec{E}(\vec{r}) = \frac{Q}{4\pi\,\epsilon_0}\,\frac{\vec{r}}{r^3}\,, \qquad E(r) = |\vec{E}(\vec{r})| = \frac{Q}{4\pi\,\epsilon_0}\,\frac{1}{r^2}$$

$$\Phi = \oint_K \vec{E}\cdot d\vec{A} = \oint_K E\,dA = \frac{Q}{4\pi\,\epsilon_0}\,\frac{1}{r^2}\oint_K dA = \frac{Q}{\epsilon_0}$$

Thus we find the following for the electric field of a point charge:

$$\Phi = \oint_K \vec{E}\cdot d\vec{A} = \frac{Q}{\epsilon_0}.$$

This we will call Gauss' law. A more general consideration shows that it actually does not at all depend on where the charge Q sits exactly and whether it is a single point charge, as long as it is enclosed by the surface we consider. Similarly, the area of that surface does not matter as long as it completely encloses Q.

Therefore we have the general form of *Gauss' law:*

$$\oint_{A_V} \vec{E}\cdot d\vec{A} = \oint_{A_V} E_n\,dA = \frac{Q_{\text{inside}}}{\epsilon_0}.$$

Q_{inside} is the electric charge completely enclosed by the surface A_V.

The total flux of an electric field through a closed surface is equal to the enclosed charge ($\times\,1/\epsilon_0$) Or, to state this more succinctly,

Charges are the sources of an electrostatic field.

While the field lines start out at positive charges (they are sources of the electric field), they end with negative charges. Negative charges are thus negative sources, i.e., they sink the field.

Since the Gauss' law is quite general (and not just valid for point charges like Coulomb's law), we can also determine the dependence of the $\vec{E}$ field for more complicated charge distributions. This will be discussed with some examples in the next subsection.

As we have already seen in fluid dynamics, we can also describe Gauss' law as a differential equation. The flux in the fluid mechanics was connected with the current density, and the conservation of matter gave us the equation of continuity for the current density. When we look at the electric flux, we see that the electric field here assumes the role of the current density, which directly leads us to the "equation of continuity" for the electric field:

$$div\vec{E} = \vec{\nabla}\cdot\vec{E} = \frac{\rho}{\epsilon_0},$$

where ρ is the density of the electric charges or charge density.

Together with the relationship between the electric potential and the electric field, $\vec{E} = -gradV = -\vec{\nabla}V$, we directly obtain a relationship between the electrostatic potential and the charge density:

$$divgradV = \nabla^2 V = -\frac{\rho}{\epsilon_0}.$$

This relationship (also called Poisson's equation) can be very useful in the determination of electrostatic potentials, as we shall see later as an example in the treatment of an ionic solution.

9.2 Electric Fields and Potentials of Specific Charge Distributions

9.2.1 The DNA Molecule: A Linear Chain

Very often, we are dealing with a linear molecular chain in which the individual segments are electrically charged. This is the case with many proteins, but *the* example is certainly the DNA molecule in which negatively charged phosphate groups are linearly stacked and thus give a homogeneously charged chain. Let us now consider what the electric field of such a charged linear chain is. As stated previously, we will do this using Gauss' law rather than directly summing up all the contributions of the point charges (this would be a lengthy and complicated calculation).

If we imagine a very long, linear wire, then the electric field must point directly away from the wire radially (if the wire is positively charged). If this were not the case, the different segments would not be interchangeable. Given this geometry of the electric field, we can choose a surface through which the field flows to calculate the flux more easily. The best choice here is the surface of a cylinder (A_C) around the wire, with the wire at its center. Then the field will always be perpendicular to the surface, which means that the scalar product of the vectors can be replaced by the product of the absolute values: $\vec{E} \cdot d\vec{A} = E dA$. Since the magnitude of the field $\vec{E}$ has to be constant along the surface of the cylinder because of symmetry, every point is always the same distance from the wire, and we obtain the following for the total flux through the surface of the cylinder:

$$\Phi = \oint_{A_C} |\vec{E}| dA = |\vec{E}| 2\pi r L = Q/\epsilon_0.$$

Here, r is the radius of the cylinder, i.e., the distance of the surface from the source, and L is the length of the cylinder. The last equation comes from the general formulation of Gauss' law that the flux has to be given by the total charge Q enclosed by the surface. For a homogeneously charged wire, the charge per length is constant, i.e., $Q/L = \lambda$ is constant. This means that we obtain the magnitude of the electric field at a distance r from the wire:

$$|\vec{E}| = \frac{\lambda}{2\pi r}.$$

That is, the field decreases inversely proportional to the distance. The direction we had already determined from symmetry points radially away from the wire (or toward the wire for negative charges).

The potential of this charge distribution is obtained by integrating the field along the distance. Since the field points away from the wire radially, the scalar product is directly given by the product of the magnitudes:

$$V(r) = -\int E dr = -\frac{\lambda}{2\pi} \int \frac{dr}{r} = -\frac{\lambda}{2\pi} \ln(r).$$

Therefore, the potential difference decreases only very slowly with distance.

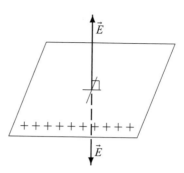

Figure 9.4   A plane with the given number of charges per unit of area.

9.2.2 Membranes: Homogeneously Charged Planes

Electrical charges are also frequently found in membranes, since membrane proteins can be charged themselves or transport charges and thus allow for ion exchange. This will be of great importance again when we look at nerve conduction later in this chapter. For now, we want to consider a cartoon model with which we can describe the charged membrane.

To this end, we consider a plane, which we assume to be arbitrarily large, on which there are the same number of charges per unit of area. Then the resulting field must be the same everywhere. Also, the field lines must be perpendicular to the plane as shown in Figure 9.4 for symmetry reasons. We will now consider the magnitude of the field strength.

The field can be obtained from the surface charge density σ, which is defined as follows:

$$\sigma = \frac{\text{Charge}}{\text{Area}} = \frac{dQ}{dA}.$$

For a homogeneously charged plane, we have $\sigma =$ const. We again use Gauss' law to determine the field strength by enclosing a part of the plane with a box A_B that has top surface area A_C and contains a charge $Q_B = \sigma A_C$. From this, we obtain the following:

$$\Phi = \oint_{A_B} E_n dA = 2 \int_{A_C} E_n dA = 2EA_C = \frac{Q_B}{\epsilon_0} = \frac{\sigma}{\epsilon_0} A_C, \quad \Rightarrow E = \frac{\sigma}{2\epsilon_0}.$$

In this calculation, we only have to consider the top surfaces (on both sides, therefore we obtain a factor of 2), because the field has to point away from the plane perpendicularly. Therefore, the field is parallel to the side surfaces of the box, $\vec{E} \perp d\vec{A}$, and the corresponding flux is zero, $\vec{E} \cdot d\vec{A} = 0$. At least this will be the case if the plate is sufficiently large. At the edge of the plate, this will no longer be the case, and the field geometry is more complicated, which we do not want to get into care about the edges.

In a membrane, however, we typically have two charged planes, which correspond to the two lipid layers of the membrane. That is, we must extend our caricature by describing the membrane as a combination of two oppositely charged homogeneous plates. The distance between the plates, d is fixed. In physics, this arrangement is called a plate capacitor. The resulting field is now restricted to the space between the plates because outside the fields of the two planes compensate each other. This is because the field of a single plate is constant

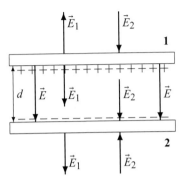

Figure 9.5 The field in a large plate capacitor.

in space, such that the opposite fields of the two oppositely charged plates will point in opposing directions, but have the same strength.

We obtain for outside and inside, respectively, as follows:

$$\vec{E}_o = \vec{E}_1 + \vec{E}_2 = 0, \ \vec{E}_i = \vec{E}_1 + \vec{E}_2 = 2\vec{E}_1 = 2\vec{E}_2,$$

$$\Rightarrow \ |\vec{E}| = \frac{\sigma}{\epsilon_0}, \vec{E} \perp \text{membrane}.$$

This can be realized in a good approximation when two plane metal plates are charged to $\pm Q$ and kept at a distance d, which is small compared to the plate diameter (see Figure 9.5). For membranes, where the thickness is very small ($d \simeq 5$ nm), this is a good description since cell membranes are several microns in diameter.

Therefore, the condition of a small distance d relative to the diameter is fulfilled very well. Apart from the edges (see Figure 9.6), the field inside the capacitor is homogeneous and constant. The surface charge density is $\sigma = Q/A$, where A is the area of the plate.

We obtain the voltage (or the potential difference) by integrating the constant field $\vec{E}$ inside the capacitor from the top plate (1) to the bottom plate (2):

$$-(V(2) - V(1)) = \int_1^2 \vec{E} \cdot d\vec{r} = E \int_1^2 dr = Ed, \quad \Rightarrow E = \frac{V(1) - V(2)}{d} \equiv \frac{V}{d}.$$

Since the electric field between the plate of a capacitor is constant and perpendicular to the plates, it is directly given by the ratio of the voltage to the plate distance. The *capacitance* C of a capacitor describes how much charge can be stored for a given voltage:

$$C \equiv \frac{Q}{V} \left(\text{Unit}: \ \text{Farad (F)} = \frac{\text{As}}{\text{V}} = \frac{\text{A}^2 \text{s}^4}{\text{m}^2 \text{kg}} \right).$$

The capacitance depends only on the geometric arrangement of a capacitor. For a plate capacitor, we have the following:

$$C = \frac{Q}{V} = \frac{\sigma A}{Ed} = \frac{\epsilon_0 E A}{Ed} = \frac{\epsilon_0 A}{d}.$$

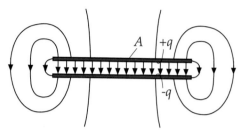

Figure 9.6 The electric field of a plate capacitor of finite size.

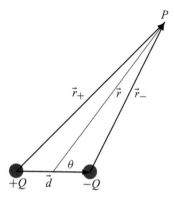

Figure 9.7 The positive and negative charges in a dipole are at slightly different distances from a given point, thus determining the field distribution of a dipole.

9.2.3 The Water Molecule: Electrical Dipoles

Many microscopically uncharged systems consist of spatially separated positive and negative charges. This applies, for example, to most atoms or molecules, if the center of gravity of the electron orbits does not coincide completely with the atomic nucleus. One prominent example of this is the water molecule, where the two hydrogen atoms are connected to the oxygen atom at a certain angle, which leads to a distribution of charges with an excess of positive charges near the hydrogen atoms and an excess of negative charges near the oxygen atom. These properties of atoms and molecules essentially determine their microscopic interactions, as we shall see in the next couple of sections. For this reason, let us examine the caricature model of such a charge distribution, the electric dipole. The idealized dipole consists of two equal-sized point charges with a reversed sign that are at a fixed distance d.

However, the field distribution no longer has a simple geometric symmetry, which is why Gauss' law cannot be applied equally easily in this case as in the two previous cases. However, we can still obtain a description of this since the dipole consists only of two point charges, where we can simply add up the contributions from each of the two charges.

At point P (see Figure 9.7), we obtain a field

$$\vec{E} = \vec{E}_+ + \vec{E}_- = \frac{Q}{4\pi \epsilon_0} \left(\frac{\vec{r}_+}{r_+^3} - \frac{\vec{r}_-}{r_-^3} \right),$$

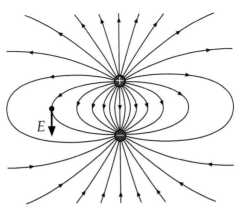

Figure 9.8 The electric field of a dipole. In the mirror plane, the field outside the dipole points in the direction from the negative to the positive charge. Between the two charges, the field shows from the positive to the negative charge. The dipole moment $\vec{p}$ is defined in the direction from the positive to the negative charge.

and a potential difference

$$V = V_+ + V_- = \frac{Q}{4\pi\epsilon_0}\left(\frac{1}{r_+} - \frac{1}{r_-}\right).$$

Far away from the dipole itself ($r_+, r_-, r \gg d$), we can Taylor-expand the two different distances using the following:

$$\vec{r}_\pm = \vec{r} \pm \frac{1}{2}\vec{d}, \qquad r_\pm = r\sqrt{1 \pm \frac{d}{r}\cos\theta + \frac{d^2}{4r^2}} \approx r\left(1 \pm \frac{d}{2r}\cos\theta\right)$$

and find

$$V(r) = \frac{Q}{4\pi\epsilon_0}\frac{d\cos\theta}{r^2}\ .$$

The resulting field is given by the gradient of the voltage and will be symmetric around the axis of the dipole (see Figure 9.8), and will be proportional to the *Dipole moment* $p \equiv Qd$. On the axis (in between the two charges), we have $\theta = 90°$, $\cos\theta = 0$, and hence

$$V(\vec{r}) = 0, \qquad \vec{E}(\vec{r}) = -\frac{Q}{4\pi\epsilon_0}\frac{\vec{d}}{r^3} \equiv -\frac{1}{2\pi\epsilon_0}\frac{\vec{p}}{r^3}.$$

The field generated far away from the dipole thus opposes the dipole moment. Although a dipole carries a total charge of zero, it does generate a field $\vec{E}$. However, as we have seen from the preceding calculations, its magnitude decreases with the cube of the distance, i.e., more rapidly than for a point charge. Because of this electric field, a dipole will exert a force on a charge or another dipole. This leads, for example, to interatomic or intermolecular forces, as will be discussed in detail later in this chapter.

In a constant electric field applied from the outside, a dipole will align itself, since the positive charge will be repelled from the field and the negative charge will be attracted to the field. This means that the dipole moment is such that the dipole moment is opposed to the applied field. If you want to put a dipole in a different configuration to the applied field, you have to perform work. The potential energy of such a dipole at an angle to the applied field is then given by the following:

$$E_{pot} = \vec{p} \cdot \vec{E}. \tag{9.1}$$

If the dipole moment and the field are antiparallel, this potential energy is minimal, as it should be for the preferred state. This potential energy will continue to interest us when we are looking at magnetic phenomena and in particular at nuclear resonance spectroscopy. Correspondingly, in the inhomogeneous electric field, there is a resulting force on the dipole given by the following:

$$\vec{F} = -\nabla \vec{p} \cdot \vec{E}.$$

The interaction between dipoles and an applied electric field also describes an important property of materials. For matter that is made up of molecules with a strong dipole moment, an applied electric field will lead to the macroscopic alignment of all of the molecules of the substance. Adding up the contributions of all of these molecular dipoles gives a strong electric field opposing the applied field. This field due to polarization of the material $\vec{E}_{pol}$ has to be added to the external field $\vec{E}_{ex}$ within the material. The total field inside is therefore given by

$$\vec{E}' = \vec{E}_{ex} + \vec{E}_{pol},$$

which is lower than the external field, since the polarization opposes the applied field. This reduction factor is called the dielectric constant of the material ϵ, which is dimensionless and only depends on the molecular properties of the material considered, such that

$$\vec{E}' = \frac{1}{\epsilon} \vec{E}_{ex}.$$

If a plate capacitor is filled with a dielectric medium, it has to be charged more by a factor of ϵ to obtain the same total field $\vec{E}'$ and hence have the same voltage V across the capacitor as one would have without the dielectric. The capacity of the capacitor Q/V therefore is increased by a factor of ϵ. Due to its strong diploe moment, water has a very high dielectric constant of roughly $\epsilon \simeq 80$. Hence the electric field in a capacitor is very strongly reduced if it is filled with water.

The fact that the polarization of the dipole moments reduces the internal field in a material is of great importance to the chemistry of aqueous solutions and thus to biology. If we consider to oppositely charged but equal ions Q^+ and Q^- in solution, their mutual attraction (and hence probability to recombine) is reduced to by a factor of ϵ:

$$|\vec{F}| = Q^- |\vec{E}'^+| = Q^- \frac{|\vec{E}^+|}{\epsilon}.$$

Because of its high value of ϵ, water is an excellent solvent (see Figure 9.9).

9.2.4 Ionic Solutions

In ionic solutions, there are electrical charges (ions) in an uncharged environment in which they are freely mobile. In addition, the thermal movement of the solution molecules can

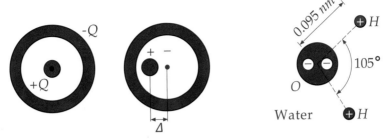

Figure 9.9 The origin of the lare dielectric constant of water lies in the strong electric dipole found in a water molecule (see right). Other atoms can also obtain an induced dipole moment by a relative shift of the nucleus to the electron clouds (see left).

excite the ions to a certain agitation, that is to say, give a distribution of the ions that does not correspond to that of static electric charges. Thus, if we obtain an unequal distribution of the positive and negative charges by thermal motion, a voltage difference results that must be described by the temperature and the charge density. Quantitatively, the charge carrier density as a function of the location $\rho(x)$ is given by the difference of the charge carrier densities of the positive and negative charges. If the positive and negative ions have the opposite charge (typically one or two elementary charges), then $\rho(x) = q_+ n_+(x) + q_- n_-(x) = e(n_+(x) - n_-(x))$. Deviations from a uniform distribution due to thermal fluctuations occur according to the Boltzmann distribution. That is, the probability of finding an ion at a position where it feels a potential difference $V(x)$ is given by $n_\pm(x) = n_0 \exp(-\frac{qV(x)}{k_B T})$, where q is the charge of the ion and n_0 is the density of ions in the solvent (uniformly distributed). That is, if there is a given ion concentration difference in a solution, such as is the case close to a cell membrane, there is a voltage $V(x)$ present in that solution, which can be calculated from the Boltzmann distribution. We thus obtain a potential difference

$$\Delta V = -k_B T/q \ln(n_2/n_1)$$

between two regions 1 and 2 containing ion densities n_1 and n_2 respectively. This potential is also called the Nernst potential. This type of potential difference typically arises between membranes where charges are separated. The most important example of this is the cell in which Ca, Na, and K ions are redistributed by ion channels, resulting in a typical potential difference between the inside and outside of the cell. So we will deal with this in Section 9.2.5 more extensively when we are looking more closely at nerve conduction. Given that the charges of ions are elementary charges and the concentrations will not be different by very many orders of magnitude, we can obtain an estimate for the voltages at play in cells from $V \simeq k_B T/e = 25$ mV at room temperature corresponding to $k_B T = 300$ K.

    If we consider an ionic solution in a dynamic sense, we have to also consider that such a redistribution of charges in an ionic solution also causes a redistribution of the sources of the electric field. We can then also consider the equilibrium distribution, or the equilibrium potential, of a single charge in an ionic solution. Let us consider one single ion in an ionic solution (to be specific, we look at one positively charged ion). The negative ions are attracted to this positive ion, and the positive ions are repelled. However, this cannot go as far as possible, because when many negative charges are in close proximity of the positive ion, the negative charges cancel the positive one and at large distances there is no more attraction or repulsion. In addition, the negative ions are moved away from the positive ion by thermal motion such that they cannot get in too close proximity of the positive ion we consider. There is thus a negatively charged cloud of ions around the positive ion, whose extent is described by $k_B T$, which screens the potential of this positive ion. We want to make this more quantitative. As we have seen, the voltage is determined by the charge carrier density, $\nabla^2 V = -\rho/\epsilon_0$. We determined this charge carrier density previously: $\rho(x) = e(n_+(x) - n_-(x))$ for singly charged ions. However, the number densities of the positive and negative charges depend on the temperature and the present potential of the one ion we consider $V(x)$ via a Boltzmann distribution: $n_\pm(x) = n_0 \exp(-\frac{\pm eV(x)}{k_B T})$. Inserting this into the description of the charge carrier density, we obtain the following:

$$\rho(x) = en_0 \left(\exp(-\frac{eV(x)}{k_B T}) - \exp(+\frac{eV(x)}{k_B T}) \right) \simeq -2en_0 \frac{eV(x)}{k_B T},$$

where we have approximated the exponentials using a Taylor expansion and used the fact that $\frac{eV(x)}{k_B T} \ll 1$. This will be the case close to the positive ion we consider. We now have a reasonably simple description of the charge carrier density depending only on the voltage and the temperature. If we insert this into the Poisson equation to determine the voltage, we obtain a differential equation for this voltage, which we have seen several times before, such that we will be able to guess the solution. In a one-dimensional problem, we obtain the following:

$$\frac{d^2 V}{dx^2} = -\frac{\rho}{\epsilon_0} = \frac{2e^2 n_0}{\epsilon_0 k_B T} V(x).$$

The spatial curvature of the voltage is thus directly proportional to the voltage itself. This will have to yield an exponential decrease of the voltage with space, i.e., $V(x) = V_0 \exp(-x/\lambda)$, where λ describes the length scale on which the voltage has decreases by a factor of $1/e$. From the specific form of the preceding Poisson equation (called the Poisson–Boltzmann equation), we directly obtain what this length scale λ is for an ionic solution:

$$\lambda = \sqrt{\frac{\epsilon_0 k_B T}{2e^2 n_0}}.$$

This means that the higher the temperature, the greater the range of the electric potential of an ion, or the higher the ion density, the shorter the range. This makes intuitive sense, because the higher the density of the ions, the more charges are there to shield the charge of a single ion, which decreases the range of the interaction. On the other hand, the higher the temperature, the more average kinetic energy the ions have, thus the larger is the cloud of shielding ions, increasing the range of interaction because the shielding is less effective. This range of an electrical potential in an ionic solution is also called the Debye screening length. If we use typical values, i.e., singly charged ions, room temperature and an ionic strength of the solution of a micromolar, we obtain a screening length of a few nanometers; thus in biological contexts, the charges of the ions will be strongly screened.

9.3 Molecular Interactions and Bonds

9.3.1 The van der Waals Interaction

We have already looked at what happens when we have molecules with a fixed dipole moment, which can be aligned in an external field and which themselves generate an electric field. As we have seen, the field of an electric dipole falls off at large distances with $E_D \propto 1/r^3$. However, there are many molecules that do not have a fixed dipole moment. In these molecules, the motion of the electrons around the nuclei is symmetrical so that, on average, not only the charges cancel, but also the dipole moments. If, however, such an atom or molecule is brought into an external electric field, a dipole moment arises, as discussed previously, by the relative displacement of the charges. This is quantified by the polarizability α, which is related to the dielectric constant of the material. The induced dipole moment $\vec{p}$ thus is given by the polarizability α and the external electric field $\vec{E}_{ex}$.

$$\vec{p} = \alpha \vec{E}_{ex}$$

This property of matter leads to one of the most important interactions of biology and chemistry, namely the van der Waals interaction, which we have already briefly introduced in Section 7.1 on continuum mechanics as one of the key players in the interactions between atoms binding solids and liquids together. We will now want to describe the spatial dependence of the van der Waals interaction quantitatively and see how it is related to the properties of an induced dipole. For this purpose, we consider two isolated atoms, both of which have a fluctuating dipole moment, which, however, is zero on average. Now consider one given moment in time, where we freeze the fluctuations. The dipole moment of atom 1, $\vec{p}_1$, generates an electric field $\vec{E}_{D1}(r)$ at the position of atom 2 (i.e., at a distance r from atom 1). This electric field induces a dipole moment within atom 2 in its own direction:

$$\vec{p}_2 = \alpha_2 \vec{E}_{D1}(r) = \alpha_2 \frac{\vec{p}_1}{2\pi\epsilon_0 r^3}.$$

This dipole moment in atom 2 now in turn generates an electric field at the position of atom 1:

$$\vec{E}_{D2} = \frac{\vec{p}_2}{2\pi\epsilon_0 r^3} = \alpha_2 \frac{\vec{p}_1}{(2\pi\epsilon_0)^2 r^6}.$$

Now we can unfreeze time, since this field is parallel to the (fluctuating) dipole moment of atom 1 at any time, since it is only induced by this fluctuating dipole moment (via the presence of atom 2). That is to say, even if the dipole moment fluctuates continuously, there is nevertheless an attraction between the two atoms via this induced dipole moment in atom 2. Thus there is a potential energy between the two atoms, which is determined by dipole moment 1 and the (induced) electric field 2 given by the following:

$$U_{21} = \vec{p}_1 \cdot \vec{E}_{D2} = \alpha_2 \frac{\vec{p}_1^2}{(2\pi\epsilon_0)^2 r^6}.$$

Here it is of great importance that the dipole moment of atom 1 appears quadratically. Because of this, even a fluctuating dipole moment with $\langle \vec{p} \rangle = 0$ will have a potential energy different from zero because it has to be that $\langle \vec{p}^2 \rangle \neq 0$. This is the same as in thermal motion of molecules, where we also obtained an energy (the temperature), even though the movements of the individual molecules all canceled each other. Thus, such van der Waals attraction occurs for all atoms no matter how they are formed, and thus it is also the most important interaction that acts upon the accumulation of many molecules. However, the attraction is relatively weak and disappears very quickly for large distances ($U \propto 1/r^6$!). Given the size of the atoms involved, one can in principle calculate their polarizability, since the size determines the maximum possible dipole moment. At typical interatomic distances of about 1 nm, one obtains a potential energy due to the van der Waals interaction of about 10^{-21} J, which is slightly less than the thermal energy $k_B T$ at room temperature. This means that while van der Waals bindings are always present, they are not very strong and therefore are easily broken up by thermal fluctuations. However, this can be counteracted by greatly reducing the distance between the molecules or increasing the number of binding sites in a large molecule. If, therefore, larger molecules, such as proteins, are to bind well together, then the distance of the atoms must be very small over the entire molecule. This can only be achieved if the geometric shape of the two molecules is spatially well matched. In other words, good enzymes actually behave stereometrically similar to the target molecule as a key fits the lock. This property of the van der Waals interaction is thus the fundamental reason for the molecular relationship between the form and the function of a biochemical substance. This can also be observed repeatedly in the determination of the structure of important biomolecules, e.g., the ion channels or transcription factors. For this reason too, structure determination of biomolecules is very important. We have already seen the physical basis of a technique for this purpose in X-ray crystallography and will shortly see another one in nuclear magnetic resonance.

9.3.2 The Bohr Model Revisited

We have already discussed that electrons can behave like waves, i.e., they can also form standing waves if they interfere with themselves. The wave of an electron that moves around an atomic nucleus very quickly will have no amplitude on average, because the phases are slightly different in each case, and if we add up many oscillations with slightly different phases we get zero. There are, however, a few specific orbits where we obtain a large amplitude, when the circumference of the orbit is just a multiple of the wavelength. This is the same as the note played by an organ pipe. Then every passage of the orbit interferes constructively with the previous ones and a standing wave forms. This is the basis of our description of the atom, the Bohr model. This standing wave has given us a quantized orbit radius, which is determined as follows:

$$r = \frac{n\hbar}{p}.$$

To get further than this in our description of the atom, we have to consider what the momentum of an electron in the atom is and how it relates to its position in the orbit radius. If we are successful, this should only depend on fundamental properties of the electron. For simplicity, we put the electron on a circular orbit. Then there is a centripetal acceleration due to the speed of orbit, which connects the momentum to the radius. In addition, we know that the force supplying this acceleration is the electrostatic attraction between the electron and the nucleus. We thus obtain the following:

$$\frac{Ze^2}{4\pi\epsilon_0 r^2} = \frac{mv^2}{r} = \frac{p^2}{mr}.$$

Here, Z is the number of protons in the nucleus and e is the elementary charge, such that Ze is the charge of the nucleus. Finally, m is the mass of the electron. We can cancel a factor of r on both sides and obtain a second relation between r and p:

$$\frac{Ze^2}{4\pi\epsilon_0 r} = \frac{p^2}{m}.$$

Thus we can now obtain the radius of the orbit of the standing wave by inserting one relation into the other. This yields the following:

$$\frac{Ze^2}{4\pi\epsilon_0 r} = \frac{(n\hbar)^2}{r^2 m},$$

where we can again get rid of on factor of r on both sides to obtain

$$\frac{Ze^2}{4\pi\epsilon_0} = \frac{(n\hbar)^2}{rm}.$$

Solving this for r gives us the size of an atom that actually only depends on fundamental properties such as the mass and charge of the electron and natural constants:

$$r = \frac{(n\hbar)^2 4\pi\epsilon_0}{Ze^2 m}.$$

If we insert the known values of the fundamental constants and consider hydrogen (i.e., $Z = 1$) in the ground state (i.e., $n = 1$), we obtain a radius for the orbit of $r = 0.52$ Å. This is also called the Bohr radius. We can insert this radius in the relation between momentum and radius and thus obtain the momentum of the electron on its orbit to be as follows:

$$p = \frac{Ze^2 m}{n\hbar 4\pi\epsilon_0}.$$

This momentum has to be related to the binding energy of the electron to the nucleus. In the end, the binding energy has to be given by the kinetic energy, since otherwise the bond would be broken for higher kinetic energy or be moved closer for higher binding energy. Therefore, $E = mv^2/2 = p^2/(2\,m)$, using $p = mv$. Then we obtain for the binding energy:

$$E = \frac{p^2}{2m} = \frac{(Ze^2)^2 m}{2(n\hbar 4\pi\epsilon_0)^2}.$$

Again using hydrogen in the ground state, this corresponds to an energy of $E = 13.6$ eV $= 2.2 \times 10{-}18 * $ J. Here we have used the unit of electronvolts (eV) as a unit of energy, which in atoms is often a natural choice leading to numerical values close to one (13.6 in the case here). An electronvolt is the energy it takes to apply a voltage of one volt to an elementary charge. Thus one electronvolt is 1 eV $= 1.6 \times 10^{-19}$ J.

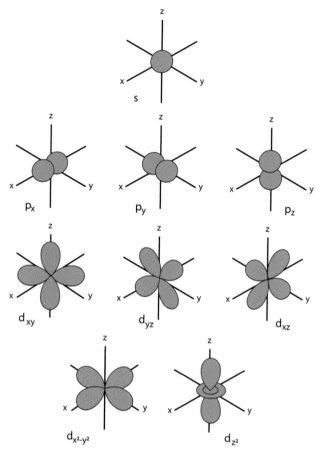

Figure 9.10 Electron probability distributions for different states in an atom. The higher excited states with larger n can have additional modes of oscillation with different local distributions of nodes and antinodes.

9.3.3 The Periodic Table of the Elements

Since two electrons cannot be in the same state by virtue of the Pauli exclusion principle, the possible states given by the Bohr model are successively filled with electrons of increasing n if there are several electrons in one atom. In addition to differences in the radius (i.e., n), the electrons can also differ in the shape of the orbit, that is, in their angular momentum (see Section 10.2), as well as in their internal angular momentum or spin. The spin $\vec{s}$ can only have two possible values, namely $|\vec{s}| = \pm\hbar/2$. The possible orbital angular momentum depends on how many antinodes there are in the standing wave of electrons around the nucleus, i.e., how large n is. Since the angular momenta are also quantized, $|\vec{L}| = l\hbar$, where l is a number that lies between 0 and n. The probability distributions of the different states, that is, the standing waves of the electrons that are formed, are represented in Figure 9.10 for different states with $l = 0, 1, 2$, which corresponds to the s-, p-, and d-shells respectively.

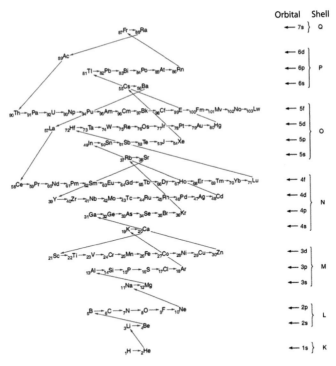

Figure 9.11 Illustration of which states belong to which shells and how they are filled in the periodic system.

These different states, which we have found in the Bohr model, are essentially occupied by the different electrons in an atom with many electrons. Changes of this scheme occur only with relatively heavy atoms due to the change in the electrostatic potential by the additional electrons. How many electrons can occupy a particular state (with a certain n) gives information on how easy it is to add additional electrons to this atom. This means that the chemical reactivity of the different atoms is well described by the energies of these states. If all of the states are occupied by a particular n, this is called a closed shell and the corresponding atom is very inert. Accordingly, these atoms are found on the very right-hand side of the periodic table and constitute the noble gases, which are very hard to bind to other atoms. How the different shells are filled is shown in Figure 9.11. This representation also determines the shape of the periodic table.

9.3.4 Covalent Bonds

If we have different atoms, which have not filled shells, and in which, e.g., the p-states are occupied, the electrons are relatively far away from the nucleus. Another atom, in which an electron is absent in a corresponding state, may then bind this electron of the first atom, while the electron is still bound to the first atom. This means that the two atoms have now effectively formed a bond, which is how covalent bonds form. The binding energies of such covalent bonds are correspondingly dependent on the magnitude of the binding energies of

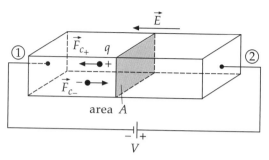

A conductor to which a voltage V is applied between points 1 and 2.

the electrons to the nucleus, that is to say, of the order of magnitude $E_B \sim 10^{-18}-10^{-19}J$ or $E_B \sim 1 - 10\ eV$ in atomic units. These are also the binding energies, which give us, e.g., the strength of substances or their optical absorption characteristics.

9.4 Electrical Currents

In electric fields, there are forces acting on charges, which lead to a movement of these charges. Such moving charges constitute an *electric current*, where electric charge is transported in space in a similar way as the fluid flows we have dealt with in Section 7.6. If the charge transport takes place in a body, one speaks of *conduction*. One important example of this is the transport of an electrical signal along the axon of a nerve cell. In order to describe how this can happen and what different possibilities nature has to produce such signals and what limits them, we will first have to lay a few foundations as to how electrical currents can be described.

9.4.1 Current and Current Density

To obtain a quantitative understanding of electric currents, consider a conductor to which we apply a voltage V between points 1 and 2 (see Figure 9.12). With this voltage, we generate an electric field $\vec{E}$, given by the following:

$$\int_1^2 \vec{E} \cdot d\vec{r} = V.$$

The charges $q_\pm$ present in the conductor then feel a Coulomb-force $\vec{F}_{C\pm} = \pm q_\pm \vec{E}$. Therefore, the mobile charge carriers in the interior of the conductor are set in motion. A current starts to flow. The current strength I is defined as the number of charges flowing through the conductor cross section A per unit of time:

$$I = \frac{dQ}{dt}.$$

Table 9.1 Typical values of electric currents in nature.		
Neve signal along an axon	10^{-7} A	0.1 μA
Current running through an electrophoresis gel	$10^{-2}-10^{-3}$ A	$1-10$ mA
Noticeable	10^{-2} A	10 mA
Deadly	$> 10^{-1}$ A	100 mA
Atrial fibrilation	$> 7 \times 10^{-1}$ A	700 mA
Appliances like television, etc.	1 A	
Current flowing in an NMR magnet	10^{2} A	0.1 kA
Lightning	3×10^{4} A	30 kA

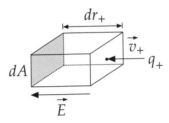

Figure 9.13 For a a positive charge q_+ to move through the cross-section dA during the time interval dt, when it is moving under the influence of the Coulomb force with a velocity v_+, its distance from dA cannot be greater than $dr_+ = v + dt$

The unit of electrical current, the *ampère* (A), is one of the base units of the international system of units. Table 9.1 shows typical values of electric currents and their effect on humans.

The current flowing through the cross-section A depends on the number of free charge carriers per unit of volume n [m^{-3}] and their charge q [As], i.e., the charge density of free carriers ρ ($= nq$) [As m^{-3}].

In order for a charge (consider a positive charge q_+ for now) to move through the cross-section dA during the time interval dt, when it is moving under the influence of the Coulomb force with a velocity v_+, its distance from dA cannot be greater than $dr_+ = v_+dt$ (see Figure 9.13).

If it is farther away, it will not reach dA within the time dt. In other words, the charges q_+ within a volume given by $dV_+^{vol}(= dA \, dr_+)$ contribute to the current by passing dA within dt:

$$dQ_+ = \rho_+ dV_+^{vol} = \rho_+ v_+ dt dA \quad \Rightarrow \quad I_+ = \frac{dQ_+}{dt} = \rho_+ v_+ dA.$$

Similar to what we found in fluid currents, the density times the velocity, $\rho_+ v_+$, constitutes the current density j_+ [A/m^2]. If we also take into account the negative charges in our consideration, we obtain the following:

$$\frac{dQ}{dt} = \frac{dQ_+}{dt} + \frac{dQ_-}{dt} = (\rho_+ v_+ + \rho_- v_-)dA = (j_+ + j_-)dA.$$

Because the negative charges will have a change in sign both in the charge density as well as in the velocity, both signs of charge will give a similar contribution to the overall current and current density:

$$\rho_- = -n_- Z_- e \,, \quad \rho_+ = n_+ Z_+ e \,, \quad \vec{v}_+ \parallel -\vec{v}_-.$$

So we obtain in general the following:

$$j = n \cdot Z \cdot e \cdot v$$

where we have to take care to use the proper signs for e and v respectively.

We have here only considered a one-dimensional current. The velocities, which are in fact vectors in three-dimensional space, thus here were basically speeds. If one wants to describe a fully three-dimensional world, we have to use velocity vectors and consider their component perpendicular to the surface of interest:

$$\frac{dQ}{dt} = \rho v_n dA = j_n dA = \rho \vec{v} \cdot d\vec{A} = \vec{j} \cdot d\vec{A} \,, \qquad d\vec{A} \equiv \hat{n} dA \,, \hat{n} \perp dA \,, |\vec{n}| = 1.$$

The total current is then given by summing up all contributions in the cross-sectional area, i.e., by carrying out the surface integral over A:

$$I = \int_A \vec{j} \cdot d\vec{A} = \int_A j_n dA = \rho \int_A v_n dA = \rho \int_A \vec{v} \cdot d\vec{A}.$$

9.4.2 Conductivity and Resistance

We would expect that the moving charges in the $\vec{E}$ field would continuously accelerate, since according to Newton's second principle, $m\vec{a} = \vec{F} = Ze\vec{E}$. However, in usual conductors, we find experimentally that the current density $\vec{j} = \rho \vec{v}$ and thus the velocity $\vec{v}$ is constant and depends on the applied field. This implies that there are friction forces acting on the moving charges that ultimately determine how the applied force (i.e., the electric field) is related to the velocity (i.e., the current density). If we assume a frictional force, with the same properties as viscous friction, such that the friction is proportional to the negative velocity, i.e., opposes it, we obtain a final, constant velocity, which is proportional to the applied field. We have already shown this in Section 6.2.4 in the case of a ball falling through the air. Applied to the electric current, this means the following:

$$\vec{j} = \sigma_L \vec{E} \qquad \text{Ohm's law in differential form}$$

The materials constant σ_L is called the electrical conductivity. It depends on the charge carrier density and their coefficient of friction. The reason for this proportionality lies in the fact that the charges are the same as we saw for terminal velocity in freefall. The charges are accelerated in the field but lose energy and momentum again and again by collisions with the atoms in the lattice or the fluid. The speed of the individual charges is highly variable (due to thermal motion). The so-called *drift velocity*, which determines the current density, only has a constant value when averaged over time and over all charges. From this version of Ohm's law and the current density above $\vec{j} = \rho \vec{v}$, we obtain the following for the drift velocity:

$$\vec{v} = \frac{\sigma \vec{E}}{\rho}.$$

A somewhat more common formulation of Ohm's law is given in terms of the total current and the applied voltage. To do this, let us think of a rectangular conductor of length l and cross-section A, where a voltage V is applied. With Ohm's law (and the definition of the current density), the total current through the conductor is given by the following:

$$I = jA = \sigma_L EA = \sigma_L \frac{V}{l} A.$$

If we further define the *resistance R* of the body in question as

$$R := \frac{l}{\sigma_L A} \qquad [R] = \mathrm{Ohm} = \Omega = \frac{\mathrm{V}}{\mathrm{A}}.$$

We obtain the common version of Ohm's law as follows:

$$V = RI.$$

The resistance of a conductor thus depends on the geometry as well as the material of the conductor. Only the conductivity is a proper materials constant that is independent of the geometry of the actual body considered. In most cases, the conductivity is still dependent on other variables, e.g., the current I itself or the temperature. This will become relevant when we will deal with the action potential shortly.

As we constantly have to apply a voltage, we have to do work on the conductor. This work is dissipated by the constant collisions of the charge carriers and is turned into disordered kinetic energy of the atoms of the conductor, i.e., into heat. A wire with a current flowing through it becomes warm. This dissipated power we want to consider more quantitatively in terms of current and voltage now. If a charge dQ flows from point 1 to point 2, the work done by the voltage source is given by the following:

$$dW_{1 \to 2} = dQ \int_1^2 \vec{E} \cdot d\vec{r} = dQ\, V.$$

The power dissipated (work performed per unit of time) then is as follows:

$$P = \frac{dW}{dt} = \frac{dQ}{dt} V = IV.$$

Therefore, to maintain a current I, a voltage V is required and thus the voltage source must supply energy.

The Joule heat generated in the resistor R per unit of time can also be written as follows:

$$P = IV = I^2 R = \frac{V^2}{R}.$$

9.4.3 Currents Flowing through Networks: Kirchhoff's Laws

Voltage sources and resistors can be connected to circuits or networks. The currents and voltages for each individual resistor can then be calculated with the help of the *Kirchhoff's laws*, which we have already encountered in Sections 7.7.2 and 7.7.3 for the vascular network and the flows of fluids.

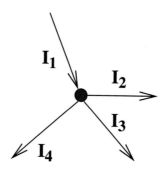

Figure 9.14 The junction rule.

The *junction rule* (see Figure 9.14), which is based on the conservation of the flowing material, i.e., the conservation of charge in case of electrical currents, states that at a branching point of a network where n conductors meet, the sum of the incoming currents is equal to the sum of the outgoing ones:

$$\sum_{k=1}^{n} I_k = 0.$$

Incoming currents are counted positive, outgoing one negative.

For a current loop, a *mesh*, the sum of all voltage sources (taking care of using the right sign) has to be equal to the sum of all the voltage drops on all resistors. Applied to the mesh shown in Figure 9.15 having ℓ voltage sources and n resistors, this gives the following:

$$\sum_{j=1}^{\ell} V_j = \sum_{k=1}^{n} V_{R,k} = I \sum_{k=1}^{n} R_k.$$

Here V_j is the voltage supplied by the jth source and $V_{R,k}$ is the voltage drop at the kth resistor. If all the resistors are known, Kirchhoff's laws for a network yield a system of equations with as many equations as there are unknown currents or voltage drops to be determined. An example of this is shown in Figure 9.16.

In addition to resistors, capacitors can be added to an electric circuit. In that case, each capacitor C will have an additional voltage drop $V_C = Q/C$. Two common types of circuits that we have already encountered in fluid networks are resistors in parallel and in series, shown in Figure 9.17.

n resistors R_k in parallel can be replaced by an effective resistor $R_\parallel$ that is given by the following:

$$I = \sum_{k=1}^{n} I_k = \sum_{k=1}^{n} \frac{V}{R_k} = V \sum_{k=1}^{n} \frac{1}{R_k} = \frac{V}{R_\parallel} \quad \Rightarrow \quad R_\parallel = \left(\sum_{k=1}^{n} \frac{1}{R_k} \right)^{-1}.$$

The inverse of the replacement resistance $R_\parallel$ is the sum of the inverse of all the individual resistances R_k.

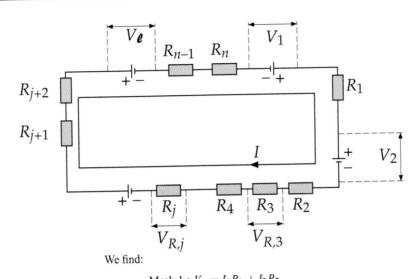

We find:

$$\text{Mesh } 1 : V_0 = I_1 R_1 + I_2 R_2$$

$$\text{Mesh } 2 : 0 = I_3 R_3 - I_2 R_2$$

$$\text{Junction} : I_1 - I_2 - I_3 = 0$$

$$\Rightarrow I_1 = V_0 \left(R_1 + \frac{R_3 R_2}{R_2 + R_3} \right)^{-1}$$

$$\Rightarrow I_2 = \frac{R_3}{R_2 + R_3} I_1, \qquad I_3 = \frac{R_2}{R_2 + R_3} I_1$$

Figure 9.15 A single mesh for illustrating the mesh rule.

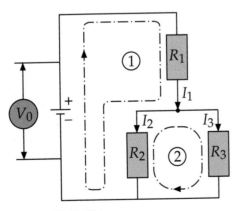

Figure 9.16 An example network for the application of Kirchhoff's laws.

n resistors R_k in series can be replaced by en effective resistor R_S that is given by the following:

$$V = IR_S = I \sum_{k=1}^{n} R_k \quad \Rightarrow \quad R_S = \sum_{k=1}^{n} R_k.$$

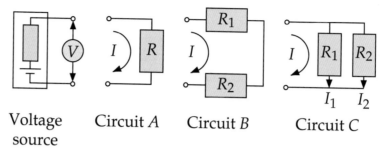

Circuit A:
$$V = IR \,, \; I = \frac{V}{R}$$

Circuit B:
$$V = IR_1 + IR_2 = IR_S \,, \; I = \frac{V}{R_S} \;\Rightarrow\; R_S = R_1 + R_2$$

Circuit C:
$$V = I_1R_1 = I_2R_2 \,, \; I = I_1 + I_2 \,, \; I = \frac{V}{R_{\parallel}} = V\Big(\frac{1}{R_1} + \frac{1}{R_2}\Big) \;\Rightarrow\; \frac{1}{R_{\parallel}} = \frac{1}{R_1} + \frac{1}{R_2}$$

Figure 9.17 Resistors in parallel (circuit C) or in series (circuit B) can be replaced by an effective total replacement resistor (circuit A).

Resistors in series are simply added.

There are similar rules for replacing capacitors in series and in parallel. However, because the voltage drop for a capacitor is proportional to $1/C$, we find that capacitors in parallel can be added directly, whereas capacitors in series have to be added inversely.

9.4.4 Electrolytes: Ionic Transport in Fluids

We have already looked at how ionic solutions can be described because of their importance in modeling biological substances. If there are voltages applied to ionic solutions, a current of negative and positive ions starts to flow and we are dealing with an electrolyte.

Let us first consider a single ion in the aqueous solution under influence of an applied electric field. There is a Coulomb force $\vec{F}_C = Ze\vec{E}$ acting on this ion of charge Ze, as well as a friction force $\vec{R} = -f\vec{v} = -6\pi\,\eta r\vec{v}$. Here, η is the viscosity of the electrolyte, which will be close to that of water. As we have discussed for the resistance, this is the same situation we have already solved in mechanics when we introduced viscous friction. We therefore know that this situation will lead to a terminal velocity for the ion given by $v_\infty = F/f$, or more specifically, when taking the full friction coefficient into account, we obtain the following:

$$v_\infty = \frac{ZeE}{6\pi\,\eta r}.$$

Once we know the terminal velocity, we can determine the conductivity as we have described previously:

$$\text{currentdensity}: \; j = \sigma_L E = q_+ n_+ v_+ + q_- n_- v_- = \frac{E}{6\pi\,\eta}\left[(Z_+e)^2\frac{n_+}{r_+} + (Z_-e)^2\frac{n_-}{r_-}\right].$$

Here, $n_\pm$ are the ion concentrations, $Z_\pm$ the charge of the ions, and $r_\pm$ their radius. The total charge has to be zero ($0 = Z_+e_+n_+ - Z_-e_-n_-$), which determines the relative concentrations of the different ions. For example, we would find for table salt (NaCl) that $Z_- = Z_+ = 1$, $n_- = n_+$. On the other hand, $CaCl_2$ would yield $Z_+ = 2, Z_- = 1, n_- = 2n_+$.

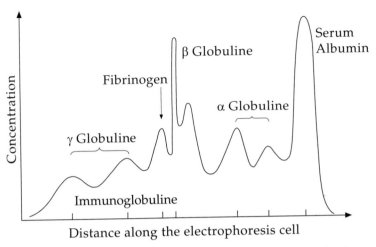

Figure 9.18 Distribution of the molecules of human blood plasma in an electrophoresis cell. Immunoglobulins, for example, protect against viral and bacterial infections.

For monovalent salts, acids, or bases with $Z_\pm = 1$ and $n_\pm = C$ (concentration), we thus find a conductivity of

$$\sigma_L = C \frac{e^2}{6\pi\eta} \left[\frac{1}{r_+} + \frac{1}{r_-} \right].$$

The conductivity of an electrolyte thus depends on the ionic radius, the valency, and the viscosity of the solvent. As long as the temperature is constant, so is the viscosity. Since the ionic radius and valence are materials properties independent of temperature, also the conductivity will be constant for constant temperature. As we have discussed in Section 8.3.5, the viscosity is strongly dependent on temperature and decreases for increasing temperature, and the conductivity of an electrolyte strongly increases with increasing temperature. The ions become more mobile. This is due to the fact that $\sigma_L \propto 1/\eta$.

The fact that the rate of drift in aqueous solutions depends on the ionic radius and the ion charge is used in electrophoresis in biomedical technology, even though we have seen in Section 8.5.5 that the separation of long-chained molecules in gel electrophoresis is also strongly influenced by the properties of the molecules and the gel. As an example, Figure 9.18 shows the concentration distribution in an electrophoresis cell for proteins in the blood plasma. The different transport speed is used to separate the constituents.

9.5 *Propagation of Nerve Signals

We now want to combine all of the building blocks we have encountered in the previous sections for the properties of charges and currents and describe the transport of an electric signal in a nerve cell. Let us first consider how we can describe the membrane of a nerve cell as a capacitor and how this can lead to a time-varying signal when the sources of charge change in time (as, for example, by ion channels).

9.5.1 Charging a Membrane

The membrane of a nerve cell separates the interior of the axon from the intracellular fluid. We can imagine the membrane as a double layer of lipid molecules that is impermeable to ions and thus allows charge separation between the interior and the exterior. Since such a membrane is essentially flat, we can describe it as the inside of a plate capacitor having a certain thickness (namely, that of the double-lipid layer of about 5 nm). For the potential of such a plate capacitor, we have already found that $V = \frac{Q}{\epsilon\epsilon_0 A}d$, where d is the thickness, A the area of the plate capacitor, and Q the total charge. At that point, we had also introduced the capacitance $C = \frac{\epsilon\epsilon_0 A}{d}$, which connects the charge on the capacitor with the applied voltage via: $Q = C \cdot V$. For the case of a membrane, we obtain a potential difference by the displacement of different charges by means of ion channels, with which we finally get a voltage difference V_0, which corresponds to the Nernst potential of the respective type of ion $V_0 = k_B T/e \ln(n_-/n_+)$. In addition, the membrane has a resistance given by $R = \rho_m A/d$, where ρ_m is the specific resistivity of the membrane. If we want to look at the temporal evolution of the charges in the membrane, we can describe a piece of membrane as a circuit shown in Figure 9.19, with a resistance R connected in parallel with a capacitance C.

Using Kirchhoff's laws, we can describe this circuit by a differential equation for the charge Q or the current I, as follows (the switch S is closed):

$$RI + \frac{Q}{C} = V_0 , \; \Rightarrow \; R\frac{dQ}{dt} + \frac{Q}{C} = V_0 , \; \Rightarrow \; R\frac{dI}{dt} + \frac{I}{C} = 0.$$

In the second and third equation, we have used the definition of the current, namely $I = dQ/dt$. In the third equation, we have used the time derivative of the first equation. This again is an equation we have encountered many times in that the time derivative is proportional to the value itself.

Written as

$$\frac{dI}{dt} = -\alpha I , \; \text{with} \; \alpha = \frac{1}{RC} ,$$

this becomes more obvious and α becomes a rate of change of the exponentially varying current given by the following:

$$I = I_0 e^{-\alpha t} , \; \Rightarrow I = \frac{E_m}{R} \exp\left(-\frac{t}{RC}\right).$$

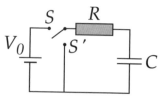

Figure 9.19 Describing a piece of membrane as a circuit.

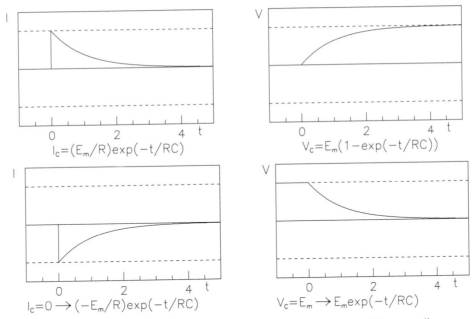

Figure 9.20 Time dependence of the current and voltage when a circuit with resistor and capacitor is switched on or off.

Here, I_0 is determined from the starting conditions stating that $t = 0$, $Q = 0$, $I = I_0$, and $RI_0 = V_0$. This can be integrated to give the charge in the capacitor:

$$Q(t) = Q_\infty \left(1 - \exp \left(-\frac{t}{RC} \right) \right), \quad Q_\infty = CV_0.$$

The charge in the capacitor Q increases after switching on up until a final value of Q_∞ is reached, when the voltage $V_C = Q_\infty/C$ has reached its maximum V_0. The current decreases exponentially from an initial value of $I_0 = V_0/R$ to zero. Once the capacitor is fully charged, there is no current flowing anymore. The characteristic time for the current to decrease $\tau \equiv RC$ is the time when the voltage is a fraction of $1/e \simeq 0.37$ from its final value. The geometry of the membrane does not influence this time scale, since using the preceding relations for capacitance and resistance, the area and thickness actually drop out: $\tau = R \cdot C = \frac{\rho_m d}{A} \cdot \frac{\epsilon \epsilon_0 A}{d} = \epsilon \epsilon_0 \rho_m$. Using typical values for a bilipid membrane, given by $\rho_m = 10^7 \ \Omega m$ and $\epsilon = 7$, we obtain a time to charge a membrane of $\tau = 0.6$ ms. This directly gives us a limit to the speed of neural processes that cannot run on time scales faster than the time to charge or discharge a membrane. In fact, we find that neural processes (e.g., firing, polarization) happen on the time scale of about 1 ms.

If we discharge the capacitor with a constant resistance R, the charge and the current are given by the following:

$$Q(t') = CV_0 \exp \left(-\frac{t'}{RC} \right), \qquad I = \frac{dQ}{dt'} = -\frac{V_0}{R} \exp \left(-\frac{t'}{RC} \right).$$

Here, the time is taken to be zero when the switch is changed from open (S) to closed (S'). If only the switch S is opened and S' remains open, the charge Q on the capacitor C remains constant. The time dependence of the current in the circuit and the voltage across the capacitor are shown in Figure 9.20.

9.5.2 The Cable Equation

In the propagation of nerve signals, not only the charging of the membrane is important, the charge signal also has to be transported along the nerve and a different nerve cell and possibly stimulate that nerve cell to fire. That is, to understand the the propagation of nerve signals, we have to look at how a voltage pulse is transported along an axon. For this purpose, we imagine the axon as a wire whose inside has a certain resistivity and is surrounded by a membrane, which consists of a capacitor and a resistance in parallel as before. We then do this along the axon in small pieces and look at how the voltage pulse is transported. The equivalent circuit diagram is shown in Figure 9.21.

We consider one piece of this axon of length dx and look at how the current flowing through the axon as well as the voltage across it depend on the position x. By virtue of the junction rule, the current at position x, $I(x)$, splits into a part flowing on along the axon and another part flowing across the membrane, i.e., $I(x) = I(x + dx) + I_m$. The current through the membrane we have already discussed has to be $I_m = \frac{V(x) - V_0}{R_m} + \frac{dQ}{dt}$, or using the relation between the charge in the capacitor and the voltage across it, we have $I_m = \frac{V(x) - V_0}{R_m} + C\frac{dV(x)}{dt}$. The area of the piece of membrane we consider is $A = 2\pi a\,dx$, which allows us to determine its resistance as $R_m = \frac{\rho_m d}{2\pi a\,dx}$ and its capacitance as $C = \frac{2\pi a\,dx\,\epsilon\epsilon_0}{d}$. With this, we obtain the following for the current through the membrane I_m:

$$I_m = \frac{(V(x) - V_0)2\pi a\,dx}{\rho_m d} + \frac{2\pi a\,dx\,\epsilon\epsilon_0}{d}\frac{dV(x)}{dt}$$

$$= \frac{2\pi a\,dx}{d}\left(\frac{(V(x) - V_0)}{\rho_m} + \epsilon\epsilon_0\frac{dV(x)}{dt}\right).$$

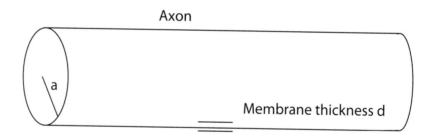

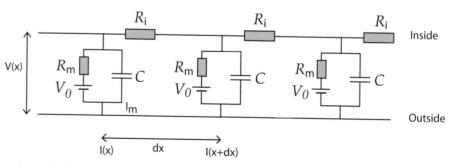

Figure 9.21 Equivalent circuit diagram of an axon.

This condition given by the junction rule can be rewritten to obtain another relation describing the current through the membrane:

$$I_m = I(x) - I(x + dx) = -\frac{dI}{dx}dx.$$

Setting both of these relations equal, we obtain an equation describing the dependence of the current $I(x)$ along the axon on the voltage $V(x)$ across the membrane:

$$-\frac{dI}{dx} = \frac{2\pi a}{d}\left(\frac{(V(x) - V_0)}{\rho_m} + \epsilon\epsilon_0\frac{dV(x)}{dt}\right).$$

Here we divided by the length dx on both sides. The current $I(x)$ is connected to the voltage $V(x)$ in another way. The resistance of the plasma inside the axon and the current $I(x)$ together describe the voltage drop along the axon. Thus we have $\Delta V(x) = V(x) - V(x + dx) = R_i I(x)$, where $R_i = \frac{\rho_i dx}{\pi a^2}$ and we have used the resistivity of the inside of the axon ρ_i. Inserting this relationship for the resistance into the equation above, we obtain the following:

$$\frac{dV(x)}{dx} = \frac{V(x + dx) - V(x)}{dx} = \frac{-\rho_i I(x)dx}{dx\pi a^2} = -\frac{\rho_i}{\pi a^2}I(x).$$

We can combine the two relations between the current and the voltage along the membrane into a single equation by again taking the derivative with respect to position of the preceding equation:

$$-\frac{dI(x)}{dx} = \frac{\pi a^2}{\rho_i}\frac{d^2 V(x)}{dx^2} = \frac{2\pi a}{d}\left(\frac{(V(x) - V_0)}{\rho_m} + \epsilon\epsilon_0\frac{dV(x)}{dt}\right)$$

or

$$\frac{a}{\rho_i}\frac{d^2 V(x)}{dx^2} = \frac{2}{d}\left(\frac{(V(x) - V_0)}{\rho_m} + \epsilon\epsilon_0\frac{dV(x)}{dt}\right).$$

We rearrange this equation such that we can see better which properties of the voltage are important:

$$\lambda^2\frac{d^2 V(x)}{dx^2} = (V(x) - V_0) + \tau\frac{dV(x)}{dt},$$

where $\tau = \rho_m\epsilon\epsilon_0$ and $\lambda = \sqrt{\frac{a\cdot d\rho_m}{2\rho_i}}$. This equation (the cable equation) reminds us of the diffusion equation, but here we describe the transport of a voltage pulse along the axon. For an intuitive understanding of this equation, let us consider the limit where the voltage does not change in time. Then the equation becomes as follows:

$$\lambda^2\frac{d^2 V(x)}{dx^2} = (V(x) - V_0).$$

As we have found several times before, this is solved by an exponential dependence of the voltage on position $V(x) = V_0 \exp(-x/\lambda)$ falling off on a characteristic length scale λ. Therefore, a voltage pulse in an axon can be transported a distance of roughly $\lambda = \sqrt{\frac{a\cdot d\rho_m}{2\rho_i}}$. With typical values for the materials constants of $\rho_i = 1\Omega m$ and geometry $a = 1$ μm, we obtain: $\lambda = 150$ μm and a propagation speed of $v = \lambda/\tau = 0.3$ m/s.

This means that a nerve pulse would not propagate very far and also very slowly. This is, however, not the case, as we know very well. After all, we can react in less than a second to an external stimulus on a limb that is roughly 1 m away. Nature therefore has solved this problem somehow. There are two possible considerations for this (and both are used in nature). The first relates to the thickness of the membrane or the insulating layer around the axon. As we have seen, the thickness of the membrane enters the characteristic length of transport. In many nerve cells (about 30%), axons are myelinated, meaning they are surrounded by cells (Schwann cells) that act as an insulating layer. The corresponding thickness is then $d \simeq 2~\mu m$, and thus we get a characteristic length of $\lambda = 7$ mm. Between the Schwann cells, the axons have voltage-sensitive sites (nodes of Ranvier) that react strongly polarizing to an external voltage. These are about 1.5 mm apart, which means that the current transport in these nerves "hops" from one node to the next (saltatory transport), which reaches the necessary speeds, since we are no longer dealing with a diffusive process of transport on long length scales. The corresponding propagation speed is in the range of 3 m/s (1.5 mm/0.5 ms), which makes typical reaction times possible.

9.5.3 The Action Potential

Now we still have the problem of the 70% unmyelinated nerve cells. Hodgkin and Huxley investigated this in detail on the axons of giant squid. With careful measurements, they could determine the precise time dependencies of the currents for Na and K ions along these axons and model how the conductivities of the different ions across the membrane has to change in response to an applied voltage. The basic principle behind the quick propagation of nerve signals is the following: in our preceding discussion, we have assumed that the resistance of the membrane remains constant and is independent of the voltage across the membrane. In their measurements, Hodgkin and Huxley have found that this does not have to be the case. The resistivity of an axon with respect to transport of Na, K, or Ca ions depends on the applied voltage. This is because the ion channels, which transport the respective ions through the membrane, have different states corresponding either to an open or closed channel. This means the resistance of the membrane ρ_m can change by several orders of magnitude depending on whether the channel is open or closed. In order to make this clear, remember the resistivity of an ionic solution, which we have found to be $\rho \propto 1/n$. Thus the larger the charge carrier density, the smaller the resistance. In the axon and the ion channels, an opening of an ion channel is similar to a strong increase in charge carrier density and thus corresponds to a strong reduction in resistivity. Since the ion channels for the different channels react differently to applied voltages, a threshold voltage results in which a strong voltage pulse can be excited. From the temporal form of the current pulses, which they have fit to a model describing the different conductances, Hodgkin and Huxley were even able to conclude how many openings the different channels have, where they have found four opening gates for K and three opening and one closing gate for Na. Forty years later, these predicted structures could be confirmed by X-ray crystallography of sodium and potassium channels (see Figure 9.22).

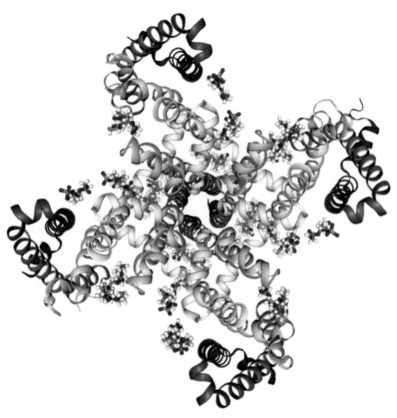

Structure of a Na-ion channel from X-ray crystallography. Data from Lenaeus et al., 2017.

Exercises

9.1   Electrostatics

An ion is accelerated due to a potential difference of 600 V. With this, it loses an energy of 1.92×10^{-17} J. What is the charge of this ion? Is this possible in real life?

9.2   Electrostatics 2

You have a salt (NaCl) solution of 0.16 M in water (the dielectric constant of water is $\varepsilon = 80$). What is the Debye screening length at a temperature of 293 K?

9.3   Electrostatics 3

Within a cell, ion channels are used to create an imbalance of sodium ions between the inside and the outside of the cell. The concentration inside is 20 mMol/L, outside it is 160 mMol/L. What potential difference on the membrane does this lead to? The membrane is at a temperature of 293 K.

9.4   Electrostatics 4

(a) A resting potential of 70(7) mV is applied to the membrane of a nerve cell. The membrane is a lipid bilayer of a thickness of about 5(1) nm. The charge concentration on the surface corresponds to the concentration of ions given by $6.0(6) \times 10^{15}$ ions/m^2. What is the dielectric constant of the membrane material?

(b) What is the error of the dielectric constant?

9.5   Electrostatics 5

(a) Consider DNA as a long, uniformly charged wire. In the spacing of a base pair, there is one elementary charge of a phosphate group. Therefore, the charge density of the charged wire is $\sigma = 1.6 \times 10^{-19}$ C / 0.34 nm. Describe the electric field emanating from this DNA molecule (strength, direction, distance dependence). Assume the DNA is kept in pure water, i.e., is not screened.

(b) What is the binding energy of such a DNA molecule when wound around a histone (with a radius of 5 nm)? For this purpose, assume that the DNA is bound to the histone at a distance of 1 nm (corresponding to the Debye screening length determined in another exercise). Also assume that the histone is charged with a single positive elementary charge that is smeared out over the boundary of the histone. Do not forget to take into account the dielectric constant of water and compare your result to the bending energy of DNA from Exercise 8.1.1.

9.6   Van der Waals interaction, dipoles

(a) Exercise 7.x describes the Lennard–Jones potential. The term with r^{-6} in this potential is usually associated with van der Waals attraction between the molecules. The prefactor M has been determined for CH$_4$ to be $M = 1.6(1) \times 10^{-77} Jm^6$. What is the polarizability of CH$_4$ if the molecule is 2.0(2)$\mathring{A}$ in size and the fluctuating dipole is created by displacing an elementary charge over this size?

(b) What is the error of this?

9.7   Dipoles

An HCl molecule has an electric dipole moment of $3.4(2) \times 10^{-30}$ Cm. The two atoms are roughly $1.0(1) \times 10^{-10}$ m apart.

(a) What is the average charge of these two atoms? Why is it not a multiple of e?

(b) What is the error of this average charge?

(c) What is the energy necessary to rotate such a molecule by 45 degrees from its equilibrium alignment in a homogeneous electric field of 2.5×10^4 V/m?

9.8   Capacitors

Determine the potential energy of a plate capacitor charged with Q_0. Determine the work needed to "push" a small charge dQ from one pole to the other. Insert the electric field E as a function of the total charge Q, then integrate the result from $Q = 0$ to $Q = Q_0$.

9.9   Capacitors 2

Figure 9.23 shows a thin cylindrical capacitor of length L as a model for an axon. Here the capacitor takes the role of the axon's membrane. The inside surface contains a charge $+Q$, the outside a charge $-Q$.

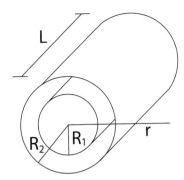

Figure 9.23 A thin cylindrical capacitor of length L as a model for an axon.

(a) Draw the corresponding field lines.
(b) Determine the electric field as a function of the distance r from the axis and plot your result.
(c) Calculate the voltage V between the two cylinders and with this the capacity of the capacitor.

9.10 Atomic physics

(a) What is the energy emitted from a hydrogen atom where the electron has changed from the state with n = 10 to the state with n = 3?
(b) What is the de Broglie wavelength of an electron in the ground state of te Bohr model of the hydrogen atom?

9.11 Atomic physics 2

(a) How does the description of the Bohr model change for higher elements in the periodic table? What is the difference between the state with $n = 1$ and the state with $n = 6$. For a numerical example, use lead (Z = 82).
(b) How does the binding energy of the Bohr model change if instead of an electron a negatively charged muon orbits the nucleus? Muons are elementary particles with basically the same properties as electrons, except that they are 200 times more massive.
(c) How does the answer in (b) change if instead of a muon an antiproton (one negative elementary charge, (2,000-fold) mass of the electron) orbits the nucleus?

9.12 Currents

A current density of $j = 0.8 A/m^2$ is stimulating a nerve membrane of a thickness of 5 nm during 1 ms. How does the voltage across the membrane change as a result of this? The dielectric constant is $\epsilon = 7$.

9.13 Currents 2

What is the current flowing when all sodium channels in a piece of muscle membrane of area 1 mm^2 are opened during a time of 10 ms? There are about 50 channels/μm^2 in muscle tissue and a single ion channel has a flux rate of 1,000 Ionen/ms.

9.14 Currents 3

What is the resistivity of buffer solution, i.e., a solution of 160 mM NaCl in water? Assume that the mobility of the ions is given by viscous friction (viscosity of water $\eta = 10^{-3}$ Pa s) and the ions are spheres of diameter $d_{Na} = 4$ Å and $d_{Cl} = 3$ Å respectively.

9.15 Currents 4

In a thunderstorm, you observe lightning of duration about 0.2 s. In order for lightning to appear, the potential difference needs to exceed 50 megawatts (MW) and you remember that the power in a flash of lightning can be around 1 gigawatts (GW) from the movie *Back to the Future*. What is the average current flowing in the lightning strike, and how much charge has been transported?

9.16 Resistors

You have a piece of clay that you roll into a cylinder of length L. You measure its electric resistance and obtain some value R. Now you take the same piece of clay and roll it into a cylinder of length $L/2$. What resistance do you measure?

Quiz Questions

9.1 Electrostatics

Two charges repel with a force given by F. What is the force if both charges are doubled?

A $F/4$
B $F/2$
C F
D $F * 2$
E $F * 4$

9.2 Electrostatics 2

A proton and an electron are placed at $x = -d$ (proton) and $x = d$ (electron) at rest and are let go. What happens?

A They are repelled.
B They meet in the middle (bei $x = 0$).
C They meet close to $x = -d$.
D They meet close to $x = d$.
E They form a hydrogen atom.

9.3 Electrostatics

A proton (charge $+1.6 \times 10^{-19}$ C, mass 2×10^{-27} kg) and an electron (charge -1.6×10^{-19} C, mass 10^{-30} kg) attract each other by electrostatic forces. Which of the following is true?

A The acceleration of the proton is larger.
B The acceleration of the electron is larger.
C The force of the proton onto the electron is larger.

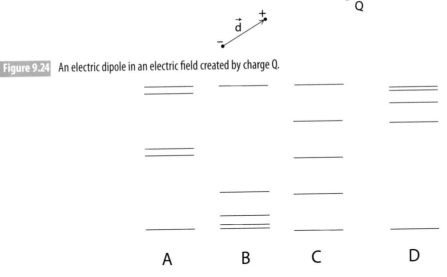

Figure 9.24 An electric dipole in an electric field created by charge Q.

A   B   C   D

Figure 9.25 Energy levels of a hydrogen atom.

D When the particles get closer, the force gets smaller.
E When the particles get closer, the velocities get smaller.

9.4 Dipole

An electric dipole is in an electric field created by charge Q (see Figure 9.24). What happens to the dipole?

A The dipole is repelled by Q.
B The dipole turns, but stays in place.
C The dipole is attracted to Q without changing its orientation.
D The dipole orients toward the charge and is attracted.

9.5 Atomic Physics: Typ A, 1P

Which of the sketches in Figure 9.25 best describes the energy levels of a hydrogen atom?

9.6 Resistance

You create two resistors out of the same material (and the same amount). One of these has a length L, and you find a resistance R. The other has a length $L/2$. What is its resistance?

A R
B $R/2$
C $2R$
D $R/4$
E $4R$

9.7 Currents

Three identical lamps are connected to a constant voltage source in a circuit, as shown in Figure 9.26. The brightness of a lamp indicates the power dissipated.

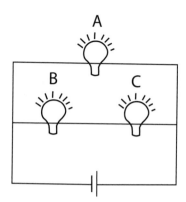

Figure 9.26 Lamps in a parallel circuit.

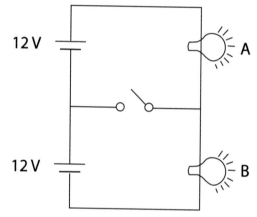

Figure 9.27 Two identical lamps put into a circuit with two constant voltage sources.

How does the brightness of lamp A compare to the combined brightness of lamps B and C?

A Four times brighter.
B Twice as bright.
C They are the same.
D Half as bright.
E A quarter of the brightness.

9.8 Currents 2

Two identical lamps are put into a circuit as shown on Figure 9.27, with two constant voltage sources. What happens if you close the switch between the two sources?

A Lamp A becomes brighter and lamp B becomes darker.
B Lamp A becomes darker and lamp B becomes brighter.
C Both lamps become brighter.
D Both lamps become darker.
E Nothing changes.
F Both lamps go out.

Magnetism

10.1 Magnetic Fields in Biology

The second large area of electromagnetism deals with magnetic fields and their description. Magnetic fields and their detection are important in various areas of biology and chemistry in a twofold way. On the one hand, different phenomena in nature, e.g., in navigation, rely heavily on the interaction with magnetic fields. On the other hand, there are different experimental methods in biology, chemistry, and medicine that are based on the properties of electric fields. We will address them briefly here and then discuss them in more detail later in this chapter.

10.1.1 Navigation

Various animals can navigate around their habitat exceptionally well; they can travel reproducibly over very long distances or find their way home. In many insects, this is achieved via their capability of perceiving the polarization of the light from the sky, as we have discussed in optics. However, in most other animals (such as migratory birds), navigation is mainly achieved via the earth's magnetic field. Also humans have navigated mainly with the help of the magnetic field (at least up to the recent availability of global positioning systems [GPS]) and long journeys were only possible by the invention of the compass needle. A very sensitive magnetic field meter is necessary for the navigation of the animals, or the differentiation of local areas around the nest, where the exact mechanisms are still not conclusively understood. What is known is that one can train pigeons or migrating birds to react to magnetic fields, such that they must have such a sensor and that they have the required sensitivity. It is also known that there are bacteria that can perceive the earth's magnetic field by having small magnetic particles (of iron oxide) in their interior that act as a compass needle. Why the particles have to have this shape and size to make an efficient receptor can be understood on the basis of the physics of the magnetic materials, which, however, is beyond the scope of this book. Such particles have also been found in other magnetically navigating animals, but not yet in migratory birds. It is also not clear how these are associated with the nervous system. Here there is still a lot for future biologists to do.

10.1.2 Spectroscopic Methods

In addition, magnetic fields are an integral part of some standard methods of particle identification in chemistry and biochemistry, namely mass spectrometry and nuclear magnetic resonance. In mass spectrometry, a combination of electric and magnetic fields is used to separate molecules (or their fragments) very sensitively to the mass of the fragments, thus allowing an identification of the substance. The sensitivity of the mass determination and relatively small amounts necessary for an accurate identification have made the method into a standard of analytical chemistry, which now also has great influence in the determination of proteins in biology. In order to know the possibilities and limitations of the method, one must, however, understand its physical basis, which we shall provide in this chapter.

An even more important experimental method in chemistry, biology, and medicine is that of nuclear magnetic resonance. Today, it is an integral part of not only medical diagnostics, but also the structural determination of macromolecules and the imaging of biological processes. Here, too, the method is based on the magnetic properties of matter, especially those of the atomic nuclei. In addition, in the treatment of the method, we will again encounter many topics from Chapter 4, e.g., resonant absorption and Fourier analysis. In order to be able to apply the method correctly, different physical principles are important. For example, there is a fundamental difference between nuclear resonance spectroscopy and imaging. The apparatus requirements for the respective methods are very different but can be understood once the underlying physics is understood.

Before discussing these two experimental methods, however, we need to look more closely at the properties of magnetic fields.

10.2 Properties of Magnetic Fields

10.2.1 Phenomenology of Static Magnetic Fields

The interaction between two magnets at a distance is described in a similar way to that of electrical charges. Hence we introduce a magnetic field $\vec{B}$, which originates from a magnetic material and interacts with the field of another magnet. These magnetic fields look similar to an electric dipole field. A magnet has two poles: the same poles repel, unequal ones attract.

The earth is also magnetic (i.e., has a magnetic field): a compass needle points north (i.e., the north pole of the compass needle points north, implying that the magnetic south pole of the earth's magnetic field lies in the arctic). The earth's magnetic field is also such a magnetic dipole field.

But in contrast to electric dipoles consisting of a positive and a negative charge that can be separated, magnetic dipoles cannot be separated. Every separation just leads to a smaller dipolar magnet. There are no magnetic charges (or monopoles), at least no such magnetic monopole has ever been found so far. Thus we can write Gauss' law for a magnetic fields as follows:

$$\oint_{A_V} \vec{B} \cdot d\vec{A} = 0.$$

The flux of a magnetic field through a closed surface is always zero. The magnetic field has neither sources nor sinks. Therefore, a magnetic field line cannot begin or stop anywhere. Magnetic field lines are always closed.

10.2.2 Ampere's Law: The Magnetic Field of a Constant Current

Magnetic fields are produced by electrical currents. Figure 10.1 shows the magnetic field near a wire carrying an electric current (i.e., containing moving charges). This field is made visible by iron filings, which, due to their magnetic properties, align like small compass needles along the magnetic field lines. The observed field lines form concentric circles around the wire.

Additionally, we find that the magnetic field increases in proportion with the current, but decreases inversely proportional to the distance, i.e., $B \propto I$ and $B \propto r^{-1}$ respectively. This is described exactly by Ampère's law.

The line integral of the magnetic field along any curve C, which completely encloses the current, is directly given by the current enclosed (see Figure 10.2). If C encloses several current carrying wire, their currents all have to be added up:

$$\oint_C \vec{B} d\vec{r} = \mu_0 \sum_i I_i.$$

In the example shown in Figure 10.3, only I_1 and I_2 have to be considered in the sum, but not I_3. The contributions are counted positively if the rotational sense of the path coincides

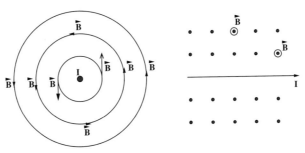

Figure 10.1   Magnetic field of a current-carrying wire; Left: section in a plane perpendicular to the wire axis; right: section in a plane containing the wire.

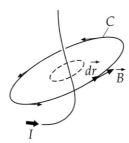

Figure 10.2   Magnetic field lines encircle electrical currents.

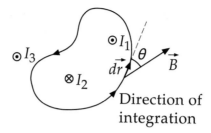

Figure 10.3 The line integral of the magnetic field is equal to only the enclosed currents.

with the direction of the current, and negative otherwise. In words, Ampère's law can be
formulated as follows:

Currents are the vortex lines of the magnetic field.

If we imagine a very short coil, that is, one circular current loop, and apply Ampère's
law, we see that this current loop produces a magnetic dipole field with a dipole moment
$p_m = I\pi R^2$, which is perpendicular to the plane of the current loop! Here, R is the radius of
the loop. Such a magnetic dipole interacts with other dipoles (be they permanent magnets
of due to currents) just as the electric dipoles did in electrostatics. This interaction is what
aligns the compass needles.

In fact, all static magnetic fields are fundamentally and microscopically generated by
electrical currents, that is, by moving electric charges. The dipole fields of the permanent
magnets we have looked at initially are generated by the superposition of the magnetic
fields of a large number of microscopic currents in the atoms of those magnets. These
are, on the one hand, the circular movements of the electrons around the nucleus, but, on
the other hand, also a kind of circular motion of the electrons around their own axis. We
can imagine that electrons have an internal angular momentum so that the rotation of the
electron's charge around the axis of rotation gives a circular current. However, this image is
flawed in several ways. For instance, as far as we know, the electron does not have a spatial
extension such that no such current would exist. But the magnetic moments are there and
measurable, so we stick to the image of an internal angular momentum and a rotation axis.
In fact, a quantum mechanical description of charged point particles such as electrons gives
the microscopic moments one observes to a quantitative accuracy of 11 significant digits, so
one can understand this. However, the image of the spinning ball gives a very similar result
to the exact quantum mechanical treatment, so we will stay at this level of description,
even though we know it has its flaws. In the image of the spinning ball, the magnitude of
the magnetic moment is given by the current caused by the rotating charge. At a frequency
of rotation of $\nu = v/(2\pi r)$, this current is given by $I = e\nu$, where e is the elementary
charge, i.e., the charge of the electron, which varies with a frequency ν, therefore directly
giving the rate of change of the charge, i.e., the current. As we have seen, this current leads
to a magnetic moment of $p_m = I\pi r^2 = e\pi\nu r^2 = evr/2$, where we have used the expression
for the frequency in the final step. The product of the rotational speed v and the distance
of rotation r is closely connected to the angular momentum of this rotation, $\vec{L} = m_e\vec{v} \times \vec{r}$,

where m_e is the mass of the electron; see Section 10.4.1. Therefore the magnetic dipole moment of a spinning charge is given by the following:

$$\vec{p}_m = \frac{e}{2m_e}\vec{L}.$$

Angular momenta are very closely related to quantum mechanics and hence magnetism is also rooted in there. Reconsidering the quantization rule of the Bohr model, we can see that angular momenta can only appear in chunks corresponding to a magnitude of $|\vec{L}| = \hbar$, where $\hbar$ is Planck's constant. So the smallest chunk of magnetic dipole moment the electron can have is given by the following:

$$p_m = \frac{e\hbar}{2m_e}.$$

An exact derivation of this is more complicated and also shows that the angular momentum of the electron (its spin) actually has the value of $\hbar/2$. Therefore, then, every single electron has a magnetic moment, but in normal materials their orientations are arbitrarily arranged so that the average magnetization is zero. If one places such a material inside a magnetic field, the dipoles are arranged along the field, as we have seen for the electric dipoles in E fields. The corresponding material property is the magnetic permeability. Just as with the electric dipoles, the material property of the permeability can determine the size of the field in a material. The individual dipoles can also interact with one another, if a part of the aligned moments creates a field, which then aligns further moments. The interaction between the dipoles, which leads to the original alignment, is again given by quantum mechanics. Given these interactions, magnetic moments can interact with a field in different ways and thus both increase or decrease the field in the material. In most materials, a lowering occurs, as was the case with the electric dipoles. The atomic dipoles are oriented against the applied magnetic field. One speaks also of diamagnetism in relation to the dielectrics. The field created by the atomic dipoles is also called the magnetization $\vec{M}$.

However, with certain materials, the magnetic moment can also be aligned parallel to the field. These are called paramagnetic and when the interaction between the dipoles is strong enough, the orientation becomes so strong that a permanent magnet can develop. In this case, one speaks of ferromagnetism.

10.2.3 The Lorentz Force

Not only do moving charges or currents create magnetic fields, they are also influenced by magnetic fields via the Lorentz-force. Therefore, this force can also be understood as an interaction between two magnetic fields. From Ampère's law, we can obtain not only the strength but also the direction of the Lorentz force. Because the field produced by the current, i.e., the moving charge is perpendicular to that movement, the force will have to be perpendicular to the movement as well. Similarly, the force will be proportional to the produced magnetic field, meaning that it will be proportional to the current or the product of charge and velocity of the charge. Likewise, it will be proportional to the applied magnetic field:

$$\vec{F}_L = q[\vec{v} \times \vec{B}].$$

Given that in this relation we see an observable force and a measurable current, we could use this as an alternative definition of the magnetic field $\vec{B}$ and we can show the presence of a magnetic field by virtue of this force onto a moving charge. This relation also gives us the units of a magnetic field relatively straightforwardly. The SI units of the field are as follows:

$$[B] = \frac{N}{Am} = \text{Tesla} = T = \frac{Vs}{m^2}.$$

The earth's magnetic field typically has a strength of 50 μT. The field of a permanent magnet is usually 0.5 T and can be 1 T for very strong neodymium magnets. A strong electromagnet reaches about 2 T and with superconducting coils up to 20 T is possible. High magnetic fields are used in medicine and biology with magnetic resonance imaging systems, in nuclear magnetic resonance (NMR) spectrometers or mass spectrometers, but also in accelerator construction for radiotherapy or in electron microscopes.

Since the Lorentz force on a moving particle is always perpendicular to the velocity, the speed always remains constant. The Lorentz force does not carry out work. It cannot change the kinetic energy, in contrast to the Coulomb force in an electric field $\vec{E}$. If a particle moves parallel to a magnetic field ($\vec{v} = \vec{v}_\parallel \parallel \vec{B}$), the vector product is zero, $\vec{v} \times \vec{B} = 0$, and hence there is no Lorentz force acting. The velocity $\vec{v}$ remains constant.

If the speed is perpendicular to the magnetic field ($\vec{v} = \vec{v}_\perp \perp \vec{B}$), there is a force and the equation of motion becomes

$$ma_\perp = m\frac{v_\perp^2}{r} = qv_\perp B.$$

The particle will move on a circular orbit with a radius r, where the plane of the orbit is perpendicular to the field. Its orbiting frequency ν, or the angular frequency $\omega = 2\pi\nu$ of the orbit, is then given by the following:

$$2\pi\nu = \omega = \frac{v_\perp}{r} = \frac{qB}{m}.$$

This is also called the cyclotron frequency, and it is noticeably independent of the particle's speed (see Figure 10.4). This means that particles with different speeds $v_\perp$ and hence different radii $r = (mv_\perp)/(qB)$, that start at the same time in point S take paths of

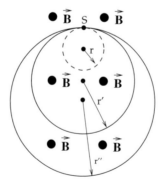

Figure 10.4 Radius of curvature of a cyclotron.

different lengths as shown, however, return to point S at the same time. This property of the motion of charged particles in a magnetic field lies at the heart of mass spectrometers, which we will discuss in the next section.

10.3 Mass Spectrometry

For analytical chemistry and biochemistry, it is often important to be able to determine the mass of molecules or their constituents. For this purpose, mass spectrometers are used. The molecules are first converted into charged ions in an ion source and then accelerated in an electric field. If the ion (charge Ze) passes through a voltage difference V, then it gathers a kinetic energy of the following:

$$\frac{m}{2}v^2 = Ze \int \vec{E}d\vec{r} = ZeV.$$

The thus accelerated ions then pass a homogeneous magnetic field, which is perpendicular to $\vec{v}$. Measuring the radius of curvature of the orbit, i.e., how far away from the source the ion returns, we can determine the mass of the ion, as long as the charge, V, and B are known:

$$r = \frac{mv}{ZeB}, \qquad \Rightarrow \qquad m = Ze\frac{r^2B^2}{2V}.$$

If the detector is moved along that distance (see Figure 10.5), one records a particle flux as a function of distance (twice the radius of curvature).

This directly corresponds to a spectral distribution of the masses of the constituents of the sample being analyzed, which can also help in elucidating its structure.

There are, however, many different types of mass spectrometers with different principles. A second example rests on the previously described cyclotron frequency, i.e., the fact that

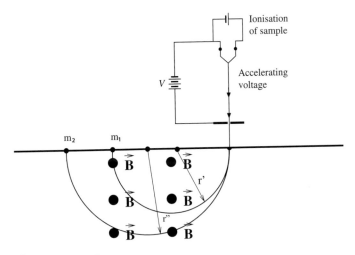

Figure 10.5 Schematic setup of a mass spectrometer.

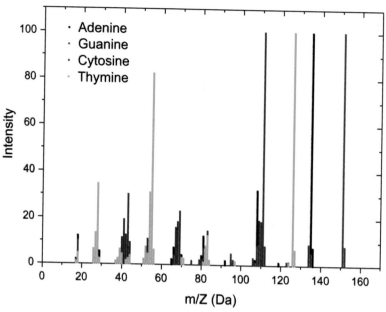

Figure 10.6 Mass spectra of the different DNA bases. Data from SDBS, 2017.

charged particles in a magnetic field produce a circular orbit with a very definite frequency. This frequency is given by $\omega = \frac{q}{m}B$. This movement of charged particles represents a current, which can be measured, and because of the cyclotron orbit, this current will oscillate with the cyclotron frequency. This depends only on the applied field and the particle mass. The period of this frequency can be easily measured, and with a known magnetic field, the mass is directly determined. However, if a spectrum of masses is present, such as shown in Figure 10.6, the determination of the frequency becomes more difficult. In this case, a Fourier analysis of the signal must be made in order to determine the different frequencies (in this case, masses). Here, the Fourier spectrum then directly corresponds to the mass spectrum.

10.4 *Nuclear Magnetic Resonance

The second important technique based on magnetic fields we want to discuss is nuclear magnetic resonance. It is based on the property of atom nuclei to have a magnetic moment, just as we have discussed previously in the case of the electrons. These are aligned by strong magnetic fields and then perform periodic oscillations in the magnetic field. There are two very different ways of applying the nuclear spin resonance: on the one hand, perhaps the somewhat more familiar tomography, which is often used as an imaging method in medical diagnostics; and, on the other hand, spectroscopy, which is of great importance for the determination of structure, as well as for analysis.

10.4.1 Angular Motion and Angular Momentum

In order to understand how nuclear spins react to magnetic fields, we first have to make a brief excursion into describing angular motions, i.e., spins or spinning tops, in order to have some intuition for the effect of external forces on spins. As we have seen in Section 3.2 on oscillations, if something moves around a circle, there needs to be an acceleration that keeps the object on its path. Furthermore, we have seen that the speed of the motion is larger the larger the radius of the orbit is. Since motion at a speed is connected to kinetic energy, a circular motion corresponds to a form of kinetic energy, rotational energy. We have also seen that a moving mass carries a certain amount of momentum. Again, this will have to be similar in rotating bodies; however, it makes sense to include the distance to the resting center of the rotation in this quantity, which is called the angular momentum, defined by the following:

$$\vec{L} = \vec{r} \times \vec{p}. \tag{10.1}$$

Since the distance to the center is a vector as is the momentum, and we want to include both of these quantities in our description, we have to use the cross-product here. If we were to use the scalar product of distance and momentum, we would obtain a quantity, which is independent of the sense of direction of the rotation, which therefore would not properly account for such a motion. Thus, if we want to describe an angular motion, we use angular momentum, which is a vector pointing in the direction perpendicular to the plane in which the object is rotating. We can rewrite the previous definition of the angular momentum by taking into account what momentum stands for, which gives $\vec{L} = m\vec{r} \times \vec{v}$, and simplify the description by considering a circular rotation. In that case, the velocity and the radius are perpendicular and their cross-product is simply given by $\vec{r} \times \vec{v} = rv\vec{e}_z$, where $\vec{e}_z$ is the unit vector pointing perpendicular to the plane of rotation. Moreover, we have seen in Section 3.5 that the speed of a circular motion is simply given by the radius and the angular frequency, i.e., $v = r\omega$, such that we obtain the following for the angular momentum:

$$L = mr^2\omega = I\omega. \tag{10.2}$$

Here we have introduced the moment of inertia I, which describes for more complicated bodies how they resist an angular motion. Basically, every point can be considered separately and all the corresponding small angular momenta have to be added up. Thus the moment of inertia of any body is given by the following:

$$I = \int r^2 dm, \tag{10.3}$$

which for instance for a sphere of radius R yields $I = \frac{2}{5}mR^2$. Thus we see that an angular momentum corresponds to an angular frequency and a moment of inertia, which depends basically on the geometry and mass of the spinning object. Therefore, a charged particle with a magnetic moment will also carry angular momentum, as we have described previously in Ampere's law.

Similar to momentum, angular momentum is conserved. You can easily experience this while ice skating: if you start rotating with your arms extended and then pull in your arms,

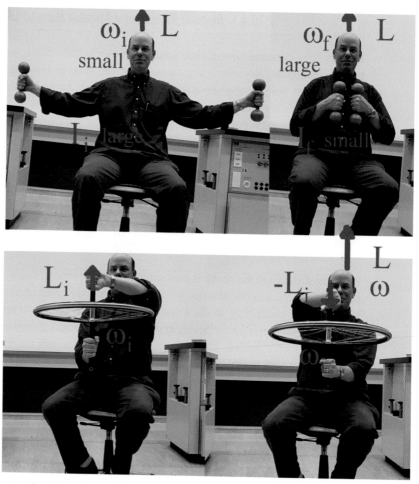

Figure 10.7 Conservation of angular momentum. Top: decreasing the moment of inertia leads to an increase in angular frequency, such that the angular momentum stays constant. This allows ice skaters to perform their fast pirouettes. Bottom: turning a spinning object changes its angular momentum. In order to conserve the total angular momentum, you have to be imparted the corresponding change. If you sit on a turning chair, you start rotating in the opposite direction.

your frequency of rotation will increase greatly. This is because the moment of inertia has decreased because you have pulled in your arms (you have decreased r) and since angular momentum ($L = I\omega$) is conserved, if I decreases, ω increases accordingly (see the top of Figure 10.7).

This conservation of angular momentum also leads to the fact that changing the orientation of spinning objects is not easy, allowing, for instance, for more stable bicycling. If I turn around a spinning object, I have inverted its sense of direction and hence the vector of angular momentum. Therefore, this change in angular momentum needs to be counteracted by an additional angular momentum on me, which leads to me rotating in the opposite direction if I sit on a turnable chair (see bottom of Figure 10.7).

The reaction of a spinning object to an external force can be more quantitatively described by considering that the force in this case does not act on the moving particle itself and that therefore there is a distance involved in this as well. This is similar to what happens in a lever, where the force times the distance counts in how far anything is rotated. The proper quantity to consider then is called a torque, given by the following:

$$\vec{\tau} = \vec{r} \times \vec{F}, \tag{10.4}$$

where again we have to use the cross-product in order to end up with a vector that can describe an orientation of turning. Now we have defined torque and angular momentum as extensions of force and momentum respectively. Since a force corresponds to a change in momentum, as we have seen in Chapter 6, a torque will also correspond to a change in angular momentum, such that

$$\vec{\tau} = \frac{d\vec{L}}{dt}. \tag{10.5}$$

Now we can consider what happens to a spinning object that is acted on with a force perpendicular to the angular momentum of the spin. This is shown in Figure 10.8, where we can see that the force downwards is perpendicular to the direction between the force and the spin. Therefore, the torque will be perpendicular to the spin and hence perpendicular to the angular momentum. Since this leads to a change in angular momentum in the direction of the torque, the angular momentum will be turned in a plane perpendicular to the applied force. The direction of the spin changes accordingly, but still stays perpendicular to the

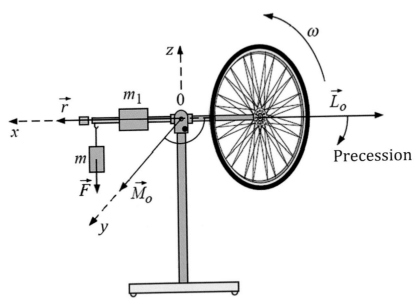

Figure 10.8 If a force is acting on a spinning object, there is a torque acting onto that object, which leads to a change in angular momentum. Given the directions of torque and angular momentum of the spin, a force downward will lead to an excursion sideways of the spin and therefore to a rotation in the plane perpendicular to the pull. This is called precession.

force, such that there is still a torque, which is still perpendicular to the spin. Therefore, the spin is turned even more, and what we obtain is that the spin direction turns around its fixation with an angular frequency of $\Omega = \frac{\tau}{L} = \frac{rF}{I\omega}$. This process is called precession.

Since magnetic moments correspond to angular momenta, these processes also happen in magnetic moments. If we apply an external field to a magnetic moment, we have already seen that what will happen is that the magnetic moment aligns with the field. This is because the field exerts a torque onto the magnetic moment given by $\vec{\tau} = \vec{p}_m \times \vec{B}$. If we treat the magnetic moment as originating from an angular momentum as described previously for an electron, such that $\vec{p}_m = \frac{e}{2m_e}\vec{L}$, we obtain a torque of $\vec{\tau} = \frac{e}{2m_e}\vec{L} \times \vec{B}$, which by definition is perpendicular to the angular momentum. Thus the spin will start to precess, and it will do so with an angular frequency of

$$\Omega = \frac{e}{2m_e}B.$$

This means that by measuring this precession frequency, we can obtain information about the applied magnetic field, which forms the basis of nuclear magnetic resonance.

10.4.2 Basic Principles

The components of the atomic nucleus, the protons and neutrons, have the same angular momentum as electrons. If one views this as originating from a spinning motion of the corresponding particle, the resulting current leads to a magnetic moment. Since the proton (as well as the neutron) is about 2,000 times heavier than the electron, this magnetic moment is smaller by the same factor. This means the energy required to turn such a magnetic moment is very small and can be done with radiowaves. However, the energy is so small that it can also be easily achieved by thermal fluctuations, which leads to an average of very few nuclear spins being aligned in an applied magnetic field. Most are disordered by thermal fluctuations. To quantify these effects, we first need to calculate the energy required for alignment. The nuclei at which nuclear spin resonance can be measured must have a resulting magnetic moment $\vec{p}_m$. For even-numbered multiples of the neutron and proton numbers, the magnetic moments of the respective building blocks can be canceled and the nucleus has no resulting magnetic moment. In such nuclei, no spin resonance can be obtained. This affects many nuclei that are very common in organic substances such as C-12, N-14, or O-16. Therefore, nuclear spin resonance in biological samples is by far the most frequently based on hydrogen nuclei (that is, on a single proton).

We have seen in electrical dipoles that they interact with an applied field. This is the same for magnetic dipoles (see Figure 10.9), and again the interaction energy is given by $\Delta E = \vec{p}_m \cdot \vec{B}$. Using the magnetic moment of the proton of $p_m = 1 \times 4 \times 10^{-26}$ J/T, we obtain an energy difference at a field of $B = 1$ T of only $\Delta E = 1 \times 4 \cdot 10^{-26}$ J. Comparing this to thermal energies of $k_B T = 4 \times 10^{-21} J$ at room temperature, we see that this energy is extremely easily excited by thermal fluctuations. The Boltzmann probability for this process is very close to one, or to be more precise, is given by $\exp(-\Delta E/k_B T) \simeq 1 - \Delta E/k_B T$. We are certainly allowed to Taylor-expand the exponential here, since $\Delta E/k_B T = 3.5 \times 10^{-6}$ and therefore very small compared to one for the values

Left: The dependency of the dipole interaction energy on the applied field. Depending on whether the magnetic moments are parallel or antiparallel to the field, a positive or negative energy is generated. To get from one state to the other, a difference energy ΔE is needed. Right: The situation of the orientation of the moments in the magnetic field. Without a field, all moments are arbitrarily aligned, whereas in a field B_0, the moments are aligned with the field predominantly parallel.

given previously. This means we can now consider quantitatively which fraction of the nuclei is ultimately aligned by the field, because the probability of the Boltzmann factor would destroy this alignment. By increasing the field or decreasing the temperature, this can be improved somewhat, but in both cases a very large change is not possible and the fraction will be of order 10^{-5}. From these considerations, we see that NMR only gives information when there are very many molecules present. However, this is not a big problem, because molecules are incredibly small and even in very small volumes, there are still many molecules present. Thus even in relatively small samples, useful measurements are possible, but single molecule experiments with NMR cannot be done.

Next, let's look at what happens with the aligned nuclear spins. As described previously, the magnetic moment, i.e., the "axis of rotation" of the proton, will carry out a precession around an applied magnetic field. No matter in which direction the nuclear moment is pointing at the beginning, the frequency of the precession is always the same and given by $\omega_L = 2m_P B/\hbar = \gamma_P B$, where γ_P is the gyromagnetic ratio and the frequency is also called the Larmor frequency. We can estimate the gyromagnetic ratio from the preceding arguments, by inserting the mass of the proton instead of the electron, i.e., $\gamma_P \sim \frac{e}{m_P}$, which yields around 100 MHz/T. For a hydrogen nucleus, one experimentally finds a value of $\gamma_P = 42$ MHz/T, where the discrepancy is due to some quantum mechanical effects that can in principle be understood. That is, with an applied field of about 3T, as is the case in modern nuclear magnetic resonance tomographs, the hydrogen nuclear spins precess with a frequency of about 120 MHz. This frequency corresponds to that of radiowaves, which is important for the further understanding of the technique.

If we imagine a precessing nuclear spin that precesses with a certain frequency, then we can change this precession by applying a time-varying magnetic field with the same frequency perpendicular to the main field. This results in the spin also being precessed around this field and thus changing its orientation. This makes it possible to determine the orientation of the spins. However, the argument from before is that, by far, not all spins will show this orientation. Such a time-varying magnetic field can be achieved, for example, by irradiating an electromagnetic wave if the latter has the correct frequency. That is, the mechanism of nuclear spin resonance is the resonant absorption of radiowaves (see the preceding frequency) by nuclear spins aligned in a strong magnetic field. This resonance is very narrow since the attenuation in the core is very small. This means that the magnetic field at the site of a nuclear spin can be determined with very high accuracy by precise measurement of the resonance frequency. This highly accurate measurement of the local magnetic field is the actual method of nuclear magnetic resonance.

There are now two methods for measuring the frequency accurately. The first is to change the frequency of irradiated radii waves and to measure the transmission of these radii waves through the sample. At the resonance frequency, absorption occurs, and the transmission becomes correspondingly smaller. If you know quite well at what frequency the resonance is, you can also apply a pulse of the correct length to align the spins perpendicular to the main field. These are then precessed with the Larmor frequency, resulting in a periodic change in the magnetization of the sample in this plane. If we now measure the temporal change of this magnetization and the Fourier transform is determined, we obtain directly a measure for the Larmor frequency and thus the local magnetic field. If several local magnetic fields are present, this method is usually more informative than that of absorption spectroscopy. Nowadays, only the second method is used.

10.4.3 Spectroscopy

We so far have seen that we can use NMR very well to measure magnetic fields inside a sample. This is now used in spectroscopy to obtain accurate information about the molecules themselves. We will now briefly look at the basic ideas. As we have seen, it is possible to measure the local magnetic field at the site of a nucleus very accurately by means of the Larmor frequency in nuclear magnetic resonance. The accuracy is actually only determined by the homogeneity of the applied field. We can estimate this by looking at how accurately we can determine a frequency. If we were to count the oscillations of the precession for a mere second, we would obtain about 100 million cycles, thus obtaining a statistical accuracy of the frequency given by one over the square root of that number, or 10^{-4}. With such accuracy, there are various influences on the local magnetic field, which may be larger and thus influence our result.

The most direct influence on the local magnetic fields are those given by the electronic properties of the atomic shell. Since the electrons in the atomic shell are indeed moving

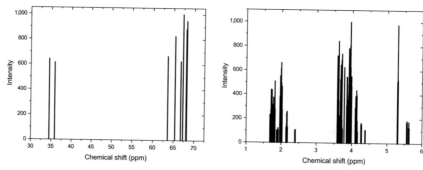

Figure 10.10 NMR spectra of deoxyribose. Left: NMR signal from C-13 nuclei; right: NMR signal from H-nuclei. Data from SDBS, 2017.

charges, they experience a Lorentz force through an applied magnetic field, which in turn changes the orbit of the electrons. This is a change in a current loop and thus a change in a magnetic field. This also affects the local field at the nucleus. This small change (typically not more than 10^{-5}) is directly measured as a chemical shift of the Larmor frequency in an NMR spectrometer and represents a fingerprint of the substance analyzed. The chemical shift is usually expressed as $\delta = \frac{\omega - \omega_{ext}}{\omega_{ext}}$ and is measured in ppm (i.e. part per million, or 10^{-6}).

As stated previously, the determination of the chemical shift requires an extremely accurate measurement of the local magnetic field, which is made possible by the frequency measurement in NMR. For this purpose, however, it is necessary to create an externally applied magnetic field that changes less than the chemical shift over the entire sample. Otherwise, an observed change in the NMR frequency could be confused with a chemical shift. The applied magnetic field therefore must be homogeneous to better than a part in a million across a sample length of at least a millimeter up to a centimeter!

For more complex molecular structures, the different chemical shifts can no longer be separated and more complicated methods are needed in which the time evolution of the nuclear spin is used further and thus correlations between different parts with similar chemical shift are distinguished. This is summarized in the techniques of multidimensional NMR spectroscopy but is beyond the scope of this book.

10.4.4 Imaging

The most well-known application of NMR is magnetic resonance imaging (MRI) (see Figure 10.11). Although it is based on the resonant absorption of radiowaves by nuclear spins in a magnetic field as in NMR spectroscopy, discussed in the preceding section, the spatial resolution is achieved by completely different means. Unlike in spectroscopy, a spatially varying magnetic field is applied in MRI. The field change in space must now be greater than the chemical shift, thus limiting the resolution. The applied radio frequency is thus only in resonance with a single layer of nuclei, and thus resonant absorption only occurs at this location. The density of nuclei of, e.g., hydrogen is then used as the imaging contrast at different positions.

The basic principle of MRI is to transform the spatial information into a spectral one. Because of the application of a gradient field, only the spins on a specific position along the z axis are susceptible to the absorption of the radiowaves of a given frequency, i.e., only these spins are flipped to the x–y plane by a $\pi/2$ pulse. But MRI gives three-dimensional images, so the positional information in the x- and y-directions also must somehow be transformed into spectral properties. After the first $\pi/2$ pulse has been applied, only one layer of spins in the x–y plane, a gradient field, is applied in the x-direction, resulting in different precession frequencies for the nuclei along the x-direction. If this is done only for a short time, it is not the frequency but the phase of the oscillation (i.e., the direction of the spin at some specific time, $t = 0$) Finally, another gradient field in the y-direction is applied, which now transforms the frequency of the oscillation into y-dependent information.

In summary, this means for an MRI experiment, the main field is a gradient field. A first $\pi/2$ pulse at a given frequency flips only the spins of a specific layer into the x–y plane.

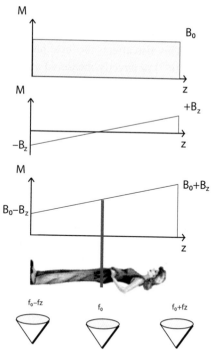

Figure 10.11 Schematic illustration of MRI. By applying a gradient field $B(z) = B_0 + \frac{\partial B}{\partial z}z$ in the main direction (z), only nuclei at a specific distance from the origin are resonantly excited by a radiopulse frequency $\omega = \gamma B(z)$. This means that only those nuclei are flipped into the x–y plane by applying a $\pi/2$ pulse. Only these nuclei thus yield an oscillating signal $M(t)$ in the magnetization. Thus, the spatial information in the z direction has been converted into frequency information of precessing nuclear spins. The determination of spatial information has therefore been transformed into a determination of a frequency spectra.

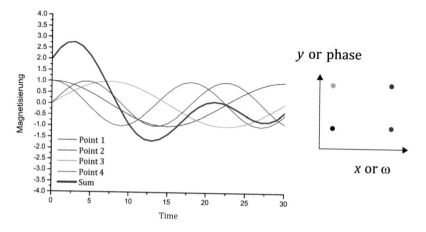

Figure 10.12 Schematic illustration of information encoding in the x–y plane in MRI. The sum of the signals from the four indicated points is measured using frequency and phase encoding. A corresponding two-dimensional Fourier transform yields the density of nuclei at the different points in the plane.

Briefly applying a gradient field in the x-direction changes the phase of the oscillation depending on the x position of the nuclei in question. Finally, applying a gradient field in the y-direction and measuring the oscillation of the magnetization in the x–y plane yields a spectrum whose frequency and phase information by a Fourier transform results in a two-dimensional density image of the layer in which the spins have been flipped by the first pulse. This is illustrated in Figure 10.12.

Exercises

10.1 Static magnetic fields

A current I is flowing through three very long and parallel wires, where the current is flowing upward in two of the wires and downward in the third. What is the magnetic field produced by these three wires at a distance r, which is much larger than the distance between the wires?

10.2 Static magnetic fields 2

A single electron has a magnetic moment of $\mu_e = 9 \times 10^{-24} J/T$. Assume that all atoms within a mole of stuff have their magnetic moments perfectly aligned. What is the magnetic field strength that you would obtain at a distance of 1 cm?

10.3 Static magnetic fields 3

(a) A long and thin coil with radius R = 0.010(1) m and length L = 0.250(1) m consists of N = 500(5) windings of a wire. A current of I = 0.10(1) A flows through the coil. What is the magnetic field in the middle of this coil? Use Ampere's law to determine this.

(b) What is the error of the magnetic field inside the coil?

10.4 Coulomb force vs. Lorentz force

Consider two beams of electrons at a distance d that are moving in parallel with speed v through a vacuum.

(a) What is the repulsive Coulomb force of one beam onto the other?

(b) What is the attractive Lorentz force of the two currents onto each other? Determine the ratio of Coulomb and Lorentz forces in this case.

(c) What is the uncertainty of this ratio if d and v are uncertain?

10.5 Lorentz force

How fast does an electron have to travel that the Lorentz force of a magnetic field of 30 T (roughly the strongest available permanent magnetic fields on earth) is the same as that exerted by a single myosin molecule in a muscle contraction, i.e., 4 pN?

10.6 Lorentz force 2

(a) Blood flow in the aorta has a speed of about $v = 0.20(5)$ m/s. The aorta is positioned inside a magnetic field of $B = 10$ T (say in an experimental MRI machine), such that the field is at an angle of 60 degrees to the aorta. How does the path of an erythrocyte in the aorta change in this situation? An erythrocyte contains about $3(1) \times 10^8$ molecules of hemoglobin and has a mass of $1.5(5) \times 10^{-13}$ kg. A hemoglobin molecule contains four ions of iron, which are doubly positively charged.

(b) What is the uncertainty of this radius of curvature?

10.7 Mass spectrometer

You want to determine the ratio of C^{12}–C^{14} in a mass spectrometer in order to carbon date an organic sample. The ions are singly and positively charged ($q = 1.6 \times 10^{-19}$ C). The mass spectrometer you use can apply a field of $B = 3$ T and a voltage of $U = 10^3$ V. You can determine the radius of curvature with an accuracy of 1 mm. How do you have to set the radius of curvature such that you can separate C^{12} and C^{14}?

10.8 Nuclear magnetic resonance

(a) In an MRI, a static, main magnetic field of $B = 5$ T is applied. For imaging, there is a gradient field with a gradient of $dB/dx = 0.5 \times 10^{-4}$ T/cm. The MRI is used to excite hydrogen spins (gyromagnetic ratio $\gamma_P = 4.2 \times 10^7 \mathrm{HzT}^{-1}$). What is the maximum frequency difference that you can determine across the head of a patient (120 cm long)?

(b) If the preceding setup has a frequency resolution of 10 Hz, what is the spatial resolution of the MRI image?

10.9 Nuclear magnetic resonance 2

Think of the proton as a charged sphere of mass $m_p = 2 \times 10^{-27}$ kg, a radius of 1 fm. What is the magnetic moment of such a proton if it rotates with an angular frequency (ω) such that its angular momentum is $m_p r_p^2 \omega/2 = \hbar/2 \simeq 0.5 \times 10^{-34}$ Js? Assume that the charge of the proton is smeared out over the equator, such that you obtain a circular current of $I = e\omega/(2\pi)$ enclosing an area of $A = \pi r_p^2$.

10.10 Nuclear magnetic resonance 3

What is the potential energy gain of the proton (magnetic moment: $\mu_P = 1.4 \times 10^{-26}$ J/T) by the alignment in a field of $B = 1$ T? At what temperature would you have a fraction of 10% of protons aligned in this field?

Quiz Questions

10.1 Static magnetic fields

A large magnet is used to attract a small iron screw. What is the force with which the screw attracts the magnet?

A Much smaller than the force of the magnet
B Much larger than the force of the magnet
C The same as the force of the magnet

10.2 Static magnetic fields 2

A sphere of radius R is placed next to (distance r) a wire carrying a current I. This current creates a magnetic field $B = \mu_0 I/(2\pi r)$. What is the magnetic flux through the surface of the sphere?

A $\mu_0 I$
B $\mu_0 I R^2/r$
C $\mu_0 I 4\pi R^2$
D zero
E $\mu_0 I/(4\pi R^2)$

10.3 Lorentz force

What is the work carried out by the Lorentz force keeping a particle on a circular orbit (charge q, speed v, field B, and radius of orbit r)?

A $2\pi q v B r$
B zero
C $q v B r$
D $q v B \pi r$

10.4 Lorentz force 2

A positively charged particle traveling with speed v arrives at the chamber of a mass spectrometer with a constant and homogeneous field at a right angle to the incoming particle's velocity. What path will the particle take in the chamber?

A A parabola
B A spiral
C A circular orbit
D A straight line

10.5 Nuclear magnetic resonance resonance

Which of the following nuclei does not give an NMR signal?

A H^1
B F^{19}
C C^{12}
D C^{13}
E Ca^{40}
F O^{16}

A Mathematical Tools

The following is a summary of some of the more important mathematical foundations of physics. They claim neither completeness nor mathematical severity. They are intended above all to repeat the mathematical apparatus from the middle school and to put it into the context of the concepts discussed in the main part of the book. The mathematical formulation is somewhat shorter than in a standard mathematical course, but this appendix is mainly there for the purpose of helping the students concentrate on the physical content of the book without being distracted by new mathematical concepts. These can be looked up later (or in advance) here. In addition, this compilation provides a summary of the mathematical concepts we will use in this book.

The Greek alphabet does not belong to mathematics, but it can be quite useful in this lecture, since Greek letters are often used as symbols for physical quantities (see Table A.1).

Table A.1 The Greek alphabet.

Alpha	α	A	(a)
Beta	β	B	(b)
Gamma	γ	Γ	(g)
Delta	δ	Δ	(d)
Epsilon	ϵ	E	(short e)
Zeta	ζ	Z	(soft z)
Eta	η	H	(long e)
Theta	θ	Θ	(th)
Iota	ι	I	(i)
Kappa	κ	K	(k)
Lambda	λ	Λ	(l)
Mu	μ	M	(m)
Nu	ν	N	(n)
Xi	ξ	Ξ	(x)
Omikron	o	O	(o)
Pi	π	Π	(p)
Rho	ρ	P	(r)
Sigma	σ	Σ	(s)
Tau	τ	T	(t)
Ypsilon	υ	Y	(y)
Phi	ϕ	Φ	(ph)
Chi	χ	X	(ch)
Psi	ψ	Ψ	(ps)
Omega	ω	Ω	(long o)

A.1 Functions and Their Derivatives

A.1.1 General Considerations

If a quantity y depends on a variable t, then y is a function of t, which is called the independent variable. One writes the following:

$$y = y(t) = f(t).$$

Examples: The position (height) of an object h changes as a function of t: $h = h(t)$. The pressure p changes as a function of volume V: $p = p(V)$. The current in an electric circuit I depends on the applied voltage V_0: $I = I(V_0)$. Figure A.1 shows such dependencies graphically.

Terms: The current as a function of voltage is a *linear* function of the type

$$y = at + b; \qquad a \text{ and } b \text{ are constant.}$$

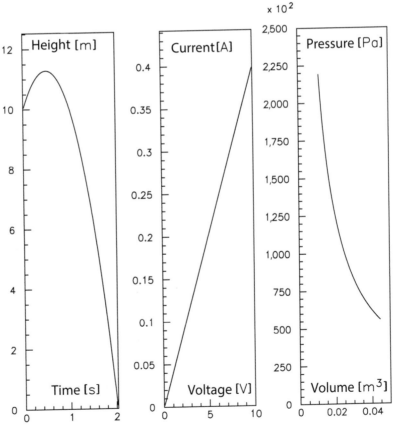

Figure A.1 Linear, quadratic, and exponential functions.

The height of a falling object is a *quadratic* function of the type

$$y = at^2 + bt + c; \qquad a, \; b, \; \text{and } c \text{ are constant.}$$

For *even* functions, we have $y(t) = y(-t)$, while *odd functions* follow $y(t) = -y(-t)$.

While functions in general can describe all kinds of relationships between two variables, there are some special functions that appear very often and that have a few properties that make them simpler to work with than others. We will now look at some of these.

A.1.2 Linear Functions, Polynomials, and Power Laws

The simplest possible function is the linear function $y(t) = at$, where a is a constant. The direct proportionality means that here y and t are intimately related and only different by a multiplicative factor. Similarly, a linear relation is given by $y(t) = at + b$, where the linear relation is shifted by a certain constant b, but otherwise remains the same. Because linear functions are very direct relations, they can be easily inverted (and stay linear relationships in this case: $t(y) = 1/ay - b/a$) and can also be easily interpolated. If two points on a linear relationship are known, the whole curve is described.

The next step in more complicated functions from the linear relationship is the polynomial. Rather than having only direct proportionalities, in a polynomial, multiples (higher powers) of the variable appear as well. This starts with quadratic equations of the form $y(t) = at^2 + bt + c$, but also includes cubic functions such as $y(t) = at^3 + bt^2 + ct + d$. The general form is $y(t) = \sum_{i=0} N a_i t^i$. The higher the number of terms (or the order of the function), the more details are encapsulated by such a function (more on this is discussed in Section A.1.6). Correspondingly, the more terms that appear in the function, the more points are needed to be known to actually describe the relationship. While a linear relation describes a straight line, a quadratic function describes a curved peak or trough and a cubic function describes a peak and a trough.

The higher orders, or powers in a polynomial, can also appear on their own and do not actually have to be a whole number. These are power laws of the form $y(t) = at^b$, which in Section 2.2.2 are shown to be very common and fundamental in quantitative descriptions of nature. In this case again, actually only two numbers describe the entire function, which is also why one can determine them using a straight line, if plotted properly, that is on a double logarithmic scale.

Power laws with even powers (or exponents) are even functions, power laws with odd exponents are odd functions (see Figure A.2).

A.1.3 Logarithms and Exponentials

In order to describe power laws properly, therefore, we have to look at the properties of logarithms. Since power laws have to do with the successive multiplication of variables, a logarithm will have to turn multiplications into sums, if we are to use them to turn power laws into linear functions. Therefore, the basic relation defining logarithms is given by $\log(x \cdot y) = \log(x) + \log(y)$. If we apply this to a power law, we find that

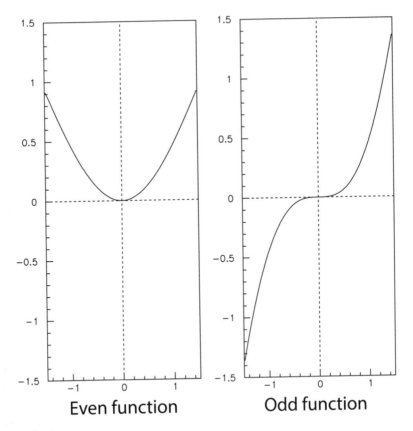

Odd and even functions.

$\log(y(t)) = \log(a) + b \cdot \log(t)$. Therefore, if $y(t)$ is a power law, plotting $\log(y)$ as a function of $\log(t)$ will give a linear relationship, where the slope is directly given by b. If we want to arrive back at the original variable from a logarithm, we have to take an exponential. That is, we are using a power law, where the exponent is not a constant, but rather the variable, i.e., $y(t) = a^t$. Here, the base of the exponent is a constant, which in principle can be any number but one or zero. What this number is will determine what type of logarithm one has to use in order to obtain back the original variable. Common bases for exponentials and logarithms are 2, 10, and Euler's number $e = \lim_{n \to \infty}(1 + \frac{1}{n})^n = \sum_{i=0} \infty \frac{1}{n!} = 2.71828\ldots..$. If the logarithm is taken with respect to Euler's number, one speaks of a natural logarithm.

Exponential functions are very commonplace in dynamics and describe the growth of capital with compound interest (successively multiplying with a constant number), but also of population growth or radioactive decay. Because the rate of change of the growth is always proportional to the present amount in such calculations, we will see that the exponential is a very important function in terms of dynamical systems and sets of differential equations. Used with base e, the rate of change is actually going to be equal to the present amount, which is why this base is usually used in physics.

A.1.4 Trigonometry

A final class of special functions, which are closely connected to the exponential if one uses complex numbers, consists of the trigonometric functions, sine, cosine, and tangent. They basically originate from the description of triangles (hence the name), but as in shown in Section 3.5, are very closely related to oscillations. Basically, this is due to the fact that these functions have angles as variables and therefore repeat themselves after the angle has run through the course of 360 degrees, or 2π, i.e., once around a full circle. As such, a sine corresponds to the y-component of a circle, while the cosine corresponds to the x-component, which directly gives the relation to their description of triangles, since the x and y-components and the radius of the circle form a triangle. From these definitions of $\sin(\alpha) = y/r$, $\cos(\alpha) = x/r$ and $\tan(\alpha) = y/x$, one can also find relations between the different trigonometric functions, such as $\sin^2(\alpha) + \cos^2(\alpha) = 1$ or $\tan(\alpha) = \frac{\sin(\alpha)}{\cos(\alpha)}$.
Other useful relations are the following:

$$\sin(\alpha \pm \beta) = \sin(\alpha)\cos(\beta) \pm \cos(\alpha)\sin(\beta)$$

$$\cos(\alpha \pm \beta) = \cos(\alpha)\cos(\beta) \mp \sin(\alpha)\sin(\beta)$$

$$\sin(\alpha) \pm \sin(\beta) = 2\sin\left(\frac{\alpha \pm \beta}{2}\right)\cos\left(\frac{\alpha \mp \beta}{2}\right)$$

$$\cos(\alpha) + \cos(\beta) = 2\cos\left(\frac{\alpha + \beta}{2}\right)\cos\left(\frac{\alpha - \beta}{2}\right)$$

$$\cos(\alpha) - \cos(\beta) = -2\sin\left(\frac{\alpha + \beta}{2}\right)\sin\left(\frac{\alpha - \beta}{2}\right)$$

In complex numbers, one finds that the two-dimensional representation of the real and imaginary numbers can also be described using sines and cosines, but also that this is closely related to the exponential in the form

$$\exp(i\alpha) = \cos(\alpha) + i\sin(\alpha),$$

which can also be used to define sine and cosine using exponentials:

$$\sin(\alpha) = \frac{1}{2i}(\exp(i\alpha) - \exp(-i\alpha))$$

$$\cos(\alpha) = \frac{1}{2}(\exp(i\alpha) + \exp(-i\alpha))$$

A.1.5 The Derivative

The derivative quantifies the rate of change of a function with the independent variable. As such, it is always used when we want to determine changes, and velocities, gradients, and angular frequencies are examples for derivatives that appear throughout the text.

Definition

The derivative of a function $y(t)$ at point $t = t_0$ is defined as follows:

$$y'(t_0) = \frac{dy}{dt}\big|_{t_0} = \lim_{t \to t_0} \frac{y(t) - y(t_0)}{t - t_0}.$$

In the limit of bringing t ever closer to t_0, both the nominator and the denominator become zero; however, usually the ratio reaches a final value that changes less and less the closer t comes to t_0. If this is the case, the function is differentiable. In physics, most functions we encounter are differentiable functions.

The quantity dy/dt is also called the *differential quotient*, and often the derivative is also written in the following way:

$$y', \frac{dy}{dt}, \dot{y},$$

where the dot is only used for the derivative of functions of time. Another way of defining the derivative that is very similar uses for a function $y(t)$ at point t the following nomenclature:

$$y = y(t); \quad y(t + \Delta t) = y(t) + \Delta y.$$

This gives the following:

$$\frac{dy}{dt} = \lim_{\Delta t \to 0} \frac{y(t + \Delta t) - y(t)}{\Delta t} = \lim_{\Delta t \to 0} \frac{\Delta y}{\Delta t}.$$

Derivatives of Some Important Functions

Derivatives of the most common functions are given in the following table:

$\frac{da}{dt} = 0$	$\frac{d}{dt}(t^a) = a t^{a-1}$
$\frac{d}{dt}(\sin t) = \cos t$	$\frac{d}{dt}(\cos t) = -\sin t$
$\frac{d}{dt}(\tan t) = \frac{1}{\cos^2 t}$	$\frac{d}{dt}(\cot t) = -\frac{1}{\sin^2 t}$
$\frac{d}{dt}(\arcsin t) = \frac{1}{\sqrt{1 - t^2}}$	$\frac{d}{dt}(\arccos t) = -\frac{1}{\sqrt{1 - t^2}}$
$\frac{d}{dt}(\arctan t) = \frac{1}{1 + t^2}$	$\frac{d}{dt}(\text{arccot}\, t) = -\frac{1}{1 + t^2}$
$\frac{d}{dt}(e^t) = e^t$	$\frac{d}{dt}(\log t) = \frac{1}{t}$

General Rules for Differentiation

There are further rules for differentiation that can be used to take the derivative of combination of the common functions:

$$\textbf{Sum rule:} \quad \frac{d}{dt}(u(t) + v(t)) = \frac{du}{dt} + \frac{dv}{dt}$$

$$\Rightarrow \text{Application:} \quad \frac{d}{dt}(u + a) = \frac{du}{dt}$$

$$\textbf{Product rule:} \quad \frac{d}{dt}(u(t) \cdot v(t)) = \frac{du}{dt} \cdot v(t) + u(t) \cdot \frac{dv}{dt}$$

$$\Rightarrow \text{Application:} \quad \frac{d}{dt}(a \cdot u) = a\frac{du}{dt}$$

$$\textbf{Chain rule:} \quad \frac{d}{dt}(u(x(t))) = \frac{du}{dx}\frac{dx}{dt}$$

$$\Rightarrow \text{Application:} \quad x = at + b, \quad \frac{d}{dt}(u(x(t))) = a\frac{du}{dx}$$

If we combine the product and the chain rules, one can derive a rule for the derivative of the fraction of two functions. This yields:

$$\textbf{Quotient} - \textbf{rule} : \quad \frac{d}{dt}\left(\frac{u(t)}{v(t)}\right) = \frac{1}{v^2}\left(\frac{du}{dt}v - \frac{dv}{dt}u\right).$$

Examples

1. $N(t) = N_0 e^{x(t)} = N_0 e^{-\lambda t}$

$$\frac{dN}{dt} = \frac{dN}{dx}\frac{dx}{dt} = N_0 e^x \frac{dx}{dt} = N_0 e^{-\lambda t}(-\lambda) = -\lambda N(t)$$

2. $f(x) = (y(x))^3 = (ax^2 + bx + c)^3$

$$\frac{df}{dx} = \frac{df}{dy}\frac{dy}{dx} = 3y^2(2ax + b) = 3(ax^2 + bx + c)^2(2ax + b)$$

3. $x(t) = x_0 \cos(\phi(t)) = x_0 \cos(\omega t + \delta)$

$$\frac{dx}{dt} = x_0 \frac{d(\cos \phi)}{d\phi}\frac{d\phi}{dt} = -x_0 \sin \phi \omega = -\omega x_0 \sin(\omega t + \delta)$$

4. $x(k) = y^{1/2} = \sqrt{y(k)} = \sqrt{ak + b}$

$$\frac{dx}{dk} = \frac{dx}{dy}\frac{dy}{dk} = \frac{1}{2}y^{-1/2}a = \frac{a}{2\sqrt{ak + b}}$$

5. $y(t) = y_0 e^{-rt} \cos(\omega t + \delta)$

$$\frac{dy}{dt} = y_0 \frac{d}{dt}\left(e^{-rt}\right)\cos(\omega t + \delta) + y_0 e^{-rt}\frac{d}{dt}\left(\cos(\omega t + \delta)\right)$$
$$= y_0 e^{-rt}\left(-r\cos(\omega t + \delta) - \omega \sin(\omega t + \delta)\right)$$

Higher Derivatives

In general, the derivative of a function is again a function of the same independent variable. Thus one can take the derivative of the derivative of a function and so on. This is usually written as follows:

$$\frac{d}{dt}\left(\frac{dy}{dt}\right) = \frac{d^2 y}{dt^2}.$$

The same rule we have stated for first derivatives are also valid for higher derivatives. Examples of higher derivatives are the curvature or the acceleration.

Example – Position: $x(t) = x_0 e^{-rt}\cos(\omega t)$

Velocity: $v(t) = \dfrac{dx}{dt} = -x_0 e^{-rt}\left(r\cos(\omega t) + \omega \sin(\omega t)\right)$

$$\frac{d^2 x}{dt^2} = -x_0 e^{-rt}(-r)\left(r\cos(\omega t) + \omega \sin(\omega t)\right) - x_0 e^{-rt}\left(-r\sin(\omega t)\omega + \omega \cos(\omega t)\omega\right)$$

Acceleration: $a(t) = \dfrac{d^2 x}{dt^2} = x_0 e^{-rt}\left(2r\omega \sin(\omega t) - (\omega^2 - r^2)\cos(\omega t)\right)$

A.1.6 Series Expansion of Functions

Certain functions (the infinitely differentiable functions) can be represented by infinite power series. Most functions appearing in physics are of this type. In addition, it is often the case that the parts of this series become increasingly small. This means that the series converges rapidly and only the first two or three terms need to be considered for a good approximation; the rest can be neglected.

The Taylor expansion of a function $f(x)$ is given by the following:

$$f(x) = f(x_0) + (x - x_0)\left(\frac{df}{dx}\right)_{x=x_0} + \frac{(x - x_0)^2}{2!}\left(\frac{d^2 f}{dx^2}\right)_{x=x_0} + \cdots,$$

or in closed form,

$$f(x) = \sum_{k=0}^{\infty} \frac{f^{(k)}(x_0)}{k!}(x - x_0)^k, \quad \text{mit } f^{(k)}(x_0) = \left(\frac{d^k f}{dx^k}\right)_{x=x_0}.$$

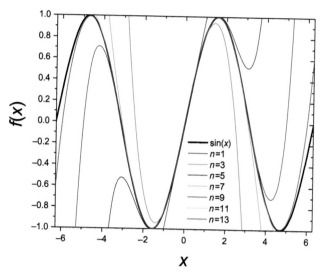

Successive Taylor expansions of the sine.

In particular, for $x_0 = 0$ and $|x| \ll 1$, we have the following:

$$(1+x)^n = 1 + nx + \frac{n(n-1)}{2}x^2 + \cdots , \qquad \frac{1}{(1+x)^n} = (1+x)^{-n}$$

$$= 1 - nx + \frac{n(n+1)}{2}x^2 + \cdots ; \log(1+x) = x - \frac{1}{2}x^2 + \frac{1}{3}x^3 + \cdots ;$$

$$\sin x = x - \frac{x^3}{3!} + \cdots , \qquad \cos x = 1 - \frac{x^2}{2!} + \cdots , \qquad \tan x = x + \frac{x^3}{3} + \cdots .$$

The example in Figure A.3 shows the Taylor expansion of a sine function with successively higher terms being taken into account. The more terms are taken into account, the larger is the region, where the Taylor expansion approximates the function itself. In a small interval, already the first or second terms are very good approximations, which is something we regularly use in our description of nature.

A.1.7 Expansion of Functions in Periodic Oscillations

Every periodic function can be treated as a sum of harmonic functions. If $u(t)$ is a function with a periodicity T, i.e., $u(t + T) = u(t)$, one can write the following:

$$u(t) = \sum_{n=0}^{\infty}(A_n \cos \omega_n t + B_n \sin \omega_n t) \qquad \text{mit } \omega_n = \frac{2\pi n}{T}$$

If the Fourier components A_n and B_n are known, $u(t)$ is uniquely determined. On the other hand, if $u(t)$ is prescibed, A_n and B_n are calculated as follows:

$$A_0 = \frac{1}{T}\int_0^T u(t)dt, \qquad\qquad B_0 = 0$$

$$A_n = \frac{2}{T} \int_0^T u(t) \cos(\omega_n t) dt, \qquad\qquad B_n = \frac{2}{T} \int_0^T u(t) \sin(\omega_n t) dt \quad (n \geq 1)$$

In the text, we have seen what happens to the Fourier components on taking the derivative, we can do the same thing when looking for the integral of a function $u(t)$ with respect to t:

$$k(t) = \int u(t) dt = \sum_{n=0}^{\infty} \int (A_n \cos(\omega_n t) + B_n \sin(\omega_n t)) dt =$$

$$= \sum_{n=0}^{\infty} \left(\frac{A_n}{\omega_n} \sin(\omega_n t) - \frac{B_n}{\omega_n} \cos(\omega_n t) \right).$$

Therefore, if we Fourier-expand $k(t)$ as

$$k(t) = \sum_{n=0}^{\infty} \int (A'_n \cos(\omega_n t) + B'_n \sin(\omega_n t)),$$

we see that $A'_n = -\frac{B_n}{\omega_n}$ and $B'_n = \frac{A_n}{\omega_n}$

The Fourier components or their graphical representation, the so-called frequency spectrum, are very well suited for characterizing complex periodic functions. We have seen that the ear does exactly that, for the magnitude of the excitation of the basil membrane at a certain place corresponds precisely to the Fourier component of excitation. The nerve signal that the ear sends out after the spatial separation of the signal corresponds exactly to a frequency spectrum of the sound. But the Fourier decomposition is also found in the method in technical instruments, such as nuclear magnetic resonance, mass spectrometry, electrocardiograms, and electroencephalograms. We have considered the functions to date as functions of the time, as they were in the preceding applications, but the same decomposition can be carried out in the place. If a local function is decomposed according to Fourier components, the frequency ω and the period T must be replaced by the wave number k and the wavelength λ:

$$u(x) = u(x + \lambda) = \sum_{n=0}^{\infty} (A_n \cos k_n x + B_n \sin k_n x) \qquad \text{mit } k_n = \frac{2\pi n}{\lambda}$$

$$A_0 = \frac{1}{\lambda} \int_0^\lambda u(x) dx, \qquad\qquad B_0 = 0$$

$$A_n = \frac{2}{\lambda} \int_0^\lambda u(x) \cos(k_n x) dx, \qquad\qquad B_n = \frac{2}{\lambda} \int_0^\lambda u(x) \sin(k_n x) dx \quad (n \geq 1)$$

Here, too, there are a large number of applications, the most prominent of which are the scattering methods, or crystallography, where a spatially periodic signal is obtained and the scatter image is a Fourier decomposition of the corresponding structure.

Figure A.4 shows the example of the Fourier decomposition of the periodic sawtooth function:

$$u(x) = 1 - 2\frac{x}{\lambda}, \quad 0 \leq x \leq \lambda; \quad u(x + \lambda) = u(x).$$

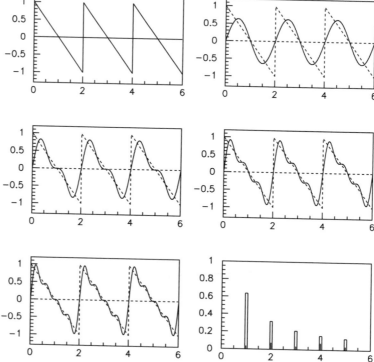

Figure A.4 The sawtooth curve shown in the first image (top-left) is decomposed into its Fourier components. The second to fifth images show the curve in comparison to the Fourier series, which is broken off after the first, second, third, and fourth term, respectively. The last image (lower-right) shows the corresponding Fourier spectrum, that is, the amplitude of the corresponding summing in the Fourier series.

The preceding equations determine the Fourier components as

$$A_n = \frac{2}{\lambda} \int_0^\lambda \left(1 - 2\frac{x}{\lambda}\right)\cos(k_n x)dx = 0, \qquad B_n = \frac{2}{\lambda}\int_0^\lambda \left(1 - 2\frac{x}{\lambda}\right)\sin(k_n x)dx = \frac{2}{\pi}\frac{1}{n},$$

which yields the following:

$$u(x) = 1 - 2\frac{x}{\lambda} = \frac{2}{\pi}\left(\sin kx + \frac{1}{2}\sin(2kx) + \frac{1}{3}\sin(3kx) + \cdots\right) = \frac{2}{\pi}\sum_{n=1}^{\infty}\frac{1}{n}\sin\left(\frac{2\pi n x}{\lambda}\right).$$

Example: Beating

As an example of a discrete Fourier transform, we are considering two oscillations with two slightly different frequencies (see Figure A.5). Thus we have $u_1(t) = u_0 \cos((\omega - \delta\omega)t)$ and $u_2(t) = u_0 \cos((\omega + \delta\omega)t)$. The total amplitude of oscillation then is the sum of the two signals: $u(t) = u_1(t) + u_2(t) = u_0(\cos((\omega - \delta\omega)t) + \cos((\omega + \delta\omega)t))$. Using the sum formula for cosines, $\cos(\alpha \pm \beta) = \cos(\alpha)\cos(\beta) \pm \sin(\alpha)\sin(\beta)$, we obtain the following:

$$u(t) = u_0(\cos(\omega t)\cos(\delta\omega t) + \sin(\omega t)\sin(\delta\omega t) + \cos(\omega t)\cos(\delta\omega t) - \sin(\omega t)\sin(\delta\omega t)).$$

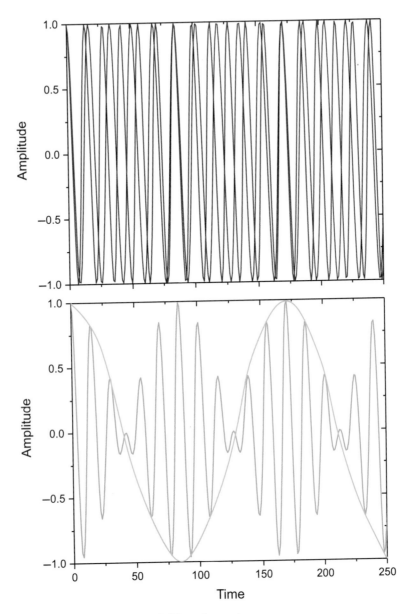

Beating: adding two harmonic oscillations with different frequencies.

The terms containing sines cancel each other and the total oscillation becomes the following:

$$u(t) = 2u_0 \cos(\delta\omega t) \cos(\omega t).$$

This means there is an oscillation with a frequency ω, but its amplitude is modulated on a time scale $1/(2\delta\omega)$, such that after a time $\pi/(2\delta\omega)$ both oscillations cancel while they amplify each other after a time $\pi/(\delta\omega)$. Given the situation that we set out describing,

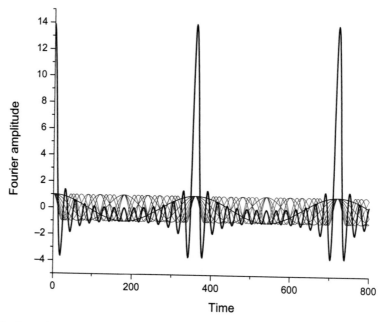

Adding higher harmonics to obtain the Fourier transform of a grid.

we know that there are two Fourier components of equal size at frequencies $\omega - \delta\omega$ and $\omega + \delta\omega$. One can view this as a maximum at ω with a width $2\delta\omega$. This means that there is an oscillation with a frequency ω that falls off on a time scale of $1/(2\delta\omega)$. However, because the beating is actually a discrete Fourier decomposition with single, narrow frequency components, the signal repeats periodically.

Example: Grid

When many regularly arranged sources come together, we have many frequencies that must be added together at the same time. If the period of the basic frequency is multiplied, all cos or sin terms have a maximum. This means that the more frequencies present, the maxima with the period that corresponds to the basic frequency are increasingly sharper. The periodicity of the signal, that is, the occurrence of very many frequencies, leads to sharp lines, and the interval between the frequencies gives the period of the signal. This is illustrated in Figure A.6, where 12 higher harmonics of a fundamental frequency are added.

This principle is also useful in crystallography, where the periodic arrangement of scatterers results in a corresponding periodic scattering signal corresponding to the Fourier transform of the scattering arrangement.

For specialists: we can also quantitatively consider the damped oscillator. We shall thus also see a method of dealing with a differential equation without actually solving it.

Reminder: the driven, damped harmonic oscillator is described by the following differential equation: $\ddot{x}(t) + \frac{1}{\tau_0}\dot{x}(t) + \omega_0^2 x(t) = K(t)$, where $K(t)$ is the external excitation or driving. Since we can write every function as a Fourier integral, we can write for the amplitude of oscillation $x(t) = \int (A(\omega)\cos(\omega t) + B(\omega)\sin(\omega t))d\omega$. Because the Fourier components do not depend on time, but only on frequency, the time derivatives are easy to take, since we only have to take the derivative of harmonic functions. We thus find $\dot{x}(t) = \int(-\omega \cdot A(\omega)\sin(\omega t) + \omega \cdot B(\omega)\cos(\omega t))d\omega$ and $\ddot{x}(t) = \int(-\omega^2 \cdot A(\omega)\cos(\omega t) - \omega^2 \cdot B(\omega)\sin(\omega t))d\omega$. If we insert these derivatives into the equation of motion of the harmonic oscillator, we no longer have to solve a differential equation, but just a normal algebraic equation in ω, to determine the frequency dependencies of the Fourier components. So we insert the Fourier transforms into the differential equation:

$$\int \left[((\omega_0^2 - \omega^2)A(\omega) + \frac{\omega}{\tau_0}B(\omega))\cos(\omega t) + ((\omega_0^2 - \omega^2)B(\omega) - \frac{\omega}{\tau_0}A(\omega))\sin(\omega t) \right]$$
$$d\omega = K(t)$$

Now we also have to describe the excitation in terms of a Fourier transform. If we drive the oscillator using a cosine, we have $K(t) = \int K\cos(\omega t)d\omega$. So we collect the terms containing sines in one equation and the terms containing cosines in another. Then we know that the terms in front of the sine have to add up to zero, whereas the terms in front of the cosines have to add up to K. So we get two equations for the two unknowns $A(\omega)$ and $B(\omega)$:

$$((\omega_0^2 - \omega^2)A(\omega) + \frac{\omega}{\tau_0}B(\omega)) = K$$

$$((\omega_0^2 - \omega^2)B(\omega) - \frac{\omega}{\tau_0}A(\omega)) = 0$$

The second equation yields

$$(\omega_0^2 - \omega^2)B(\omega) = \frac{\omega}{\tau_0}A(\omega)$$

or

$$B(\omega) = \frac{\omega}{\tau_0(\omega_0^2 - \omega^2)}A(\omega),$$

which we can insert in the first equation:

$$(\omega_0^2 - \omega^2)A(\omega) + \frac{\omega^2}{\tau_0^2(\omega_0^2 - \omega^2)}A(\omega) = K$$

or

$$\left[\frac{(\omega_0^2 - \omega^2)^2 + \omega^2/\tau_0^2}{(\omega_0^2 - \omega^2)} \right] A(\omega) = K,$$

which yields the following:

$$A(\omega) = K \left[\frac{(\omega_0^2 - \omega^2)}{(\omega_0^2 - \omega^2)^2 + \omega^2/\tau_0^2} \right]$$

$$B(\omega) = K \left[\frac{\omega}{\tau_0(\omega_0^2 - \omega^2)} \frac{(\omega_0^2 - \omega^2)}{(\omega_0^2 - \omega^2)^2 + \omega^2/\tau_0^2} \right] = K \left[\frac{\omega/\tau_0}{(\omega_0^2 - \omega^2)^2 + \omega^2/\tau_0^2} \right]$$

The magnitude of the amplitude is given by the sum of the squares of A and B, i.e., $|x_0(\omega)| = \sqrt{A^2(\omega) + B^2(\omega)}$, or the following if we insert our previous result:

$$|x_0(\omega)| = K \sqrt{\left[\frac{(\omega_0^2 - \omega^2)^2}{((\omega_0^2 - \omega^2)^2 + \omega^2/\tau_0^2)^2} + \frac{(\omega/\tau_0)^2}{((\omega_0^2 - \omega^2)^2 + \omega^2/\tau_0^2)^2} \right]}$$

$$= K \sqrt{\left[\frac{(\omega_0^2 - \omega^2)^2 + (\omega/\tau_0)^2}{((\omega_0^2 - \omega^2)^2 + \omega^2/\tau_0^2)^2} \right]} = K \left[\frac{1}{\sqrt{(\omega_0^2 - \omega^2)^2 + \omega^2/\tau_0^2}} \right].$$

This is exactly the resonance curve, which we had found by solving the differential equation in Section 4.2. Thus, we also see that the Fourier transform actually describes the resonance curve, as can be imagined physically. So we see that the resonance curve (or Fourier transform) describes the oscillating system. The maximum of the curve corresponds to the frequency of the oscillation, and the width of the curve corresponds to the temporal fall of the oscillation.

A.2 Vectors

In physics, we often deal with quantities that are not only determined by their *magnitude* (in some suitable units) but also by their *direction*. These are called *vectors*.

Examples of this are the velocity, the acceleration, and electric and magnetic fields.

Nomenclature : $\vec{v} =$ vector $|\vec{v}| = v =$ Absolute value (or length) of the vector

Quantities that only describe a magnitude, such as mass, time, volume, or temperature, are called *scalars*.

Two vectors $\vec{a}$ and $\vec{b}$ are equal, i.e., $\vec{a} = \vec{b}$,

- If the magnitudes of $\vec{a}$ and $\vec{b}$ are equal $(a = b)$, and
- If $\vec{a}$ is parallel to $\vec{b}$ and points in the same direction $(\vec{a} \parallel \vec{b})$.

This definition of the equality of two vectors is true for *free* vectors, which can be moved around without changing their direction at will. In physics, there can be *fixed* vectors that are connected to a point of action, such as forces acting on an extended body. The rules for these vectors in terms of what we discuss in the following still hold when taking the fixed position in space into account.

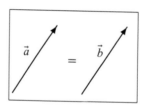

Figure A.7 Two equal vectors.

A.2.1 Adding and Subtracting Vectors

Multiplication with a Scalar

$\alpha \cdot \vec{a} = \vec{b}$ is a vector whose magnitude is $b = \alpha a$. It points in the same direction as $\vec{a}$, if $\alpha > 0$ and in the opposite direction if $\alpha < 0$. For $\alpha = -1$, we have the situation depicted in Figure A.8.

Figure A.8 Two opposite vectors.

Addition

Vectors are added according to the parallelogram or vector triangle (vector polygon for multiple vectors) to obtain the sum or *resulting* vector (see Figure A.9).

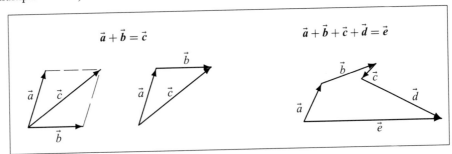

Figure A.9 Addition of vectors.

Subtraction

$$\vec{a} - \vec{b} = \vec{a} + (-1)\,\vec{b} = \vec{c}$$

or

$$\vec{a} = \vec{c} + \vec{b}$$

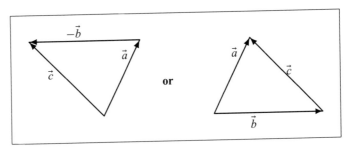

Figure A.10 Subtraction of two vectors.

A.2.2 Representation of a Vector

The description of a vector in three-dimensional space needs a combination of three numbers to represent the vector in a system of coordinates. In general, one chooses a system where the different coordinates are orthogonal (Cartesian coordinate system). The vector $\vec{a}$ is then given by the components in the three orthogonal directions, x, y, and z, i.e., a_x, a_y, a_z. On the other hand, one can use the magnitude of the vector $|\vec{a}|$ and two angles ϕ and θ describing the latitude and longitude in which the vector points (see Figure A.11).

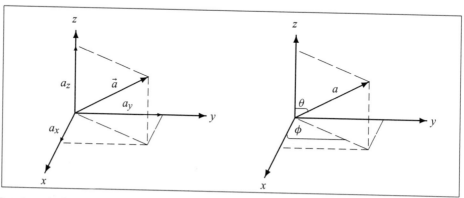

Figure A.11 Cartesian and polar representation of a three-dimensional vector.

In this case,

$$a_x = a \sin \theta \cos \phi , \qquad a_y = a \sin \theta \sin \phi , \qquad a_z = a \cos \theta ,$$

$$a \equiv |\vec{a}| = \sqrt{a_x^2 + a_y^2 + a_z^2} , \qquad \cos \theta = \frac{a_z}{a} , \qquad \tan \phi = \frac{a_y}{a_z} .$$

For working in Cartesian coordinates (see Figure A.12), one defines *unit vectors* for each of the three spatial dimensions:

$$|\hat{i}| = |\hat{j}| = |\hat{k}| = 1.$$

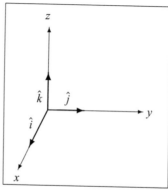

Figure A.12 Defining unit vectors for Cartesian coordinates.

The vector is then written as

$$\vec{a} = a_x\hat{i} + a_y\hat{j} + a_z\hat{k} \equiv \begin{pmatrix} a_x \\ a_y \\ a_z \end{pmatrix}.$$

The numbers a_x, a_y, and a_z are called the *components* of the vector in the three coordinate directions.

In such a component representation, the rules for *equality* of two vectors become the following:

$$\vec{a} = \vec{b}, \quad \text{wenn} \quad a_x = b_x, \ a_y = b_y, \ a_z = b_z.$$

A vector equation therefore corresponds to a list of three scalar equations. Similarly, the rules for *addition* and *subtraction* of two vectors become the following:

$$\vec{a} \pm \vec{b} = (a_x \pm b_x)\hat{i} + (a_y \pm b_y)\hat{j} + (a_z \pm b_z)\hat{k} \equiv \begin{pmatrix} a_x \pm b_x \\ a_y \pm b_y \\ a_z \pm b_z \end{pmatrix}.$$

Finally, the multiplication with a scalar is as follows:

$$\alpha\,\vec{a} = \alpha\,a_x\hat{i} + \alpha\,a_y\hat{j} + \alpha\,a_z\hat{k} \equiv \begin{pmatrix} \alpha\,a_x \\ \alpha\,a_y \\ \alpha\,a_z \end{pmatrix}.$$

A.2.3 Differentiating Vectors

A vector can be a function of another vector or of a scalar. The latter case, which we want to look at in more detail, has been important in mechanics and kinematics, where vectors such as position, velocity, and acceleration depend on time. When a vector changes as a function of time, either its length (**a**) its direction (**b**), or both (**c**) can change. We will study these three cases separately. The change of the vector $\vec{a}(t)$ during time dt is called $d\vec{a}$. The derivative of $\vec{a}$ with respect to t we call $\vec{b}$, which is given by the following:

$$\vec{a}(t) = \vec{a}, \qquad \vec{a}(t + dt) = \vec{a} + d\vec{a}, \qquad \vec{b} = \frac{d\vec{a}}{dt}.$$

(a) If only the *magnitude* of the vector $\vec{a}$ changes, then $d\vec{a}$ and hence $\vec{b}$ points in the same direction as, i.e., is parallel to, $\vec{a}$ (see Figure A.13). If $\hat{a}_\parallel$ denotes a unit vector parallel to $\vec{a}$, we have the following:

$$\vec{b} \equiv \frac{d\vec{a}}{dt} \parallel \vec{a}, \quad \Rightarrow \quad \vec{b} = \frac{da}{dt}\hat{a}_\parallel, \ \text{ mit } \hat{a}_\parallel = \frac{\vec{a}}{|\vec{a}|} = \frac{\vec{a}}{a}.$$

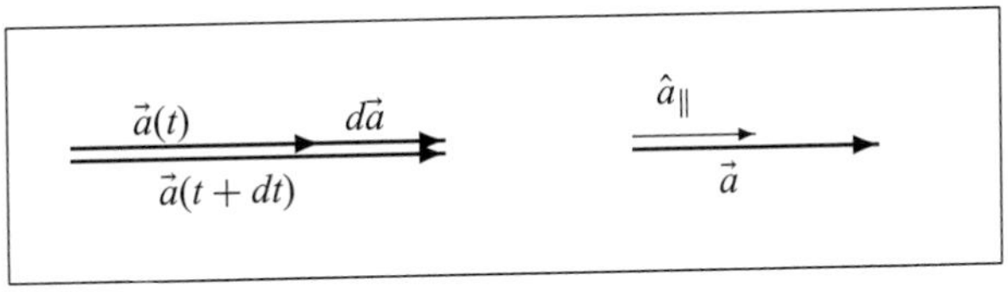

Figure A.13 Differentiating vectors when the magnitude of a vector changes.

(b) If the vector $\vec{a}$ only changes *direction*, $d\vec{a}$ and hence $\vec{b}$ are *perpendicular* to $\vec{a}$. If we denote the change in direction with $d\phi$, and $\hat{a}_\perp$ is a unit vector perpendicular to $\vec{a}$, we have the following:

$$\vec{b} = \frac{d\vec{a}}{dt} \perp \vec{a}, \qquad \tan d\phi \approx \sin d\phi \approx d\phi = \frac{|d\vec{a}|}{|\vec{a}|} = \frac{da}{a}, \qquad \Rightarrow$$

$$\vec{b} = \frac{da}{dt}\hat{a}_\perp = a\frac{d\phi}{dt}\hat{a}_\perp.$$

Figure A.14 shows clearly that $d\vec{a}$ does not have to equal $d|\vec{a}|$, since $d|\vec{a}|$ has to be zero.

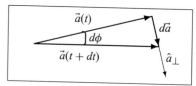

Differentiating vectors if a vector only changes direction.

(c) In general, both the magnitude and the direction can change (see Figure A.15). In this case, one finds the following:

$$\vec{b} = \frac{d\vec{a}}{dt} = \frac{da}{dt}\hat{a}_\| + a\frac{d\phi}{dt}\hat{a}_\perp$$

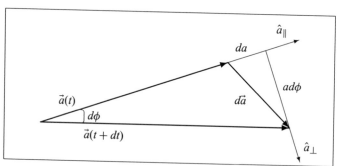

Differentiating vectors when both magnitude and direction change.

Not using these geometrical considerations, one can also simply use directional components that are each viewed as function to take the derivative:

$$\vec{b} \equiv \frac{d\vec{a}}{dt} = \frac{da_x}{dt}\hat{i} + \frac{da_y}{dt}\hat{j} + \frac{da_z}{dt}\hat{k} \equiv \begin{pmatrix} \frac{da_x}{dt} \\ \frac{da_y}{dt} \\ \frac{da_z}{dt} \end{pmatrix}.$$

For the magnitude of the vector $\vec{b}$, one finds the following:

$$b = \sqrt{\left(\frac{da_x}{dt}\right)^2 + \left(\frac{da_y}{dt}\right)^2 + \left(\frac{da_z}{dt}\right)^2} \,.$$

A.2.4 Multiplying Vectors

The Scalar Product

The scalar product of two vectors (see Figure A.16) is defined as follows:

$$\vec{a} \cdot \vec{b} = |\vec{a}||\vec{b}| \cos \phi = ab \cos \phi \,.$$

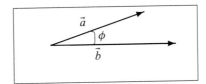

Figure A.16 The scalar product of two vectors.

The following rules need to be obeyed:

$$\vec{a} \cdot \vec{b} = \vec{b} \cdot \vec{a} \quad \text{(commutative)},$$
$$\vec{a} \cdot (\vec{b} + \vec{c}) = \vec{a} \cdot \vec{b} + \vec{a} \cdot \vec{c} \quad \text{(distributive)},$$
$$\vec{a} \cdot \vec{a} = a^2 = |\vec{a}|^2 = \vec{a}^2;$$
$$\vec{a} \cdot \vec{b} = 0, \quad \text{if } \phi = 90°, \text{ i.e., } \vec{a} \perp \vec{b}.$$

An example is $\vec{a} = \vec{b} + \vec{c}$ (see Figure A.17). Then we have the following:

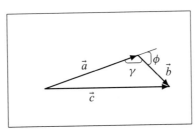

Figure A.17 The scalar product and the cosine theorem.

$$(\vec{a} + \vec{b})^2 = (\vec{a} + \vec{b}) \cdot (\vec{a} + \vec{b}) = a^2 + b^2 + 2\vec{a} \cdot \vec{b} = c^2 \,,$$
$$c^2 = a^2 + b^2 + 2ab \cos \phi = a^2 + b^2 - 2ab \cos \gamma \,.$$

This is the cosine theorem in geometry.

In components, one finds the following:

$$\vec{a} \cdot \vec{b} = \begin{pmatrix} a_x \\ a_y \\ a_z \end{pmatrix} \cdot \begin{pmatrix} b_x \\ b_y \\ b_z \end{pmatrix} = (a_x b_x + a_y b_y + a_z b_z) \,, \quad \text{and}$$

$$\vec{a} \cdot \hat{i} = a_x , \quad \vec{a} \cdot \hat{j} = a_y , \quad \vec{a} \cdot \hat{k} = a_z.$$

The Cross-Product

The cross-product of two vectors is defined as follows:

$$\vec{a} \times \vec{b} = \vec{c}, \qquad |\vec{c}| = |\vec{a}||\vec{b}| \sin \phi, \qquad \vec{c} \perp \vec{a}, \vec{b}.$$

The vectors $\vec{a}$, $\vec{b}$, and $\vec{c}$ form a right-handed system (see Figure A.18). The magnitude of vector $\vec{c}$ corresponds to the parallelogram spanned by the two vectors $\vec{a}$ and $\vec{b}$ or twice the area of the triangle with the sides $\vec{a}$ and $\vec{b}$.

The following rules have to be obeyed:

$$\vec{a} \times \vec{b} = -\vec{b} \times \vec{a} \quad \text{(not commutative)},$$
$$\vec{a} \times (\vec{b} + \vec{c}) = \vec{a} \times \vec{b} + \vec{a} \times \vec{c} \quad \text{(distributive)},$$
$$\vec{a} \times \vec{a} = 0;$$
$$\vec{a} \times \vec{b} = |\vec{a}||\vec{b}| = ab , \quad \text{if } \phi = 90°, \text{ i.e., } \vec{a} \perp \vec{b},$$
$$\vec{a} \times \vec{b} = 0, \quad \text{if } \phi = 0°, 180°, \text{ i.e., } \vec{a} \parallel \pm\vec{b}.$$

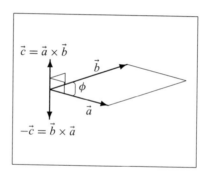

Figure A.18 The cross-product.

Examples for applying the rules on unit vectors include the following:

$$\hat{i} \times \hat{i} = \hat{j} \times \hat{j} = \hat{k} \times \hat{k} = 0 , \qquad \hat{i} \times \hat{j} = \hat{k}, \ \hat{j} \times \hat{k} = \hat{i}, \ \hat{k} \times \hat{i} = \hat{j}.$$

In components, one finds the following:

$$\vec{a} \times \vec{b} = \begin{pmatrix} a_x \\ a_y \\ a_z \end{pmatrix} \times \begin{pmatrix} b_x \\ b_y \\ b_z \end{pmatrix} = \begin{pmatrix} a_y b_z - a_z b_y \\ a_z b_x - a_x b_z \\ a_x b_y - a_y b_x \end{pmatrix}.$$

Solutions to Quizzes

Chapter 1

1. A
2. C, D
3. A
4. A

Chapter 2

1. C
2. D
3. E
4. B
5. D
6. C
7. E
8. A, D, E, F, H, K

Chapter 3

1. B
2. B
3. B
4. A, B, C

Chapter 4

1. A, C, E
2. C, E
3. A, D
4. B
5. D
6. B
7. C
8. A

9. C
10. B
11. D
12. A, D, E, F
13. A, E, F
14. A, B, C, E
15. F
16. A, D, E
17. B, D
18. C, D, E, F
19. D

Chapter 5

1. E
2. B
3. A
4. B
5. A, B, D, F
6. A, B, C, E, F
7. B
8. C
9. A
10. C
11. D
12. C
13. E
14. C, D
15. B

Chapter 6

1. B
2. C
3. C
4. B, D, F
5. A
6. C
7. D
8. B
9. C
10. C

Chapter 7

1. C, D
2. A
3. C, D
4. D
5. C
6. B
7. A

Chapter 8

1. B, D, E
2. A
3. D
4. A
5. A
6. D

Chapter 9

1. E
2. C, E
3. B
4. D
5. D
6. D
7. B
8. E

Chapter 10

1. C
2. D
3. B
4. B
5. C, E, F

References

Aegerter-Wilmsen, T., Aegerter, C.M., and Bisseling, T. 2005. Model for the Robust Establishment of Precise Proportions in the Early Drosophila Embryo. *J. Theor. Biol.*, **234**, 13–19.

Bates, M., Huang, B., Dempsey, G.T., and Zhuang, X. 2007. Multicolor Super-Resolution Imaging with Photo-Switchable Fluorescent Probes. *Science*, **317**, 1749–1753.

Davey, C.A., Sargent, D.F., Luger, K., Maeder, A.W., and Richmond, T.J. 2003. Solvent Mediated Interactions in the Structure of the Nucleosome Core Particle at 1.9 A Resolution. *J. Mol. Biol.*, **319**, 1097–1113.

Detlefsen, D.J., Thanabal, V., Pecoraro, V.L., and Wagner, G. 1991. Solution Structure of Fe(II) Cytochrome c551 from *Pseudomonas aeruginosa* as Determined by Two-Dimensional 1H NMR. *Biochemistry*, **30**, 9040–9046.

Dodds, P.S., Rothman, D.H., and Weitz, J.S. 2001. Re-examination of the "3/4-Law" of Metabolism. *J. Theor. Biol.*, **209**, 9–27.

Drew, H.R., Wing, R.M., Takano, T., Broka, C., Tanak, S., Itakura, K., and Dickerson, R.E. 1981. Structure of a B-DNA Dodecamer: Conformation and Dynamics. *Proc. Natl. Acad. Sci. USA*, **78**, 2179–2183.

Filmetrics. 2017. Reflectance Spectra were Calculated with the Reflectance Calculator by Filmetrics. www.filmetrics.com/reflectance-calculator.

Finer, J.F., Simmons, R.M., and Spudich, J.A. 1994. Single Mysosin Molecule Dynamics: Piconewton Forces and Nanometre Steps. *Nature*, **368**, 113–119.

Franklin, R.E., and Gosling, R.G. 1953. Molecular Configuration in Sodium Thymonucleate. *Nature*, **171**, 740–741.

Gierer, A., and Meinhardt, H. 1972. A Theory of Biological Pattern Formation. *Kybernetik*, **12**, 30.

Göttfert, F., Wurm, C.A., Mueller, V., Berning, S., Cordes, V.C., Honigmann, A., and Hell, S.W. 2013. Coaligned Dual-Channel STED Nanoscopy and Molecular Diffusion Analysis at 20 nm Resolution. *Biophysical J.*, **105**, L01–L03.

Graceffa, P., and Dominguez, R. 2003. Crystal Structure of Monomeric Actin in the ATP State. Structural Basis of Nucleotide-Dependent Actin Dynamics. *J. Biol. Chem.*, **278**, 34172–34180.

Heide, G. 1968. Flugsteuerung durch nicht-fibrilläre Flugmuskeln bei der Schmeißfliege Calliphora. *Zeitschrift für vergleichende Physiologie*, **59**, 456–460.

Käs, J., Strey, H.H., and Sackmann, E. 1994. Direct Imaging of Reptation for Semiflexible Actin Filaments. *Nature*, **368**, 226–229.

Kiang, N.Y.S. 1980. Processing of Speech by the Auditory Nervous System. *Journal of the Acoustic Society of America*, **68**, 830.

Kleiber, M. 1947. Body Size and Metabolic Rate. *Physiological Reviews*, **27**, 11–541.

LaBarbera, M. 1990. Principles of Design of Fluid Transport Systems in Zoology. *Science*, **249**, 992–1000.

Lenaeus, M.J., Gamal El-Din, T.M., Ing, C., Ramanadane, K., Pomes, R., Zheng, N., and Catterall, W.A. 2017. Structures of Closed and Open States of a Voltage-Gated Sodium Channel. *Proc. Natl. Acad. Sci. U.S.A.*, **114**, E3051–E3060.

MacLulich, D.A. 1937. Fluctuations in the Number of the Varying Hare (*Lepus americanus*). University of Toronto Press, Toronto.

Nettles, J.H., Li, H., Cornett, B., Krahn, J.M., Snyder, J.P., and Downing, K.H. 2004. The Binding Mode of Epothilone A on Alpha,beta-Tubulin by Electron Crystallography. *Science*, **305**, 866–869.

Pfau, H.K. 1986. Untersuchungen zur Konstruktion, Funktion und Evolution des Flugapparates der Libellen (Insecta, Odonata). *Tijdschrift voor Entomologie*, **129**, 35–123.

Potyrailo, R.A., Ghiradella, H., Vertiatchikh, A., Dovidenko, K., Cournoyer, J.R., and Olson, E. 2007. Morpho Butterfly Wing Scales Demonstrate Highly Selective Vapour Response. *Nature Photonics*, **1**, 123–128.

Prahl, Scott. 1998. (scott.prahl@oit.edu) Using Data from W.B. Gratzer and N. Kollias. Medical Research Council Labs, Holly Hill, London; and Wellman Laboratories, Harvard Medical School, Boston, MA. https://omlc.org/spectra/hemoglobin/summary.html.

Sanderson, J. 2000. The Theory of Contrast Control in the Microscope. *Quekett Journal of Microscopy*, **38**, 617–627.

SDBS. 2017. *Spectral Database for Organic Compounds*.

Smith, S.B., Cui, Y., and Bustamante, C. 1996. Overstretching B-DNA: The Elastic Response of Individual Double-Stranded and Single-Stranded DNA Molecules. *Science*, **271**, 795–799.

Strain, H.H., Thomas, M.R., and Katz, J.J. 1963. Spectral Absorption Properties of Ordinary and Fully Deuteriated Chlorophylls a and b. *Biochim. Biophys. Acta*, **75**, 306–311.

Suder, F. 1994. Verhaltensphysiologische und neuroanatomische Untersuchungen zur Organisation des Flugsystems von Manduca sexta. PhD thesis, University of Cologne.

Thermo-Fischer. 2017. *Spectral Data from Thermo-Fisher Spectravisualizer*.

Turing, A. 1952. The Chemical Basis of Morphogenesis. *Philos. Trans. R. soc. London B.*, **237**, 37–72.

Webb, D. 1996. Maximum Walking Speed and Lower Limb Length in Hominids. *Am. J. Phys. Anthropol.*, **101**, 515–525.

Wehner, R. 1982. Himmelsnavigation bei Insekten: Neurobiologie und Verhalten. *Neujahrsblatt Naturforschende Gesellschaft Zürich*, **182**, 1–132.

Wehner, R. 1994. Himmelsbild und Kompassauge – Neurobiologie eines Navigationssystems. *Verhandlungen der Deutsch Zoologischen Gesellschaft*, **87.2**, 9–37.

Wehner, R. 2000. Panting Desert Ants – and How They Navigate. *Schweiz. Med. Wochenschr.*, **130**, 258–263.

Wehner, R., and Wehner, S. 1990. Insect Navigation: Use of Maps or Ariadne's Thread. *Ethol. Ecol. Evol.*, **2**, 27–48.

Index

absorption
  colour, 204
  radiation, 208
  resonant, 204, 414
acceleration, 65
acceleration sensor, 78
accuracy, 22
actin, 220, 345
action potential, 394
age of the earth, 209, 330
alveoli, 279
Ampere's law, 403
aneurism, 281
angular momentum, 409
area moment of inertia, 344

bell curve, 321
bending, 49, 271
bicoid, 324
Bohr model, 137, 378
Boltzmann distribution, 314, 328, 339, 352, 375, 412
bond, 203, 235
  acto-myosin, 247
  chemical, 13
  covalent, 221, 381
  energy, 203
  hydrogen, 138, 342
  interatomic, 236
  ionic, 264
bone
  growth regulation, 50
  scaling, 9
  size, 48
  stress–strain curve, 271
Brownian motion, 313

capacitor, 371, 386
capillary, 300, 324
Cataglyphis, 4, 173
Celsius, 315
charge
  electric, 363
  elementary, 363
circular motion, 80

collision
  elastic, 241
  inelastic, 242
color, 2, 151, 193, 201, 204, 205
compression, 269, 271
conduction, 331
  electrical, 9, 382
  heat, 9, 329, 332, 351
conductivity, 384
conservation
  energy, 250
  mass, 285
  momentum, 239
contraction
  lateral, 268
  muscle, 220
convection, 324, 350
coordinate system, 62
coordinates
  Cartesian, 77
Coulomb's law, 364
current, 284, 331
  density, 284, 332, 383
  diffusion, 287, 328
  drift, 328
  electrical, 8, 382, 403
  fluid, 284
  loop, 404
curvature, 427
  radius of, 182, 271, 278, 343, 407

damping, 88
Darwin, 19, 210, 331
decibel scale, 113
deformation
  bending, 262
  plastic, 262
  shear, 262
  tension, 262
derivative, 424
diffraction
  droplet, 159
  grating, 164
  hole, 168
  limit, 170, 191, 207

diffraction (cont.)
  slit, 167
  x-ray, 11, 164
diffusion, 322, 350
  Fick's law, 323
diffusivity, 323
dimensional analysis, 39
dipole
  induced, 377
  magnetic, 402
dipole moment, 373
direction sensor, 218
dispersion, 192, 200
distribution
  Boltzmann, 328, 375, 412
  Gauss, 321
  Gaussian, 29
  probability, 339
  speed, 327
DNA, 13, 14, 369
  information, 340
  packing, 346
  persistence length, 349
  stretching, 341
Doppler effect, 138
Drosophila, 12, 324, 356

Einstein relation, 326
elastic
  collision, 241
  constant, 265
  deformation, 261
  modulus, 266
  shear, 274
  strain, 265
  stress, 264
  tension, 264
electric
  charge, 363
electrolyte, 388
electron, 11, 98, 136, 151, 171, 203,
    230, 363, 372, 378, 404, 408
electrophoresis, 14, 229, 239,
    348, 389
energy, 244
  conservation, 250
  elastic, 268
  internal, 333
  kinetic, 248
  potential, 248
  thermal, 319
entropy, 336
  Boltzmann, 338
  force, 339
  information, 340
  spring, 344
equation
  cable, 393

diffusion, 287, 322
  ideal gas, 317
  motion, 227
  Navier–Stokes, 287, 292
  of continuity, 284
  Poisson, 368
equilibrium, 311
  thermal, 315
equipartition theorem, 320
error
  absolute, 31
  measurement, 25
  propagation, 32
  relative, 31
  single measurement, 30
  statistical, 27
  systematic, 27
estimate, 22
exponential, 423

falsification, 19, 49
Fick's law, 323
field
  dipole, 372
  electric, 14, 229, 364
  electromagnetic, 99, 151, 402
  flow, 283
  gravitational, 249
  magnetic, 15, 401
  polarizability, 377
  polarization, 374
  vector, 78
  velocity, 283
flow
  laminar, 284
  Poiseuille, 294
  turbulent, 293
fluctuation
  dipole, 377
flux
  electric, 366
  fluid, 284
focal length, 183
force
  action, 224
  entropic, 339
  friction, 232
  impact, 243
  Lorentz, 405
  normal, 232
  reaction, 222
  sensor, 218
Fourier expansion, 428
Fourier series, 74
Fourier–transform, 11, 116, 408, 433
  grid, 158
  sphere, 159
freefall, 39, 229

frequency, 109
  cyclotron, 406
  Larmor, 413
  precession, 412
  radio, 414
friction
  inertial, 226
  sliding, 234
  viscous, 236, 291, 384

Gauss
  distribution, 29
  error propagation, 34
  law of electrostatics, 368
  law of magnetism, 402

harmonic functions, 424
hearing
  frequency dependence, 113
  mechanism, 115
  threshold, 112
heat, 313, 333
  specific, 335
Hooke's law, 265
hydrogen bond, 138

ideal gas, 316
image
  mirror, 179
  real, 178, 186
  virtual, 178, 187
imaging
  magnetic resonance, 13, 415
  optical, 178, 188
  x-ray, 12
impact, 243
inertia, 224
information, 340
inner ear, 78, 218
instability, 352
interference, 3, 8, 136
  constructive, 134, 160
  destructive, 134
  double slit, 134, 156
  thin film, 4, 160
ionic solution, 374
irreversibility, 313

kelvin, 315
Kirchhoff's laws, 299, 385, 390

lens, 182, 190
logarithmic scale, 43, 423
  double, 43

macroscopic description, 260, 310
magnification, 192

mass spectrometry, 15
materials properties, 261
mean square displacement, 314
measurement error, 21
membrane, 370, 375, 390
  axon, 392
  basilar, 67
  drum, 129
  rubber, 278
metabolic rate, 51
microscopic description, 310
microscopy
  confocal, 11, 196
  dark field, 195
  DIC, 196
  electron, 2
  fluorescence, 194, 207
  imaging, 178
  light, 11, 191
  phase contrast, 196
  super-resolution, 11, 207
microtubule, 345
mirror
  image, 179
  spherical, 181
mobility, 326
  electrophoretic, 389
modulus
  compression, 269
  shear, 274
  Young's, 266
moment
  magnetic, 404
moment of inertia
  area, 273, 344
  mass, 409
momentum, 239
  angular, 404, 409
Morpho menelaus, 2, 159
morphogen, 207, 324, 353
motion
  equation of, 227
myosin, 220

navigation, 4, 401
nerve, 389
  firing rate, 226
nonlinear dynamics, 85
nuclear magnetic resonance
  imaging, 13, 415
  spectroscopy, 12, 414
nuclear magnetic resonance, 408

Ohm's law, 297, 331, 384
optical tweezer, 221
oscillation
  amplitude, 67
  circular motion, 80

oscillation (cont.)
  coupled, 92
  damped, 88
  forced, 99
  harmonic, 70
  period, 67
  resonant, 100

pattern formation, 353
period
  oscillation, 67
  pendulum, 40
periodic table, 381
persistence length, 344
pigment, 2
Poisson ratio, 268
polarizability, 377
polarization, 170
  birefringence, 173, 174
  field, 374
  scattering, 7, 173, 196
  sky, 6, 173
  spin, 413
population dynamics, 83
potential
  action, 394
  electric, 365
power law, 44, 422
precession, 412
predator-prey systems, 83
pressure, 316
  Bernoulli, 289
  dynamic, 289
  Laplace, 277
  microscopic, 317
  static, 269
proton, 138, 208, 363, 379, 412

radiation, 208
  alpha, 208
  beta, 209
  gamma, 209
  natural, 209
radius of curvature, 182, 271, 278, 343, 407
rainbow, 17, 198
random walk, 321, 326, 343
reflection, 152, 176
refraction, 153, 176
refractive index, 152, 176, 192, 200
reproducibility, 21
reptation, 348
resistance
  electric, 385
  fluid flow, 297
  thermal, 331

resistor
  electric, 385
  heat, 331
resonance, 100
  damped, 102
  excitation of basilar membrane, 115
  phase shift, 105
Reynolds number, 46, 238, 284, 292, 326

scaling
  allometric, 9
  blood flow, 301
  bone size, 49
  flight speed, 47
  metabolic rate, 51
scaling law, 42
screening, 376
SI units, 36
significance, 29
slope, 425
sonar, 125
spectrum, 74
  Fourier, 408
  mass, 15, 408
  nuclear magnetic resonance, 12, 415
  resonance, 102
  visible light, 5, 193, 204, 205
speed, 65
  flight, 42
  flow, 301
spin, 409
stability analysis, 354
standard deviation, 29, 321
stiction, 233
strain, 265
  shear, 274
stream lines, 284
stress
  normal, 264
  yield, 271
stress-birefringence, 174
stress–strain curve, 271
stretching, 271
surface tension, 275

Taylor expansion, 23, 34, 93, 267, 339, 353, 354, 373, 376, 412, 427
temperature, 315
  microscopic, 319
tension, 264
  surface, 275
torque, 411
transport, 331

uncertainty, 21
  measurement, 25

principle, 136, 341
  statistical, 27
  systematic, 27

van der Waals interaction, 262, 377
vector, 76, 217, 434
  addition, 435
  derivative, 437
  field, 78
  multiplication, 439
  subtraction, 435
velocity, 64
viscosity, 45, 236, 291, 389
voltage, 365

walking speed, 41
wave
  acoustic, 112
  amplitude, 107

energy transport, 111
equation, 108
frequency, 109, 139
harmonic, 108
intensity, 111
period, 109
reflection, 119
standing wave, 126, 137, 378
string, 107
superposition, 119
transmission, 119
velocity, 110
wavelength, 3, 109, 128, 130, 136, 139,
  157, 161, 167, 191, 378
work, 244, 333
  expansion, 268

yield stress, 9, 271